Human Brain Responses and Functional Brain Networks across the Lifespan

Human Brain Responses and Functional Brain Networks across the Lifespan

Guest Editors

Birgit Mathes
Canan Başar-Eroğlu

MDPI

Basel • Beijing • Wuhan • Barcelona • Belgrade • Novi Sad • Cluj • Manchester

Guest Editors
Birgit Mathes
Human and Health Sciences
University of Bremen
Bremen
Germany

Canan Başar-Eroğlu
Faculty of Arts and Science
Izmir University of
Economics
Izmir
Turkey

Editorial Office
MDPI AG
Grosspeteranlage 5
4052 Basel, Switzerland

This is a reprint of the Special Issue, published open access by the journal *Brain Sciences* (ISSN 2076-3425), freely accessible at: https://www.mdpi.com/journal/brainsci/special_issues/N92V4D7SXI.

For citation purposes, cite each article independently as indicated on the article page online and as indicated below:

Lastname, A.A.; Lastname, B.B. Article Title. *Journal Name* **Year**, *Volume Number*, Page Range.

ISBN 978-3-7258-5145-4 (Hbk)
ISBN 978-3-7258-5146-1 (PDF)
https://doi.org/10.3390/books978-3-7258-5146-1

Contents

About the Editors

Birgit Mathes

Birgit Mathes studied psychology, with a minor in biology and sociology, at the University of Bremen (Germany) and the City University in London (United Kingdom). She gathered research experience in psychology, neurobiology, psychiatry, and neurosciences at the University of Bremen and the University of Melbourne (Australia) and was a guest scientist at Istanbul Kültür Üniversitesi (Turkey). She later became the head of the research lab and Bremen group of the Bremen Initiative to foster early childhood development (BRISE). She currently works as a professor in the faculty of human and health sciences at the University of Bremen.

Canan Başar-Eroğlu

Canan Başar-Eroğlu studied human biology and received her PhD in biophysics from Hacettepe University in Ankara, Turkey. Her post-doctoral career brought her to the Institute of Physiology at the Medical University of Lübeck, Germany, where she became an assistant professor. She subsequently worked at the University of Bremen, Germany. She served as head of the EEG Research Group and as a professor at the Institute of Psychology and the Cognition Research and Center for Cognitive Sciences. She later advanced to the head of the Institute of Psychology and Cognition Research. Currently, she is a professor of psychology at Izmir University of Economics in Turkey.

Preface

Life-long adaptations of human brain networks allow for responses to life-period-specific challenges. Measuring brain responses in real time using electrophysiology enriches our understanding of changes in behavior and cognitive function across the lifespan. This reprint of the Special Issue "Human Brain Responses and Functional Brain Networks across the Lifespan" compiles the research of recognized experts in the field, all with a longstanding history in discovering the functions of the brain through non-invasive methods.

Contributions cover age groups ranging from 4 months to 86 years, including healthy participants as well as patients with neuropsychiatric and neurodegenerative disorders. Three contributions focused on environmental influences on brain function and address the under-representation of socioeconomically disadvantaged populations in neuroscience research. Experimental paradigms include resting EEG; passive stimulation protocols; and active behavioral tasks examining attentional states, language comprehension, and cognitive control. Analytical methods encompass the time–frequency analysis of oscillatory dynamics, source localization, event-related potential (ERP) measures, and the link between EEG/ERP and behavioral performance.

The findings indicate that EEG and other neurophysiological measures provide relevant data on age-related changes in brain function and their associations with cognition, risk factors, life experiences, and mental health. They also demonstrate that research in human neurophysiology will profit from the advantages of combined EEG strategies.

We hope to inspire the reader with new understandings and ideas within this important field of research.

Birgit Mathes and Canan Başar-Eroğlu
Guest Editors

brain sciences

MDPI

Editorial

Human Brain Responses and Functional Brain Networks Across the Lifespan

Birgit Mathes [1,*] and Canan Başar-Eroğlu [2]

1 Faculty of Human and Health Sciences, University of Bremen, 28359 Bremen, Germany
2 Faculty of Arts and Science, Izmir University of Economics, 35330 Izmir, Turkey; canan.basar@izmirekonomi.edu.tr
* Correspondence: birgit.mathes@uni-bremen.de

1. Introduction

Measuring brain responses in real time using electrophysiology enriches our understanding of changes in behavior and cognitive function across the lifespan [1]. Life-long adaptations of human brain networks allow responding to life-period-specific challenges. This enables learning and improvement of new abilities during development and compensational processes during aging [2–4]. Research investigating human brain responses across the lifespan further allows us to understand risks of emerging mental health or neurodegenerative disorders [5–7]. Recent studies increasingly mirror how brain development and function are influenced by social experiences and socioeconomic status, i.e., the social and economic position of a person or group within society, typically measured by income, education, and occupation [8–11].

Adaptations in human brain networks through life can be characterized by spontaneous and event-related activity. Characterizing oscillatory EEG measures specifically enables determining age-related changes in the temporal coordination of multiple neural activation patterns and their integration within functional neural networks [12]. Both spontaneous and event-related measures allow the implementation of similar or even identical routines in data acquisition and analysis, thus supporting a focus on age-related changes. Abilities of life-long importance, including attention and response inhibition, can be investigated with well-established paradigms, such as the oddball or Go/NoGo task across age [13–17].

The aim of this Special Issue was to bring together a broad range of EEG and other neuroscientific studies to better understand the mechanisms and functions of brain changes through the lifespan. In the following, you will find a short description of all contributions in the order of age groups, ranging from 4 months to 86 years.

2. Overview of Published Articles

1. Neural responses related to social influences and cognitive abilities in infants and toddlers

Infancy is characterized by fast learning of basic and life-long important abilities as well as increased neuroplasticity, meaning the adaptation of brain structure and function due to learning experiences [18,19]. Living circumstances may thus have a lasting effect on brain maturation. Its impact on cognition may root in infancy and increase over the life course. Peykarjou et al. (2024) demonstrate that fast visual characterization of living and nonliving objects is already developed in 4-month-old children. Tarullo et al. (2024)

https://doi.org/10.3390/brainsci15040402

and Wienke and Mathes (2024) investigate how early living circumstances may affect early neural development.

- Tarullo et al. (2024) undertook a challenging longitudinal resting-state EEG study in rural South Africa. They aimed to gain better insights into the determinants of early brain maturation at 7, 17, and 36 months of age from children growing up with restricted resources. Their results indicated that larger head size at birth predicted smaller gamma and larger theta power, particularly at 7 months of age. This finding, when looking at high-SES literature, counterintuitively indicates delayed maturation in children with a presumably better fetal environment. Tarullo and colleagues argue that high-SES children, in comparison to children experiencing more environmental prenatal challenges, may experience a prolonged time window of increased neural plasticity, an important mechanism for life-long learning and cognitive performance. Taken together, their studies underline the importance of reflecting social context in developmental neuroscience.
- In a comprehensive study, Peykarjou et al. (2024) demonstrate that 4-month-old infants already elicit a brain response indicative of categorizing visual stimuli into "living" and "non-living". Their elegant task design enables Peykarjou et al. (2024) to study fast categorization abilities and their neural correlates similarly in infants of 4, 7, and 11 months, as well as in 6–7-year-old children and young adults. In all groups, categorization responses emerged at posterior-occipital and frontal leads and were driven mostly by global Gestalts and to a lesser degree by low-level visual features. The early emergence and similar, though not identical, neural categorization response across age points towards the life-long importance to distinct living, and potentially interacting, from nonliving objects.
- The EEG study of Wienke and Mathes (2024) examined how the brain responses of infants aged 6 to 14 months are influenced by their families' socioeconomic background. During a passive auditory oddball paradigm, the results showed an immature N2-P3a complex, with the N2 amplitude being more positive for deviants than for standards. This effect was more pronounced in infants from families with lower parental education and with a recent migration history, especially when burdens coincide. This suggests that early neural development is affected by socioeconomic and cultural challenges. The observed deviations may have an impact on attention and learning processes early on.

2. Neural responses related to social influences and cognitive abilities during childhood

This Special Issue includes three reviews, all approaching changing neural networks during childhood with a different focus. Karakaş (2024) summarizes EEG studies on attentional processes and emphasizes studying brain oscillations. Accordingly, Ünsal et al. (2024) provide an update on functions of brain oscillations during neurodevelopment. The review of Schneider et al. (2024) ties into the discussion of how the environment and socioeconomic position of a child correlates with brain function.

- The review of Karakaş (2024) spans between the roots and underlying concepts of EEG research, how these relate to up-to-date neurodevelopmental studies on attention in children and directions for future research. Event-related potentials (ERPs) are commonly used to investigate early and late stages of attentional processing in children. ERP latencies indicate slower processing speed in children than adults, while the overall shape, topography, and amplitude of ERPs indicate qualitative differences and, thus, the developmental reorganization of neural networks between childhood and adulthood. Investigating brain oscillation may allow a better understanding of spatial and temporal integration within parallel processing neural networks during childhood. Developmental changes in spontaneous oscillations are discussed as the building blocks of event-related brain oscillations (EROs) and the superposition of event-related oscillations as the building blocks of ERPs; thus, Karakaş (2024) stresses the necessity to better understand their inter-relation in neurodevelopmental research. She further demonstrates the importance of understanding clinical applications of attentional processing deficits originating in childhood.

- Schneider et al. (2024) provide a comprehensive and future-directed review illustrating environmental risk factors on brain maturation and cognitive functioning. Low maternal education and low family income are often correlated but may affect the brain of the maturing child via distinct pathways. Less diverse linguistic input, often a result of low educational attainment, affects the development of the language processing system. Low income may lead to increased everyday stress, which affects brain development globally and may have detrimental effects for the child's cognitive control and executive functions. The authors further propose that cultural context, including family size, parental beliefs, and living area, need to be considered more thoroughly in future research.
- Ünsal et al. (2024) review the state of knowledge regarding event-related brain oscillations between infancy and childhood. Following a thorough literature search, they summarize functional correlates as well as age-related changes in event-related delta, theta, alpha, beta, gamma, and mu oscillations across 86 studies. Theta oscillations have been researched most often so far, as they play a role in perception, attention, and memory. The reviewed studies demonstrate that in children, similar to adults, oscillatory responses of all frequency bands are important for cognitive functions. The authors emphasize the fruitful approach of understanding brain development through an age-related view on brain oscillations. They conclude that future studies should strongly focus on the interplay between brain responses in different frequency bands and further integrate advanced neuroimaging and EEG techniques.

3. Neural responses related to language abilities during adolescence

- Behboudi et al. (2023) illustrate the maturation of ongoing gamma oscillations in early adolescents (8 to 15 years) elicited during sentence processing. Broadly increased ongoing gamma activation was observed in younger participants. Older participants showed a more localized activation pattern at left central and posterior regions and smaller amplitude values than younger participants. In concert with a previous study of their group [20], the authors argue that theta and gamma networks elicited during sentence processing still mature at the onset of adolescence. Their work may be taken as an example of how oscillatory neural networks become better defined during adolescence, which may specifically reduce the need for word-by-word integration for sentence comprehension.

4. Structural and Functional Brain Networks during young adulthood

Young adulthood seems to be a period of relative stability in brain function and underlying structure. Subtle age-related changes may, however, be found. Fehr et al. (2024) and Rizzo et al. (2024) analyze brain oscillations during young adulthood. Salisbury et al. (2024) use MRI data to investigate if the onset of a mental illness during young adulthood accelerates early aging of the brain.

- Fehr, T. et al. (2024) present a thorough signal- and source-based analysis of age-related changes in resting-state EEG during the third decade of life. EEG power remained relatively stable across age in all major oscillatory frequency bands (delta, theta, alpha, beta, and gamma). Network activation pattern, as indicated by the succession of activations in modeled sources of the EEG signal, changed across age. The findings comprised different and broadly distributed sources for different frequency bands. The authors conclude that age-related changes in the third decade are subtle and affect how information is processed (as indicated by network activation pattern) rather than source strength (as indicated by power).

- Rizzo et al. (2024) explore neural oscillatory dynamics underlying reduced one-sided pain perception utilizing the mirror-induced feedback illusion. Watching tapping one's right finger in a mirror while the left finger is unmoved and hidden elicits an illusion of the left finger moving. During the illusion, painful versus painless electrical stimulation of the hidden finger induced increased desynchronization of alpha activity. This trend was reversed in two control conditions, indicating that the illusion affected cortical activation. The authors argue that the illusion changed internal body perception and allocation of attentional resources to the painful stimulus during both anticipation and experience of pain.
- Salisbury et al. (2024) present structural imaging data from healthy young adults and patients approximately 27 days (median) after their first clinical contact due to a suspected psychotic disorder. Previous studies reported signs of early aging in patients with established schizophrenia and progressive gray matter loss when transitioning into a first psychosis [21,22]. Power analysis of the authors make it, however, unlikely that early aging of the brain is found early during the course of a psychotic illness. They report older predicted brain age in patients with poorer emotional expressivity, i.e., in relation to the extent of negative symptoms, as well as with lower premorbid IQ, i.e., in relation to life-long and pre-illness acquisition of knowledge.

5. Changes in neural responses and cognition across childhood to late adulthood

Two cross-sectional studies investigate changes in brain function covering a broad age range. Barry et al. (2024) looks at pre-stimulus EEG and post-stimulus ERPs during response inhibition and execution in children, young and old adults. Yener et al. (2024) utilize a visual oddball paradigm to investigate the effects of age and gender on event-related delta and theta oscillations during young, middle, and late adulthood.

- Barry et al. (2024) thoroughly explore the links between pre-stimulus EEG, post-stimulus ERPs, and performance during Go and NoGo processing in children (8–12 years) as well as young (18–24 years) and older adults (59–74 years). Children showed the lowest and young adults showed the highest performance. In accordance, young adults elicited the simplest, indicating the most efficient, pattern of links between EEG/ERP and performance measures. The links between EEG/ERPs and performance were more complex in children and older adults. This result may be indicative of less effective modulation of attentional and vigilance states and of reduced cognitive control. Brain maturation may lead to better distinction between Go and NoGo processing.
- Yener et al. (2024) utilize a classical visual oddball paradigm and analyze amplitude differences in delta and theta oscillations across adulthood (18–86 years) and gender. Their rationale was that development of Alzheimer's disease is twice as likely in women than men and slow oscillations elicited by targets relate to abnormalities in cognitive performance. Their results indicate that decline of theta amplitudes with age is steeper in women, leading to increased theta amplitudes in women compared to men until middle adulthood. These results may relate to early signs of dementia being more likely in women than in men. The authors call for gender-stratified and gender-adjusted analysis in future studies on aging to overcome women's under-

representation in research and better adjustment of treatment strategies according to gender.

3. Conclusions

Contributions in this Special Issue covered age groups ranging from infancy to old age. Studies focused on age factors in healthy participants and patients with psychotic symptoms. Three contributions focused on environmental influences on brain function and overcoming under-representation of people with socioeconomic challenges in research. Paradigms included resting EEG, passive stimulation, and active behavioral paradigms, with a focus on attentional states, language comprehension, and cognitive control. Multifold analytical methods were applied, looking at time frequency components of the oscillatory dynamics of the brain, source localization, and ERP measures as well as the link between EEG/ERP and behavioral performance. Thus, EEG measures provide important and various opportunities to measure life-long changes in brain function, its relation to cognition, risk factors, and the course of mental illnesses.

One of the presented studies included longitudinal data and another MRI data. Future direction of lifespan research would profit from more longitudinal studies, ideally covering longer age ranges and a combination of methods to integrate information from brain structure and neural dynamics. Mobile EEG studies would allow for brain dynamics to be investigated in real-life situations. We further stressed that increasing diversity in research and understanding how life experiences and living circumstances relate to brain neuroplasticity remain important to reach better representability and to improve prevention and support strategies and medical treatment. These considerations call for collaboration between research institutions and research disciplines, data sharing, and long-term studies that allow for comprehensive recruitment and multiple testing. Future research therefore faces both challenges and opportunities to gain comprehensive insights into brain dynamics and their real-life implications across the lifespan.

Author Contributions: Both authors contributed equally to the editorial. All authors have read and agreed to the published version of the manuscript.

Funding: BM was funded by the German Federal Ministry of Education and Research (grant no. 01NV2001A–I).

Acknowledgments: We are thankful to all authors who contributed to our Special Issue with their valuable work.

Conflicts of Interest: The author declares no conflicts of interest.

List of Contributions:

1. Barry, R.J.; De Blasio, F.M.; Clarke, A.R.; Duda, A.T.; Munford, B.S. Age-Related Differences in Prestimulus EEG Affect ERPs and Behaviour in the Equiprobable Go/NoGo Task. *Brain Sci.* **2024**, *14*, 868. https://doi.org/10.3390/brainsci14090868.
2. Behboudi, M.H.; Castro, S.; Chalamalasetty, P.; Maguire, M.J. Development of Gamma Oscillation during Sentence Processing in Early Adolescence: Insights into the Maturation of Semantic Processing. *Brain Sci.* **2023**, *13*, 1639. https://doi.org/10.3390/brainsci13121639.
3. Fehr, T.; Mehrens, S.; Haag, M.C.; Amelung, A.; Gloy, K. Changes in Spatiotemporal Dynamics of Default Network Oscillations between 19 and 29 Years of Age. *Brain Sci.* **2024**, *14*, 671. https://doi.org/10.3390/brainsci14070671.
4. Karakaş, S. A Review of Childhood Developmental Changes in Attention as Indexed in the Electrical Activity of the Brain. *Brain Sci.* **2024**, *14*, 458. https://doi.org/10.3390/brainsci14050458.
5. Peykarjou, S.; Hoehl, S.; Pauen, S. The Development of Global-Level Categorization: Frequency Tagging EEG Responses. *Brain Sci.* **2024**, *14*, 541. https://doi.org/10.3390/brainsci14060541.

6. Rizzo, M.; Petrini, L.; Del Percio, C.; Arendt-Nielsen, L.; Babiloni, C. Neurophysiological Oscillatory Mechanisms Underlying the Effect of Mirror Visual Feedback-Induced Illusion of Hand Movements on Nociception and Cortical Activation. *Brain Sci.* **2024**, *14*, 696. https://doi.org/10.3390/brainsci14070696.
7. Salisbury, D.F.; Wulf, B.M.; Seebold, D.; Coffman, B.A.; Curtis, M.T.; Karim, H.T. Predicted Brain Age in First-Episode Psychosis: Association with Inexpressivity. *Brain Sci.* **2024**, *14*, 532. https://doi.org/10.3390/brainsci14060532.
8. Schneider, J.M.; Behboudi, M.H.; Maguire, M.J. The Necessity of Taking Culture and Context into Account When Studying the Relationship between Socioeconomic Status and Brain Development. *Brain Sci.* **2024**, *14*, 392.
9. Tarullo, A.R.; Evans, D.; Coetzee, L.; Lopera-Perez, D.C.; Brady, S.P.; Gabard-Durnam, L.J.; Fink, G.; Hamer, D.H.; Yousafzai, A.K.; Rockers, P.C. Neonatal Physical Growth Predicts Electroencephalography Power in Rural South African Children. *Brain Sci.* **2024**, *14*, 552. https://doi.org/10.3390/brainsci14060552.
10. Ünsal, E.; Duygun, R.; Yemeniciler, İ.; Bingöl, E.; Ceran, Ö.; Güntekin, B. From Infancy to Childhood: A Comprehensive Review of Event- and Task-Related Brain Oscillations. *Brain Sci.* **2024**, *14*, 837.
11. Wienke, A.S.; Mathes, B. Socioeconomic Inequalities Affect Brain Responses of Infants Growing Up in Germany. *Brain Sci.* **2024**, *14*, 560. https://doi.org/10.3390/brainsci14060560.
12. Yener, G.; Kıyı, I.; Düzenli-Öztürk, S.; Yerlikaya, D. Age-Related Aspects of Sex Differences in Event-Related Brain Oscillatory Responses: A Turkish Study. *Brain Sci.* **2024**, *14*, 567. https://doi.org/10.3390/brainsci.

References

1. Babiloni, C. The Dark Side of Alzheimer's Disease: Neglected Physiological Biomarkers of Brain Hyperexcitability and Abnormal Consciousness Level. *J. Alzheimer's Dis.* **2022**, *88*, 801–807. [CrossRef]
2. Janacsek, K.; Fiser, J.; Nemeth, D. The Best Time to Acquire New Skills: Age-related Differences in Implicit Sequence Learning across the Human Lifespan. *Dev. Sci.* **2012**, *15*, 496–505. [CrossRef]
3. Cabeza, R.; Albert, M.; Belleville, S.; Craik, F.I.M.; Duarte, A.; Grady, C.L.; Lindenberger, U.; Nyberg, L.; Park, D.C.; Reuter-Lorenz, P.A.; et al. Maintenance, Reserve and Compensation: The Cognitive Neuroscience of Healthy Ageing. *Nat. Rev. Neurosci.* **2018**, *19*, 701–710. [CrossRef] [PubMed]
4. Pauen, S.; Bechtel-Kuehne, S. How Toddlers Acquire and Transfer Tool Knowledge: Developmental Changes and the Role of Executive Functions. *Child. Dev.* **2016**, *87*, 1233–1249. [CrossRef] [PubMed]
5. Başar, E.; Schmiedt-Fehr, C.; Mathes, B.; Femir, B.; Emek-Savaş, D.D.; Tülay, E.; Tan, D.; Düzgün, A.; Güntekin, B.; Özerdem, A.; et al. What Does the Broken Brain Say to the Neuroscientist? Oscillations and Connectivity in Schizophrenia, Alzheimer's Disease, and Bipolar Disorder. *Int. J. Psychophysiol.* **2016**, *103*, 135–148. [CrossRef]
6. Haigh, S.M.; Coffman, B.A.; Salisbury, D.F. Mismatch Negativity in First-Episode Schizophrenia. *Clin. EEG Neurosci.* **2017**, *48*, 3–10. [CrossRef] [PubMed]
7. Başar-Eroğlu, C.; Küçük, K.M.; Rürup, L.; Schmiedt-Fehr, C.; Mathes, B. Oscillatory Activities in Multiple Frequency Bands in Patients with Schizophrenia During Motion Perception. *Clin. EEG Neurosci.* **2023**, *54*, 349–358. [CrossRef]
8. Yerlikaya, D.; Hünerli-Gündüz, D.; Fide, E.; Özbek, Y.; Kıyı, İ.; Öztura, İ.; Yener, G.G. The Reliability of P300 and the Influence of Age, Gender and Education Variables in a 50 Years and Older Normative Sample. *Int. J. Psychophysiol.* **2022**, *181*, 1–13. [CrossRef]
9. Troller-Renfree, S.V.; Costanzo, M.A.; Duncan, G.J.; Magnuson, K.; Gennetian, L.A.; Yoshikawa, H.; Halpern-Meekin, S.; Fox, N.A.; Noble, K.G. The Impact of a Poverty Reduction Intervention on Infant Brain Activity. *Proc. Natl. Acad. Sci. USA* **2022**, *119*, e2115649119. [CrossRef]
10. Tarullo, A.R.; Tuladhar, C.T.; Kao, K.; Drury, E.B.; Meyer, J. Cortisol and Socioeconomic Status in Early Childhood: A Multidimensional Assessment. *Dev. Psychopathol.* **2020**, *32*, 1876–1887. [CrossRef]
11. Çelik, S.; Yıldırım, E.; Güntekin, B. Reduced Resting and Task-Related Alpha Activity in Mine Workers: Implications for Occupational Health and Neurodegenerative Risk. *Int. J. Psychophysiol.* **2025**, *207*, 112466. [CrossRef]
12. Başar, E. Oscillations in "Brain–Body–Mind"—A Holistic View Including the Autonomous System. *Brain Res.* **2008**, *1235*, 2–11. [CrossRef] [PubMed]
13. Karakaş, S. A Comparative Review of the Psychophysiology of Attention in Children with and without Attention Deficit Hyperactivity Disorder. *Int. J. Psychophysiol.* **2022**, *177*, 43–60. [CrossRef]

14. Barry, R.J.; De Blasio, F.M.; Cave, A.E. Development of Children's Performance and ERP Components in the Equiprobable Go/NoGo Task. *Int. J. Psychophysiol.* **2022**, *171*, 12–19. [CrossRef] [PubMed]
15. Müller, V.; Jirsa, V.; Perdikis, D.; Sleimen-Malkoun, R.; von Oertzen, T.; Lindenberger, U. Lifespan Changes in Network Structure and Network Topology Dynamics During Rest and Auditory Oddball Performance. *Front. Aging Neurosci.* **2019**, *11*, 138. [CrossRef]
16. Schmiedt-Fehr, C.; Mathes, B.; Kedilaya, S.; Krauss, J.; Basar-Eroglu, C. Aging Differentially Affects Alpha and Beta Sensorimotor Rhythms in a Go/Nogo Task. *Clin. Neurophysiol.* **2016**, *127*, 3234–3242. [CrossRef]
17. Wienke, A.S.; Basar-Eroglu, C.; Schmiedt-Fehr, C.; Mathes, B. Novelty N2-P3a Complex and Theta Oscillations Reflect Improving Neural Coordination Within Frontal Brain Networks During Adolescence. *Front. Behav. Neurosci.* **2018**, *12*, 218. [CrossRef] [PubMed]
18. Pauen, S.; Peykarjou, S. Infants' Fast Neural Categorization of Artificial Objects: The Impact of Stimulus and Task Characteristics. *Dev. Psychol.* **2023**, *59*, 1496–1510. [CrossRef]
19. Noble, K.G.; Hart, E.R.; Sperber, J.F. Socioeconomic Disparities and Neuroplasticity: Moving toward Adaptation, Intersectionality, and Inclusion. *Am. Psychol.* **2021**, *76*, 1486–1495. [CrossRef]
20. Maguire, M.J.; Schneider, J.M.; Melamed, T.C.; Ralph, Y.K.; Poudel, S.; Raval, V.M.; Mikhail, D.; Abel, A.D. Temporal and Topographical Changes in Theta Power between Middle Childhood and Adolescence during Sentence Comprehension. *Dev. Cogn. Neurosci.* **2022**, *53*, 101056. [CrossRef]
21. Pantelis, C.; Velakoulis, D.; McGorry, P.D.; Wood, S.J.; Suckling, J.; Phillips, L.J.; Yung, A.R.; Bullmore, E.T.; Brewer, W.; Soulsby, B.; et al. Neuroanatomical Abnormalities before and after Onset of Psychosis: A Cross-Sectional and Longitudinal MRI Comparison. *Lancet* **2003**, *361*, 281–288. [CrossRef] [PubMed]
22. Ballester, P.L.; Romano, M.T.; de Azevedo Cardoso, T.; Hassel, S.; Strother, S.C.; Kennedy, S.H.; Frey, B.N. Brain Age in Mood and Psychotic Disorders: A Systematic Review and Meta-analysis. *Acta Psychiatr. Scand* **2022**, *145*, 42–55. [CrossRef] [PubMed]

brain sciences

MDPI

Article

Age-Related Differences in Prestimulus EEG Affect ERPs and Behaviour in the Equiprobable Go/NoGo Task

Robert J. Barry *, Frances M. De Blasio, Adam R. Clarke, Alexander T. Duda and Beckett S. Munford

Brain & Behaviour Research Institute, School of Psychology, University of Wollongong, Wollongong, NSW 2522, Australia; francesd@uow.edu.au (F.M.D.B.); aclarke@uow.edu.au (A.R.C.); aduda@uow.edu.au (A.T.D.); rm798@uowmail.edu.au (B.S.M.)

* Correspondence: rbarry@uow.edu.au

Abstract: Detailed studies of the equiprobable auditory Go/NoGo task have allowed for the development of a sequential-processing model of the perceptual and cognitive processes involved. These processes are reflected in various components differentiating the Go and NoGo event-related potentials (ERPs). It has long been established that electroencephalography (EEG) changes through normal lifespan development. It is also known that ERPs and behaviour in the equiprobable auditory Go/NoGo task change from children to young adults, and again in older adults. Here, we provide a novel examination of links between in-task prestimulus EEG, poststimulus ERPs, and behaviour in three gender-matched groups: children (8–12 years), young adults (18–24 years), and older adults (59–74 years). We used a frequency Principal Component Analysis (f-PCA) to estimate prestimulus EEG components and a temporal Principal Component Analysis (t-PCA) to separately estimate poststimulus ERP Go and NoGo components in each age group to avoid misallocation of variance. The links between EEG components, ERP components, and behavioural measures differed markedly between the groups. The young adults performed best and accomplished this with the simplest EEG–ERP–behaviour brain dynamics pattern. The children performed worst, and this was reflected in the most complex brain dynamics pattern. The older adults showed some reduction in performance, reflected in an EEG–ERP–behaviour pattern with intermediate complexity between those of the children and young adults. These novel brain dynamics patterns hold promise for future developmental research.

Keywords: prestimulus EEG components; poststimulus ERP components; behaviour; temporal and frequency PCA; developmental brain dynamics

1. Introduction

The equiprobable Go/NoGo task is an interesting paradigm that bridges the active auditory Oddball and Go/NoGo tasks, which are traditional two-choice reaction time tasks. In the former, the target stimulus is presented with low probability (commonly 20%)—making it the unusual or "oddball" stimulus. In the latter, the target stimulus has high probability (commonly 80%)—making its response prepotent and requiring its inhibition for the correct withholding of a response to the rarer NoGo stimulus. The equiprobable task presents each stimulus at 50% probability and reduces the demand for active inhibition. It also benefits exploration of the competing Go/NoGo processing chains by providing equal numbers of stimulus–response pairings. As a simple two-choice reaction time task, it can be used with participants of a wide range of ages across the lifespan [1].

With neurotypical adults, Barry and De Blasio [2] developed a processing schema linking event-related potential (ERP) components, derived via temporal Principal Component Analysis (t-PCA), to the perceptual and cognitive processing involved in completing the equiprobable auditory Go/NoGo task. In general, the initial stimulus evaluation processing stages (marked by the early P1/N1 components) are similar in both the Go and NoGo processing chains. When the stimuli are identified as Go or NoGo, marked by P2/N2 components, the processing chains diverge, as the Go imperative initiates the active response (marked by P3b) and the NoGo processing (marked by P3a) winds down without active responding. Subsequent components may reflect evaluative processing of that decision. This general schema has now been confirmed in young and older adults [1] and explored in children [3,4]. It has also been explored in relation to development between children and adults [5] and in younger and older children [6]. A current version of the schema is shown in Figure 1.

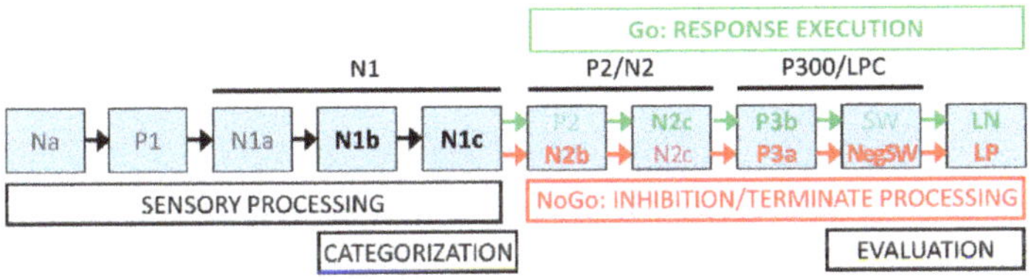

Figure 1. A processing schema for the equiprobable auditory Go/NoGo task. Sensory processing of the auditory stimulus is marked by the early P1 and N1 subcomponents as information is extracted to identify it as Go or NoGo. Categorisation is marked by P2/N2 components, beginning separate processing chains for Go (marked by P3b and the RT response) or NoGo (marked by P3a and termination of processing). Both chains may be followed by one or more slow waves (SWs) indicating evaluation of the stimulus/response pairing. Component labels are faded to indicate those not always seen. LPC = late positive complex; NegSW = negative SW; LN = late negativity; LP = late positivity.

The development of an EEG over the lifespan has been well documented, with largely consistent results regardless of measure (power or amplitude). This is characterised by a reduction in slow wave activity below 10 Hz (delta, theta, and low alpha), with an increase in faster waveforms (high alpha and beta) from childhood to adulthood [7–12]. Healthy older adults generally show reduced delta and theta, some reduction in alpha, and increased beta activity [13–16]. These studies have explored development in the resting state (with eyes open and/or closed), without any task involved. We have previously noted that the EEG power/frequency distribution differs from rest to task. Karamacoska et al. [17,18] reported that the band powers in delta, theta, alpha, and beta all increased in young adults from the resting, eyes-open condition to prestimulus activity during the equiprobable Go/NoGo task, although we are unaware of similar investigations in child or older adult populations. We are also unaware of developmental information on EEG *during* the Go/NoGo or similar cognitive tasks across the lifespan. However, we have observed age-related global reductions in prestimulus delta and theta [19], as well as focal reductions in alpha and increases in beta [20] in healthy older adults compared with young adults within the equiprobable auditory Go/NoGo task, broadly consistent with the changes reported in the resting state.

In this study, we expand upon our earlier work by examining ERP components and behavioural efficiency from a developmental perspective across three groups: neurotypical children, young adults, and older adults. The novel aspect of this study is our investigation of the relationships between the ongoing intrinsic EEG activity present at stimulus onset and the subsequent ERP components observed in the task. Genesis of the early exogenous ERP components has been related to phase locking in the ongoing EEG, while the later endogenous components have been related to differential evoked brain activity in response to Go vs. NoGo stimuli [21]. These generative processes indicate the important role that EEG in the immediate prestimulus period plays in determining the ERP components and task performance. We examine EEG from the relatively quiet prestimulus period in the equiprobable Go/NoGo task, avoiding contamination from the major event-related changes occurring in the poststimulus period. We add to the novelty of this study by applying a data-driven approach to EEG measurement, using a frequency Principal Components Analysis (f-PCA) to extract the natural frequency components [15]. f-PCA is a simple extension of t-PCA, in which the ERP waveform (voltage vs. time) is replaced by the EEG frequency spectrum (voltage vs. frequency). The other parameters (electrodes, conditions, and participants) are identical. The data sets involved are analogous, differing only in variable labels; these labels do not impact the mathematics involved, and the PCA methods are identical. f-PCA is implemented here with improvements from our earlier proof-of-concept study within the brain dynamics context [22]. The extracted components are data-driven and avoid issues of arbitrary frequency band limits common in the literature [15].

The current exploratory study maps the links between the prestimulus EEG activity immediately before Go and NoGo stimulus onset, the Go and NoGo ERP components subsequently occurring, and the associated behavioural outcomes within this task. We examine these linkages in three groups, children, young adults, and older adults, to obtain a first look at developmental changes across the lifespan in the brain dynamics involved in this cognitive task.

2. Materials and Methods

2.1. Participants

The current study assessed 60 self-reported healthy participants, drawn evenly from three age groups, child, young adult, and older adult; the group characteristics are reported in Table 1. The participants were initially recruited and participated voluntarily in separate studies, and their raw archival data were re-analysed here with a new processing pipeline that was applied consistently across all participants. Informed consent was obtained from each volunteer (adult samples) or their parent/guardian (child sample), and the protocols were in accordance with the Declaration of Helsinki and were approved by the joint University of Wollongong/South East Sydney and Illawarra Area Health Service Human Research Ethics Committee.

Table 1. Participant group characteristics.

		Group		
Characteristic	Measure	Child	Young	Older
Sample size	*n*	20	20	20
Biological sex	Female:Male	15:5	15:5	15:5
Handedness	Right:Left	17:3	20:0	20:0
Age (years)	Range	8–12	18–24	59–74
	M (*SD*)	9.75 (1.33)	19.90 (1.62)	67.75 (4.45)

All participants self-reported their handedness, abstinence from caffeine and tobacco for a minimum of 2.5 h prior to their testing session, and their health status. Participants who reported prior head trauma resulting in unconsciousness, epileptic seizures, and prior or current psychiatric illness and/or psychoactive drug use were excluded, as were those found to have high error rates and those with high levels of EEG artefacts. In addition, the child sample was screened for learning disabilities and low IQ scores (<85), and the older adults were screened for possible cognitive impairment (Rowland Universal Dementia Assessment Scale [RUDAS] score $\leq$ 22; [23]). As seen in Table 1, care was taken to minimise confounds of biological sex and handedness, resulting in a final sample size of 20 participants per group.

2.2. Task and Procedure

Following informed consent and screening, participants were fitted with electrophysiological recording equipment and sat in a quiet and dimly lit testing room. Continuous EEG was recorded while participants completed a battery of tasks, some of which differed between the groups, although all participants completed at least two blocks of an equiprobable auditory Go/NoGo task, the task assessed in this study. The stimuli were 1000 Hz and

1500 Hz tones, each of 50 ms duration; these were inclusive of a 5 ms rise/fall and played at 60 dB SPL in the child and young adult samples but included a rise/fall of 15 ms and were played at 70 dB SPL in the older adult sample to ensure tone discriminability due to the link between age-related hearing decline and cognition [24,25]. All participants received the stimuli with a fixed onset asynchrony (SOA) of 1100 ms, and each of the assessed blocks contained 75 Go and 75 NoGo tones that were shuffled to obtain a random presentation order that varied between participants. The tone assigned as the Go stimulus was held consistent between the blocks and was counterbalanced across participants within the child and young adult samples; however, this designation alternated between the blocks within the older adult sample, and the tone assigned as the first Go stimulus was counterbalanced within this group. Participants were instructed to respond to the designated Go tone with a button-press response using their dominant hand, and brief breaks were offered between the blocks to minimise fatigue.

2.3. Electrophysiological Recording

For all participants, EEG data were recorded from 19 scalp sites (Fp1, Fp2, F7, F3, Fz, F4, F8, T7, C3, Cz, C4, T8, P7, P3, Pz, P4, P8, O1, and O2), and electro-ocular (EOG) data were recorded from four facial electrodes positioned above/below the left eye to capture vertical activity and beyond the outer canthus of each eye to capture horizontal activity. These data were recorded at 512 Hz via a 16-bit A/D system (AMLAB II) with a linked ear reference for the child (data from 0.03 to 35 Hz) and young adult (data from 0.15 to 30 Hz) groups. The older adult data (DC to 30 Hz) were recorded at 1000 Hz using a Neuroscan Synamps 2 amplifier with Neuroscan Acquire software (Compumedics, Version 4.3.1), with the left ear serving as the online reference and the right ear recorded as a separate channel. All electrodes were tin, and care was taken to keep impedances below 10 kΩ.

2.4. Data Quantification

All data processing was completed in MATLAB (The Mathworks; Version 9.7.0.1190202, R2019b), using EEGLAB (version 2022.0) [26], the ERP PCA Toolkit (version 2.891) [27], and custom scripts, with the processing pipeline applied uniformly across the groups unless otherwise specified.

The raw continuous EEG data for each participant were first bandpass-filtered (0.1–30 Hz, zero-phase shift) to ensure uniformity in their spectral range, and the older adult data were re-referenced to the digital equivalent of linked ears. Non-overlapping epochs were extracted at −500 to 600 ms relative to stimulus onset for valid Go and NoGo trials. Valid Go trials were those with a button-press response within the 100–600 ms poststimulus period, and valid NoGo trials were those with no response recorded within the 1100 ms SOA period following NoGo stimulus onset; these criteria were used to extract the error rates for behavioural analysis, with the total Go errors further subdivided into errors of omission (i.e., no response recorded within SOA period), fast RT errors (i.e., a Go trial with RT $\leq$ 100 ms), and slow RT errors (i.e., a Go trial with RT $\geq$ 600 ms, thus overlapping the prestimulus period of the subsequent trial).

The epochs were manually visualised in EEGLAB, and those with identifiable flatlines and dropouts in any of the four EOG channels were rejected. The Gratton et al. [28] eye movement correction procedure was then implemented using the MATLAB function gratton_emcp.m (retrieved from https://github.com/kylemath/MathewsonMatlabTools/blob/master/EEG_analysis/gratton_emcp.m; accessed on 21 December 2023). Following EOG correction, the epochs were again manually visualised; epochs with residual EOG artefacts were rejected, and poor-quality scalp channels were interpolated using EEGLAB's spherical interpolation function. Although this may cause "ghosting" issues from rank reduction in an Independent Component Analysis (ICA) [29], it does not pose a problem for our PCAs here because we selected only components above a variance threshold of 1.5%. No more than 3 of the 19 channels (15.8%) were interpolated for any individual participant, and the interpolation data are summarised in Table 2 for each group. The EEG data for the older adults were then down-sampled to 512 Hz, to match the sample rate in the child and young groups.

Table 2. Data characteristics for each participant group.

Data		Group		
		Child	Young	Older
Channel Interpolations	Range	0–3	0–3	0–3
	M (SD)	1.30 (1.08)	1.35 (1.04)	1.10 (1.02)
Assessed epochs in each condition	Range	66–134	79–149	81–146
	M (SD)	102.85 (18.10)	121.60 (16.31)	127.35 (17.85)

The EOG-corrected epochs for all participants and groups were baseline-corrected (−100 to 0 ms), and a three-step automatic artefact rejection procedure was run similar to [30]; this process rejected epochs detected as having extreme amplitudes > 150 μV, voltage jumps > 50 μV between datapoints, and those with a maximum voltage change < 0.05 μV in any 100 ms period. Finally, the remaining trials with a button-press response within their prestimulus period were identified and rejected to remove contamination of the prestimulus spectra. To maintain the signal-to-noise ratio within subjects between conditions, the number of accepted epochs was matched between Go and NoGo. This number was set (within subject) by the condition with the lower number of accepted trials, and the corresponding number of trials was then randomly selected for analysis in the alternate condition. That is, if there were 61 Go and 68 NoGo epochs, all of the Go epochs were assessed and 61 epochs were randomly selected for analysis in NoGo. Table 2 presents a summary of the number of epochs assessed in this study for each group. From the assessed Go trials, the mean reaction time (RT) and RT variability (standard deviation in RT) data were extracted for each participant and stored with the Go and NoGo error rate data for later behavioural analysis.

To quantify the natural prestimulus spectral components, the prestimulus period (−500 to 0 ms) of each accepted epoch was extracted and DC-corrected, followed by the application of a 10% Hanning window. The windowed data were zero-padded to 512 points and submitted to Discrete Fourier Transformation (DFT), yielding single-sided prestimulus spectral amplitudes with 1 Hz resolution after corrections for the use of the Hanning window (correction factor of 1.05) and zero-padding (correction factor of 2). The DC-30 Hz within-subject mean spectra were then derived separately for Go and NoGo and submitted to f-PCA using the ERP PCA Toolkit; separate PCAs were conducted for each group (child, young, and older), and each included the prestimulus spectra for Go and NoGo given the equiprobable nature of the task and the observed consistency in the spectra between conditions. For each of the three PCAs, the input consisted of 760 cases (20 participants × 19 channels × 2 conditions) and 31 variables (DC-30 Hz at 1 Hz resolution), providing a case-to-variable ratio of 24.5. Following the recommendations in [15], PCAs were conducted with the covariance matrix, Kaiser normalisation, and unrestricted Promax rotation of 31 components. Frequency components accounting for ≥1.5% variance were considered for analysis, and the within-subject global mean peak amplitude was extracted across the Go and NoGo conditions for those identified.

To quantify the ERP components, the ERP period (−100 to 600 ms) was extracted from the EOG-corrected epochs identified for analysis, and the mean Go and NoGo ERPs were derived within subjects. The average ERPs were half-sampled to 256 Hz and submitted to separate t-PCAs for each group (child, young, and older) and condition (Go and NoGo). For each of the six PCAs, the input comprised 380 cases (20 participants × 19 channels × 1 condition) and 179 variables (−100 to 600 ms at 3.9 ms resolution), providing a case-to-variable ratio of 2.1. PCAs were conducted with the covariance matrix, Kaiser normalisation, and unrestricted Varimax rotation of 179 components following the recommendations in [31]. Temporal components accounting for ≥1.5% variance were considered for analysis, and the within-subject global mean peak amplitude was extracted separately for each Go and NoGo component identified.

2.5. Statistical Analysis

All analyses were undertaken in JASP (version 0.18.3) [32]. Spearman's rank-order correlations (rho, ρ) were conducted separately for each group, and each assessed the relationships between the global mean peak amplitudes for each identified prestimulus f-PCA EEG component, t-PCA ERP component, and behavioural measures. All correlations had 18 degrees of freedom and were assessed with two-tailed significance. An alpha level of 0.05 was used to detect significant associations, and due to the exploratory nature of this investigation, near significant effects ($0.05 < p \leq 0.10$) are also reported. The 95% confidence intervals (CIs) are reported for the Spearman correlations [33].

3. Results

3.1. Child Group

Behaviourally, the child group had an average of 6.3% commission errors in NoGo and a 13.5% total error rate in Go, which comprised 3.1% omission errors, 0.5% fast RT errors, and 9.9% slow RT errors. Their mean RT was 389 ms (SD = 38), and their RT variability was 85 ms (SD = 16).

The child group prestimulus EEG spectra at the midline sites (Fz, Cz, and Pz) are shown in Figure 2A. There is a clear parietal peak at 2 Hz in the delta band and a marked parietal peak at 9 Hz in the alpha band, with activity declining over the beta range and becoming frontal. The prestimulus activities before the Go and NoGo stimuli were very similar, with the absolute maximum amplitude difference across frequency and sites being 0.286 μV. These data were submitted to a combined f-PCA, which yielded two delta components (D1 at 1 Hz and D2 at 2 Hz), one alpha component (A1 at 9 Hz), and three beta components (B1 at 14 Hz, B2 at 18 Hz, and B3 at 24 Hz), each carrying at least 1.5% variance, together carrying 92.7% of the spectral variance. Their scaled factor loadings and topographical headmaps are shown in Figure 2B.

GM ERPs from the child group are shown in Figure 3A, with Go ERPs at the midline sites (Fz, Cz, and Pz) on the left and NoGo on the right. Clear P1 and N1 components are apparent in both sets, followed by a large frontal N2, with marked morphological Go/NoGo differences most apparent from approximately 250 ms. The Go ERP f-PCA yielded eight components carrying more than 1.5% variance each, for a total of 94.5% of the ERP variance. These were identified as Na, P1/N1a, and N1b, followed by N1c/P2, N2b, N2c, P3b, and SW. Their scaled factor loadings, component information (variance and peak latency), and topographic headmaps are shown in Figure 3B. The NoGo ERP f-PCA also yielded eight components carrying at least 1.5% variance each, for 90.4% of the total variance. These were identified as a P1/N1a, N1b, and N1c/P2, followed by N2b, N2c, P3a, P3b, and SW. Their scaled factor loadings, component information (variance and peak latency), and topographic headmaps are shown in Figure 3C.

Child Global Amplitude Correlation Analyses

Spearman correlations indicated that Go Na was somewhat inversely affected by A1 (ρ = 0.41, p = 0.07, CI = [−0.06, 0.73]) and B1 (ρ = 0.42, p = 0.07, CI = [−0.05, 0.74]); that is, when the A1 and B1 amplitudes were larger, the Na amplitudes tended to be more positive (i.e., smaller). The global positivity of the Go P1/N1a component was significantly affected directly by D1 (ρ = 0.55, p = 0.01, CI = [0.11, 0.81]), inversely by B2 (ρ = −0.46, p = 0.04, CI = [−0.76, 0.00]), and somewhat affected inversely by B3 (ρ = −0.41, p = 0.08, CI = [−0.73, 0.06]). Go N1c/P2 positivity was inversely affected by D2 (ρ = −0.44, p = 0.05, CI = [−0.75, 0.02]) and A1 (ρ = −0.58, p = 0.01, CI = [−0.83, −0.14]) and somewhat by B1 (ρ = −0.39, p = 0.09, CI = [−0.72, 0.08]). Go N2b negativity was somewhat affected inversely by A1 (ρ = 0.43, p = 0.06, CI = [−0.03, 0.74]) and directly by B2 (ρ = −0.42, p = 0.07, CI = [−0.74, 0.05]). Go P3b positivity was inversely affected by D2 (ρ = −0.45, p = 0.05,

CI = [−0.76, 0.01]), while Go SW positivity was directly affected by D2 ($\rho = 0.45$, $p = 0.05$, CI = [−0.01, 0.76]) and B1 ($\rho = 0.58$, $p = 0.01$, CI = [0.15, 0.83]). Negativity in the NoGo N1b was somewhat directly affected by B3 ($\rho = -0.39$, $p = 0.09$, CI = [−0.72, 0.08]), N2b was inversely affected by A1 ($\rho = 0.65$, $p = 0.01$, CI = [0.25, 0.86]), and N2c was directly affected by B2 ($\rho = -0.48$, $p = 0.03$, CI = [−0.77, −0.02]). Positivity in the NoGo P3b was somewhat directly affected by D2 ($\rho = 0.41$, $p = 0.08$, CI = [−0.06, 0.73]), while the SW positivity was inversely affected by D1 ($\rho = -0.54$, $p = 0.02$, CI = [−0.80, −0.09]) and directly by D2 ($\rho = 0.55$, $p = 0.01$, CI = [0.11, 0.81]).

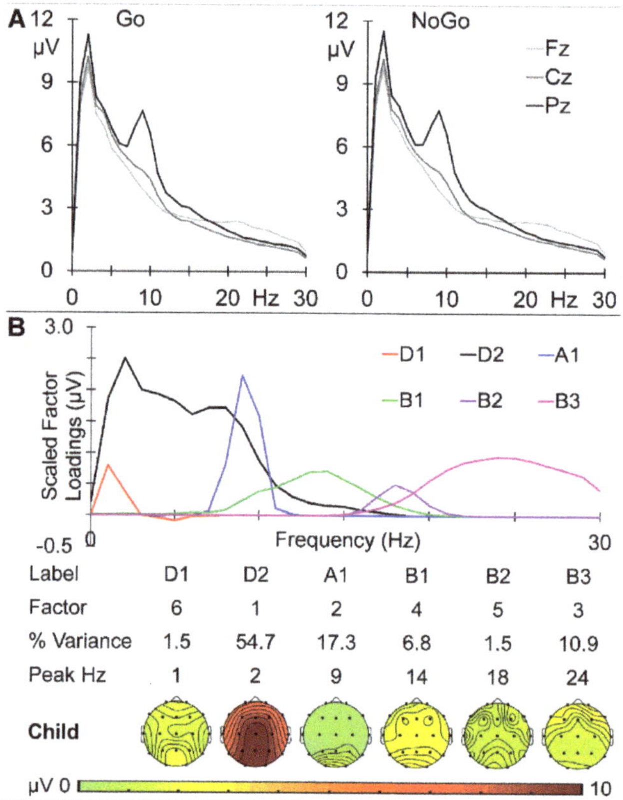

Figure 2. Child prestimulus spectra. (**A**) Amplitude as a function of frequency at the midline sites before Go (**left**) and NoGo (**right**) stimuli. (**B**) f-PCA outputs showing the scaled factor loadings as a function of frequency above the components in temporal order, together with the % variance carried and their peak frequency. Below this information is their topographic distribution with voltage scale.

Child behaviour measures showed six links with their prestimulus EEG components. Go RT variability was somewhat affected inversely by A1 ($\rho = -0.41$, $p = 0.07$, CI = [−0.73, 0.06]). Go omissions and Go total errors were somewhat affected directly by D2 ($\rho = 0.42$, $p = 0.07$, CI = [−0.05, 0.74] and $\rho = 0.41$, $p = 0.07$, CI = [−0.06, 0.73], respectively). NoGo commissions were significantly affected directly by D2 ($\rho = 0.51$, $p = 0.02$, CI = [0.06, 0.79]) and B3 ($\rho = 0.45$, $p = 0.05$, CI = [−0.01, 0.76]) and somewhat affected inversely by D1 ($\rho = -0.42$, $p = 0.06$, CI = [−0.74, 0.05]).

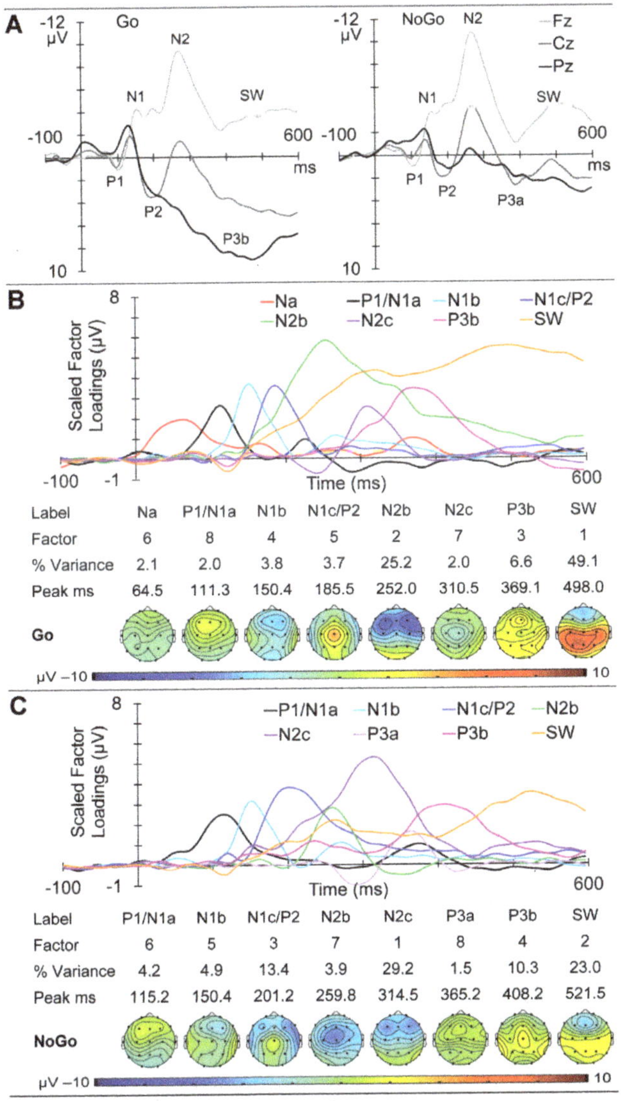

Figure 3. Child ERP results. (**A**) Morphology of the Go and NoGo responses at the midline sites as a function of time from stimulus onset. (**B**) t-PCA outcomes for Go: scaled factor loadings are shown above the component information and component headmaps. (**C**) NoGo component results.

ERP–behaviour links were also apparent. Go RT variability was somewhat directly associated with Go N2b ($\rho = -0.40$, $p = 0.08$, CI = [−0.72, 0.07]) and inversely with Go N2c ($\rho = 0.38$, $p = 0.10$, CI = [−0.09, 0.71]) and Go P3b ($\rho = -0.42$, $p = 0.07$, CI = [−0.73, 0.05]). Go omission errors were inversely related to Go P3b ($\rho = -0.59$, $p = 0.01$, CI = [−0.83, −0.16]) and somewhat so to Go P1/N1a ($\rho = -0.42$, $p = 0.07$, CI = [−0.74, 0.05]). Slow RT errors and total Go errors were each inversely related to Go P3b ($\rho = -0.52$, $p = 0.02$, CI = [−0.79, −0.07] and $\rho = -0.62$, $p = 0.01$, CI = [−0.85, −0.21], respectively).

Mean RT was inversely related to NoGo N2b negativity ($\rho = 0.45$, $p = 0.05$, CI = [−0.01, 0.76]) and somewhat inversely related to NoGo P3a positivity ($\rho = -0.43$, $p = 0.06$, CI = [−0.74, 0.04]), while RT variability was somewhat directly related to NoGo N2b negativity ($\rho = -0.41$,

p = 0.07, CI = [−0.73, 0.06]). Go omissions were somewhat directly related to NoGo N2c negativity (ρ = −0.40, p = 0.08, CI = [−0.72, 0.07]) and inversely to P3a positivity (ρ = −0.40, p = 0.08, CI = [−0.72, 0.07]). Fast RT errors were directly related to NoGo N2c negativity (ρ = −0.51, p = 0.02, CI = [−0.79, −0.06]), while slow RT errors and Go total errors were inversely related to NoGo P3a (ρ = −0.50, p = 0.03, CI = [−0.78, −0.04] and ρ = −0.48, p = 0.03, CI = [−0.77, −0.02], respectively). NoGo commission errors were somewhat directly related to NoGo P3b (ρ = 0.39, p = 0.09, CI = [−0.08, 0.72]) and showed a significant inverse association with Go P3b (ρ = −0.54, p = 0.01, CI = [−0.80, −0.09]) and near-significant direct associations with the negativity in Go Na (ρ = −0.40, p = 0.08, CI = [−0.73, 0.07]) and Go N2c (ρ = −0.44, p = 0.05, CI = [−0.75, 0.03]). These links are shown in Figure 4.

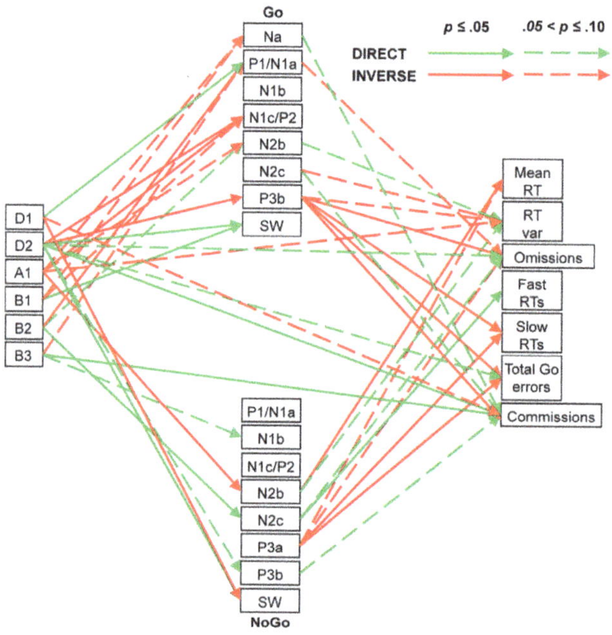

Figure 4. Child dynamics summary showing links between prestimulus frequency components (**left**), ERP components (centre: Go above NoGo), and behavioural measures (**right**).

3.2. Young Adult Group

The young group averaged 1.3% NoGo commission errors and a total Go error rate of 6.9%, reflecting 2.1% Go omission errors, 0.1% fast RT errors, and 4.6% slow RT errors. Their mean RT was 352 ms (SD = 57), and their RT variability was 71 ms (SD = 13).

The GM young group prestimulus EEG spectra at the midline sites are shown in Figure 5A. There is a clear centroparietal peak at 2 Hz in the delta band and a marked parietal peak at 9 Hz in the alpha band, with activity declining over the beta range. The prestimulus activities before the Go and NoGo stimuli were very similar, with the absolute maximum amplitude difference across sites and frequency being 0.203 μV. A combined f-PCA yielded one delta component (D1 at 2 Hz), one theta component (T1 at 6 Hz), four alpha components (A1 at 8 Hz, A2 at 9 Hz, A3 at 10 Hz, and A4 at 11 Hz), and two beta components (B1 at 16 Hz and B2 at 25 Hz), each carrying at least 1.5% variance and together carrying 94.6% of the spectral variance. Their scaled factor loadings and topographic headmaps are shown in Figure 5B.

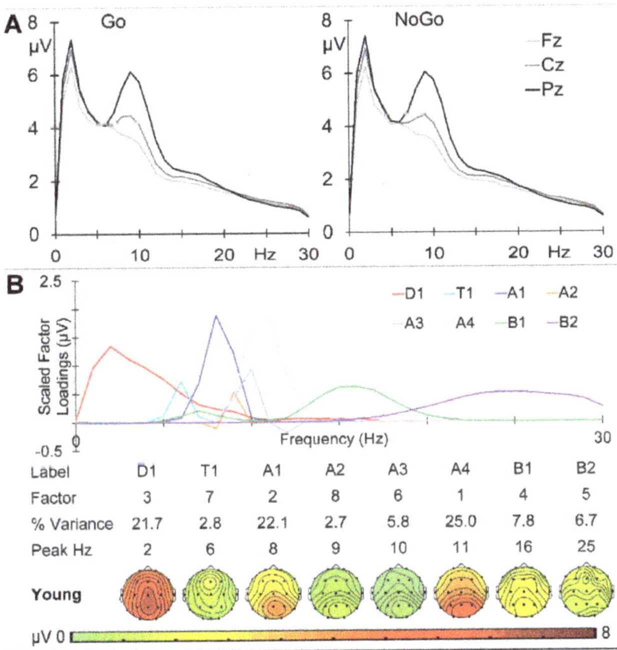

Figure 5. Young adult prestimulus spectra. (**A**) Amplitude at the midline sites before Go (**left**) and NoGo (**right**) stimuli. (**B**) f-PCA outputs with the scaled factor loadings above the components in temporal order, together with their % variance, peak frequency, and topographic headmaps.

GM ERPs from the young group are shown in Figure 6A, with Go ERPs at the midline sites on the left and NoGo on the right. Clear P1 and N1 components are apparent in both sets, with a larger Go P1 and later NoGo N1. Subsequent Go/NoGo morphology differences are apparent from approximately 150 ms. The Go ERPs yielded five components carrying more than 1.5% variance each, for a total of 90.3% of the ERP variance. These were identified as N1b, followed by N1c, P2/N2c, P3b, and SW. Their scaled factor loadings, component information, and topographic headmaps are shown in Figure 6B. The NoGo ERPs yielded seven components carrying more than 1.5% variance each, for a total of 88.8% of the overall variance. These were identified as a P1, N1a, and N1b, followed by a P2/N2b, N2c, P3a, and SW. Their scaled factor loadings, component information, and topographic headmaps are shown in Figure 6C.

Young Global Amplitude Correlation Analyses

In the young group, Spearman correlations indicated that Go P2/N2c positivity was somewhat inversely affected by A4 ($\rho = -0.42$, $p = 0.07$, CI = [−0.74, 0.05]), while Go SW positivity was directly affected by D1 ($\rho = 0.45$, $p = 0.05$, CI = [−0.01, 0.75]) and somewhat affected by A1 ($\rho = 0.39$, $p = 0.09$, CI = [−0.08, 0.72]). NoGo P1 was inversely affected by T1 ($\rho = -0.40$, $p = 0.08$, CI = [−0.73, 0.07]), while NoGo P2/N2b positivity was directly affected by A2 ($\rho = 0.47$, $p = 0.04$, CI = [0.01, 0.77]) and A3 ($\rho = 0.44$, $p = 0.05$, CI = [−0.02, 0.75]).

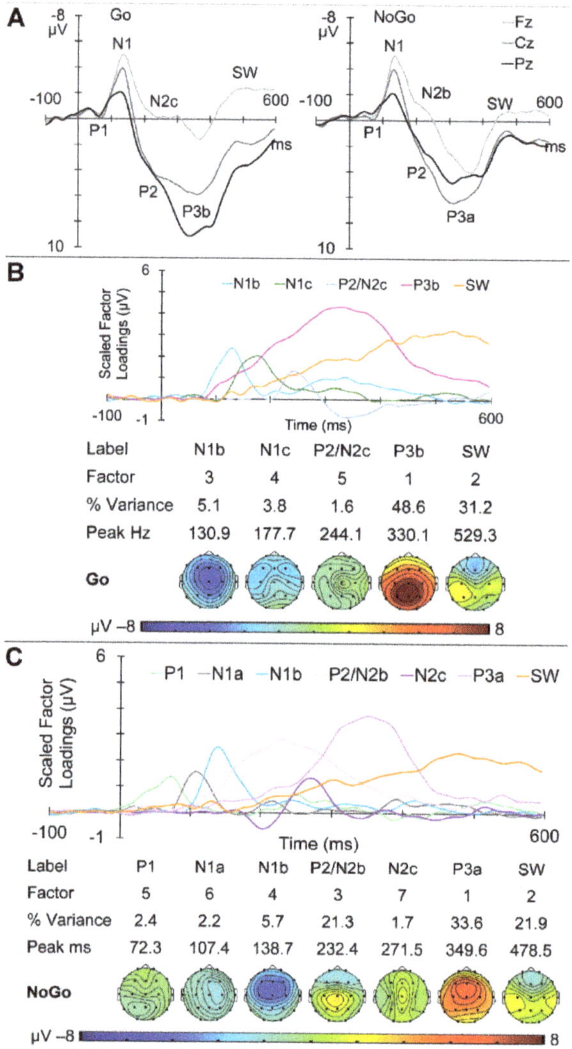

Figure 6. Young adult ERP results. (**A**) Go and NoGo response morphology at the midline sites as a function of time from stimulus onset. (**B**) t-PCA outcomes for Go: scaled factor loadings above the component information and component headmaps. (**C**) NoGo component results.

In terms of EEG–behaviour links, Go RT variability was directly affected by D1 ($\rho = 0.47$, $p = 0.04$, CI = [0.01, 0.77]) and somewhat also by T1 ($\rho = 0.40$, $p = 0.08$, CI = [−0.07, 0.73]). While none of the Go behavioural measures showed associations with the Go ERP components, NoGo commission errors were directly linked to NoGo N2c ($\rho = -0.45$, $p = 0.05$, CI = [−0.76, 0.01]). Go RT variability was directly associated with NoGo N2c ($\rho = -0.58$, $p = 0.01$, CI = [−0.82, −0.14]), and inverse associations were seen between NoGo P3a positivity and Go fast RT errors ($\rho = -0.45$, $p = 0.05$, CI = [−0.75, 0.02]), Go slow RT errors ($\rho = -0.59$, $p = 0.01$, CI = [−0.83, −0.16]), and Go total errors ($\rho = -0.46$, $p = 0.04$, CI = [−0.76, 0.00]). These links are shown in Figure 7.

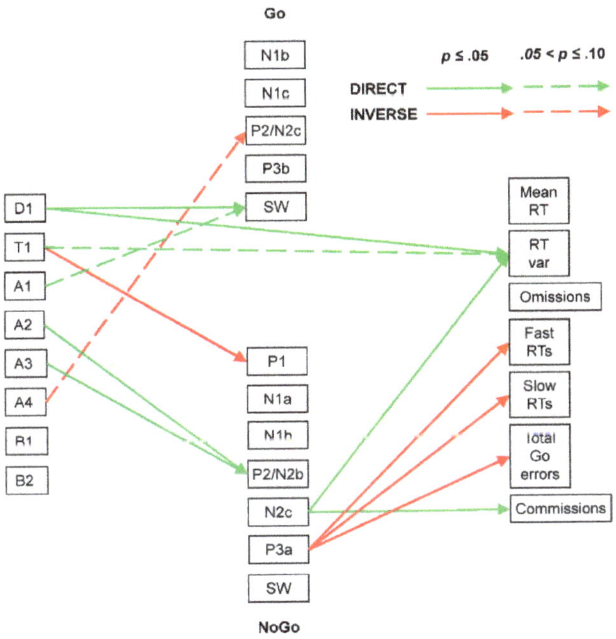

Figure 7. Young adult brain dynamics summary linking prestimulus frequency components (**left**), ERP components (centre: Go above NoGo), and behavioural measures (**right**). Note the simpler summary than that for the child group.

3.3. Older Group

The older group averaged 3.4% NoGo commission errors and a total of 8.2% Go errors, comprising 2.2% Go omission errors, 0.1% fast RT errors, and 5.9% slow RT errors. Their mean RT was 366 ms (SD = 38), with a RT variability of 83 ms (SD = 11).

The GM older group prestimulus EEG spectra at the midline sites are shown in Figure 8A. There is a clear peak at 2 Hz in the delta band and a marked parietal peak at 9 Hz in the alpha band, with activity declining over the beta range. The prestimulus activities before Go and NoGo stimuli were very similar, with the absolute maximum amplitude difference across frequency and sites being 0.175 µV. A combined f-PCA yielded one delta component (D1 at 2 Hz), two alpha components (A1 at 8 Hz and A2 at 10 Hz), and three beta components (B1 at 14 Hz, B2 at 20 Hz, and B3 at 25 Hz), each carrying at least 1.5% variance. Together these carried 90.3% of the spectral variance. Their scaled factor loadings, component information, and topographical headmaps are shown in Figure 8B.

GM ERPs from the older group are shown in Figure 9A, with Go ERPs at the midline sites on the left and NoGo on the right. Clear P1 and N1 components are apparent in both sets, with subsequent morphological Go/NoGo differences apparent from approximately 250 ms. The Go ERPs yielded seven components carrying more than 1.5% variance each, for a total of 89.2% of the ERP variance. These were identified as P1, N1a, and N1b, followed by P2, P3a, P3b, and SW. Their scaled factor loadings and topographic headmaps are shown in Figure 9B. The NoGo ERPs yielded six components carrying more than 1.5% variance each, for a total of 87.1% of the total variance. These were identified as P1, N1a/b, and N1c, followed by P2, P3a, and SW. Their scaled factor loadings, component information, and topographic headmaps are shown in Figure 9C.

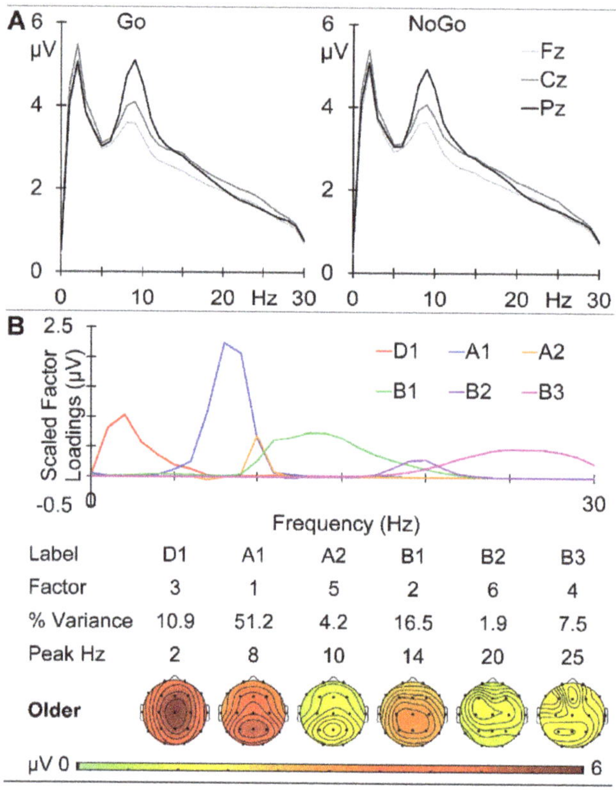

Figure 8. Older adult prestimulus spectra. (**A**) Prestimulus amplitude spectra at the midline sites before Go (**left**) and NoGo (**right**). (**B**) f-PCA outputs with the scaled factor loadings above the components in temporal order, together with their % variance, peak frequency and topographic headmaps.

Older Global Amplitude Correlation Analyses

Spearman correlations in the older group indicated that Go N1a negativity was directly affected somewhat by A2 ($\rho = -0.42$, $p = 0.07$, CI = [−0.74, 0.05]) and significantly by B1 ($\rho = -0.57$, $p = 0.01$, CI = [−0.82, −0.13]), while Go P3b positivity was somewhat inversely affected by D1 ($\rho = -0.42$, $p = 0.07$, CI = [−0.74, 0.05]). NoGo P1 positivity was inversely affected by A1 ($\rho = -0.51$, $p = 0.02$, CI = [−0.79, −0.06]), A2 ($\rho = -0.45$, $p = 0.05$, CI = [−0.75, 0.01]), B1 ($\rho = -0.48$, $p = 0.04$, CI = [−0.77, −0.02]), and somewhat by B2 ($\rho = -0.39$, $p = 0.09$, CI = [−0.72, 0.08]). NoGo N1a/b negativity was directly affected somewhat by A2 ($\rho = -0.40$, $p = 0.08$, CI = [−0.72, 0.07]) and significantly by B1 ($\rho = -0.47$, $p = 0.04$, CI = [−0.77, −0.01]). NoGo SW positivity was somewhat inversely affected by A2 ($\rho = -0.39$, $p = 0.09$, CI = [−0.72, 0.08]) and B1 ($\rho = -0.44$, $p = 0.06$, CI = [−0.75, 0.03]).

In the only older group's EEG–behaviour link, Go RT variability was directly affected by B3 ($\rho = 0.51$, $p = 0.02$, CI = [0.06, 0.79]).

In terms of ERP–behaviour links, near-significant inverse associations were seen between the Go SW positivity and both Go mean RT ($\rho = -0.42$, $p = 0.06$, CI = [−0.74, 0.05]) and total Go errors ($\rho = -0.38$, $p = 0.10$, CI = [−0.71, 0.09]) and between Go fast RT errors and Go N1b negativity ($\rho = 0.38$, $p = 0.06$, CI = [−0.09, 0.71]). NoGo commission errors were inversely linked to NoGo P2 positivity ($\rho = -0.52$, $p = 0.02$, CI = [−0.79, −0.07]) and showed some direct association with Go SW positivity ($\rho = 0.37$, $p = 0.10$, CI = [−0.10, 0.71]). Go mean RT was inversely associated with NoGo P3a ($\rho = -0.53$, $p = 0.02$, CI = [−0.80, −0.08]), while Go omissions were somewhat inversely associated with NoGo N1c negativity

($\rho = 0.42$, $p = 0.06$, CI = [−0.05, 0.74]), and Go fast RT errors were somewhat inversely associated with NoGo N1c negativity ($\rho = 0.38$, $p = 0.10$, CI = [−0.09, 0.71]) and NoGo P2 positivity ($\rho = -0.38$, $p = 0.10$, CI = [−0.71, 0.09]). Go slow RT errors and Go total errors were each inversely associated with NoGo P3a (each $\rho = -0.71$, $p = 0.01$, CI = [−0.89, −0.34]). These links are shown in Figure 10.

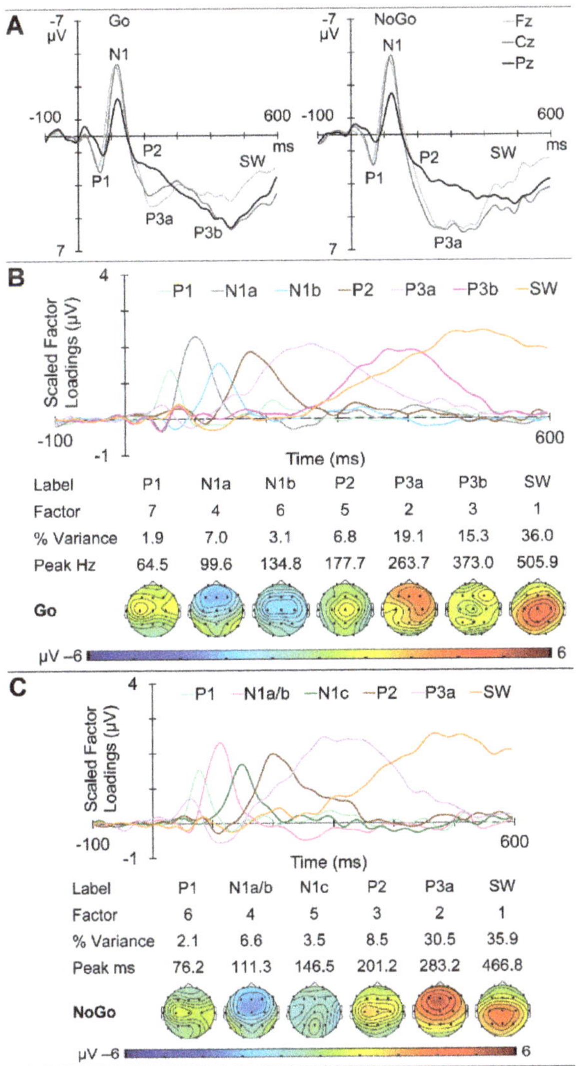

Figure 9. Older adult ERP results. (**A**) Go and NoGo response morphology at the midline sites as a function of time from stimulus onset. (**B**) t-PCA outcomes for Go: scaled factor loadings above the component information and component headmaps. (**C**) Results for NoGo components.

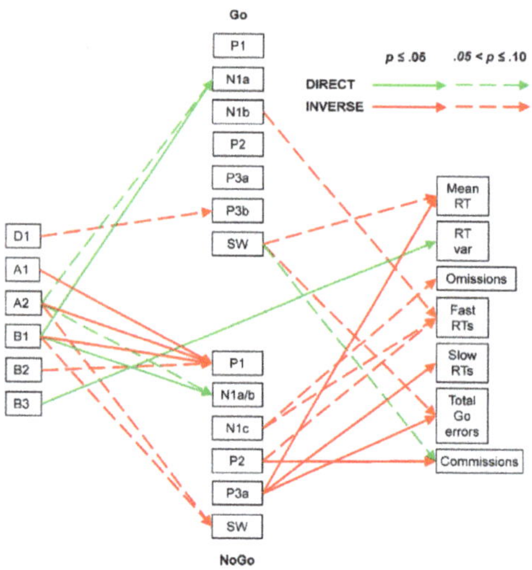

Figure 10. Older adult brain dynamics summary linking prestimulus frequency components (**left**), ERP components (centre: Go above NoGo), and behavioural measures (**right**). Note the increase in complexity compared with that in the young group's summary.

4. Discussion

This was an exploratory developmental study focussed on the brain dynamics involved in a fixed-interstimulus interval equiprobable auditory Go/NoGo task. We obtained novel prestimulus EEG measures *during* the task by using f-PCA to decompose EEG spectra obtained from the 500 ms before each stimulus onset, uncontaminated by poststimulus responses (i.e., ERPs or behaviour). t-PCA was then used to decompose the separate Go and NoGo ERP waveforms. Spearman correlations sought links between the prestimulus EEG components, ERP components, and behavioural measures. These analyses were carried out in separate groups of children, young adults, and older adults.

The sequential processing schema shown in Figure 1 was developed to help conceptualise the processing stages involved in this equiprobable auditory Go/NoGo task, and their links to ERP markers. The schema first covers the initial stages of sensory processing, during which the stimulus comes to be categorised as Go or NoGo. Subsequently, the processing splits to some extent, with the processing of the Go stimulus aimed at rapid response production and execution, and the NoGo processing aimed at inhibiting any such active response. Both these chains are followed by an evaluation of the performance in relation to the stimulus categorisation and preparation for the next stimulus. We remind readers that these stages are the simplified component peaks or "highlights" of the *parallel processing* indicated by the overlapping timing of the scaled ERP factor loadings plotted as amplitude vs. time (e.g., see Figure 3B,C). The ERP components found in each of our groups were broadly compatible with the component stages in Figure 1 and our previous reports on children [4,6], young adults [31], and older adults [1]. However, the schema has developed over the years, and previous reports have been based on differing and developing EEG/ERP processing steps and PCA techniques. The current three-group data sets have been processed using the same EEG/ERP pipeline for comparability, and our current optimised f-PCA and t-PCA techniques, allowing a coherent examination of developmental brain dynamics across the lifespan.

4.1. Changes with Age

4.1.1. EEG

The EEG data examined here were extracted from 500 ms immediately prestimulus in an equiprobable Go/NoGo task and are relatively novel in reflecting task rather than rest conditions. The dominant low-frequency activity (primarily delta) was frontal in the children and young adults (see Figures 2 and 5) but central in the older adults (Figure 8); delta amplitude reduced from some 11 to 7 to 5 µV with age. In contrast, the 9 Hz alpha peak remained parietal but reduced from approximately 7 to 6 to 5 µV over this age range, while beta showed little change over age. This systematic age reduction across the spectrum is compatible with resting-state expectations [13,16] and reflects the expected greater reduction in slow than fast oscillations. The f-PCA components broadly reflected these changes.

4.1.2. ERPs

The general ERP morphology was similar across the age groups, with P1 followed by the N1, P2, N2, P3, and SW components; NoGo showed a larger central P3a compared with parietal Go P3b. The major stand-out is the extremely large frontal N2 apparent in children, which appears larger in NoGo than Go. The parietal/central Go/NoGo P3b/P3a differences appear to reduce with age. The t-PCA results largely reflected these expected components and matched previous age-related components (e.g., in children and young adults [5] and in young and older adults [1]).

4.1.3. Behaviour

As expected, children performed relatively poorly in the task compared with the young adults, and the older adults showed some decline in performance (see [1,4]). The young adult sample showed faster mean RT and greater consistency in their responding (as indicated by their lower RT variability) relative to the child and older adult samples. This U-shaped pattern in response speed is typical of the developmental and ageing literatures [34,35] and may be attributable to the implementation of multiple strategies in this task and/or changes in brain morphology [35].

4.2. Brain Dynamics

4.2.1. Child

Behaviourally, the child group had the slowest RTs; greatest RT variability; and highest number of Go omissions, fast RT errors, slow RT errors, total RT Go errors, and NoGo commission errors. These data confirm our previous results that children find this task difficult and require active inhibition when processing the NoGo stimulus [6], which is most apparent in their large dominant N2b ERP component (see Figure 3).

The summary of our child findings (Figure 4) indicates that this difficulty is reflected in a complex array of EEG/ERP/behaviour links. The child Go sensory processing ERP components reflect both facilitation and inhibition links with delta, alpha, and beta components, where corresponding NoGo sensory components are somewhat facilitated only by beta. Go and NoGo categorisation component markers are also substantially different in their links to EEG components, with only the Go N1c/P2 positivity being linked to delta, alpha, and beta activity. In the subsequent Go processing chain, N2b is unexpected, given its traditional link with P3a, which was not large enough in the Go chain to be extracted here. Go P3b was significantly reduced by delta activity, while delta and beta activity directly facilitated the evaluatory Go SW. In the NoGo processing chain, the inhibitory marker N2b was inversely affected by alpha; the unexpected N2c/P3b components were, respectively, increased by beta and higher delta, perhaps reflecting similar impacts of higher and lower levels of attention/vigilance. NoGo SW was enhanced by higher delta and reduced by lower delta, perhaps suggesting a link between outcome evaluation, the innervation of the unexpected N2c and P3b, and lower levels of attention/vigilance.

There were many apparent links between the Go and NoGo ERP components and behaviour. The most important of these in Go linked P3b inversely with omissions, RT variability, and slow RT errors, illustrating the expected role of P3b in this condition of the task. Moreover, the unexpected NoGo P3b showed some direct association with NoGo commission errors, reiterating the link between P3b and response execution (both valid as in Go and invalid as in NoGo). The unexpected NoGo N2c linked directly to fast RT errors, pointing to the non-functional role of this component, along with P3b, in child NoGo processing. The N2b component showed (near significant) direct associations with Go RT variability *irrespective* of its stimulus condition; this component is typically seen in the NoGo processing stream in conjunction with P3a, and its links here with RT variability likely reflect immature development of these stimulus-specific linkages. Surprisingly, NoGo P3a was associated with reduced mean RT, as well as omission, slow RT, and total Go errors.

Interestingly, prestimulus EEG components themselves were also linked to behavioural outcomes in children. High delta was directly linked with Go omissions and more so with NoGo commission errors, suggestive of the effects of reduced attention on the task, while high beta was also directly linked to NoGo commission errors, perhaps suggestive of over-activation and reduced inhibitory control.

4.2.2. Young Adult

The young group had the best behavioural performance, with the shortest RT, smallest RT variability, fewest Go omissions, fewest slow RT errors, fewest total Go errors, and fewest NoGo commission errors.

The summary of their EEG/ERP/behaviour links (Figure 7) is also the simplest, suggesting an efficient implementation of brain dynamics in this paradigm. Only their theta activity affected sensory processing, with an inhibitory link between T1 and NoGo P1. Alpha (A4) inversely affected Go P2/N2c and facilitated (A2 and A3) NoGo P2/N2b, suggesting the role of focussed attention in the young categorisation of the Go/NoGo stimuli. In the NoGo processing chain, the unexpected N2c (the usual precursor of P3b) directly linked to more RT variability and commission errors, aspects of impaired NoGo processing. Surprisingly, NoGo P3a was linked to fewer fast and slow RT errors, and total Go errors.

Prestimulus EEG activity also impacted behavioural measures. Lower frequencies (D1 and T1) affected RT variability (i.e., with more low-frequency EEG, RT was more variable). This link may reflect the impact of lapses in attentional focus.

4.2.3. Older Adult

The older group had somewhat poorer behavioural performance than the young group (mean RT, RT variability, omission errors, slow RT errors, and NoGo commission errors), suggesting a reduction in performance with increasing age in adults.

Their EEG/ERP/behaviour summary (Figure 10) also indicates more complex brain dynamics than in the young group. EEG–ERP links in the Go processing stream showed that A2 and B1 directly linked to both Go N1a and NoGo N1a/b, confirming older participants' attention/vigilance involvement in the sensory-processing stage of the task. Alpha (A1 and A2) and beta (B1 and B2) prestimulus EEG components also inversely linked to NoGo P1, suggesting that similar EEG processes facilitating sensory processing in general reduced the earlier P1 stage in NoGo processing. The functional role of P1 in this paradigm is not yet well understood, but its reduction might facilitate greater N1a negativity (and associated processing) in the NoGo processing chain, perhaps due to the utilisation of a proactive (cf. reactive) cognitive control approach, or a wider mix of proactive and reactive control mechanisms across the task [20]. No prestimulus EEG components appear linked to the Go/NoGo categorisation markers in the older group, but Go P3b and NoGo SW appear inversely linked to D1 and A2/B1, respectively. This suggests that loss of attentional focus (D1) might impact Go responding (P3b), and attentive/vigilant states (A2/B1) might lead to better NoGo performance, not requiring extensive evaluation (SW).

Enhancement of Go sensory processing (N1b) was associated with fewer fast RT errors, while that of NoGo sensory processing (N1c) surprisingly appeared linked to reduced omissions and fewer fast RT errors. Go SW was linked with reduced mean RT and less total Go errors but, surprisingly, more commission errors. In the NoGo processing stream, the P2 categorisation marker was linked with reduced fast RTs and less commission errors, while the subsequent NoGo P3a was associated with lower mean RT and fewer slow RTs and total Go errors.

Only one prestimulus EEG component appeared linked to behavioural measures in the older group: high beta (B3) was directly linked to RT variability, suggesting that variation in high-activation EEG was associated with increased response variability. Here, the increased beta activity may reflect a reduction in dopaminergic function [36], which likely impacted cognitive motivation and the implementation of control strategies within this participant sample and task [37].

4.3. The Processing Schema across Ages

Our processing schema illustrated in Figure 1 generally accommodated the Go/NoGo ERP t-PCA outcomes from the three age groups and supported our expectations of similar sensory-processing and categorisation stages, leading to separate Go and NoGo processing stages and ending in similar evaluative processing, as reflected in the SW component. However, in all three groups, NoGo P3a amplitude was linked to improvements in Go behavioural measures, suggesting that this aspect should be included in future iterations of the schema. Perhaps P3a inhibition and the SW evaluation mark processing common to both Go and NoGo stimuli after the RT response has been activated or inhibited, respectively.

Deviations from the schema in the child group included Go N2b, and NoGo N2c and P3b, suggestive of poorly developed differentiation of Go versus NoGo processing and reflecting their greater difficulty and poorer performance in the task. In the older group, the appearance of Go P3a also deviated from expectations, although this has been found in some of our prior work [20]. This could reflect a "brake" on the RT production associated with the P3b and might reflect a strategic difference where the older group concentrated on accuracy rather than speed, as suggested in [1].

4.4. Limitations

Although our groups are matched on sex and handedness, they are relatively small (each $N = 20$), and this reduces confidence in the present results. However, our aim here was to present a preliminary study that could supply ideas and suggestions for future research in the developmental brain dynamics sphere. The novel EEG/ERP/behaviour summaries from each age group reflect achievement of our aim and provide many interesting linkages that should spur research on numerous fronts.

Another limitation is that the brief (50 ms duration) "beep" stimuli presented to the older group were 70 dB (cf. 60 dB) and included rise/fall times of 15 ms compared with the 5 ms used with the other groups. These changes were designed to ensure the clear audibility and differentiability of the Go and NoGo stimuli for all participants, recognising the link between age-related hearing decline and cognition [24,25]. The alternative to using identical stimuli could be explored in future studies but would require testing of hearing levels and may severely restrict the number of available older participants.

5. Conclusions

This exploratory study has shown developmental changes from children to younger adults and older adults in novel prestimulus EEG f-PCA components, ERP t-PCA components, and behaviour based on performance in an equiprobable auditory Go/NoGo task. Our framework was a t-PCA-based processing schema, which largely accommodated the ERP components and behavioural data, although the consistent unexpected connections between NoGo P3a and Go behaviours suggest the need for further development. Links between prestimulus EEG components, ERP components, and behaviour were obtained using simple Spearman correlations. These generated three age-specific summary plots that illustrate the wide range of brain dynamics involved in performing this task, and their variation throughout the lifespan.

In addition to indicating the value of the present approach, these summary plots suggest fruitful areas to explore in future studies. Such studies should include a revised processing schema better accommodating the NoGo P3a and its links to behavioural measures. More importantly, future studies should use larger samples than those available for this study in order to enhance our confidence in the links reported here. Perhaps further studies could vary age in large samples to better examine longitudinal aspects of development. Also, more-focussed studies examining portions of the EEG/ERP/behaviour ranges, such as focussing on part of the Go or NoGo processing chain, could be enlightening. Such limited studies could use the present data to generate hypotheses to be formally assessed with more powerful statistical techniques, such as multiple regression analysis. Although this exploratory study illustrates the value of obtaining extensive data and understanding from one paradigm, such as the auditory equiprobable Go/NoGo task used here, there would also be value in exploring this approach with different tasks. However, a stepwise approach, varying one parameter at a time (such as the interstimulus interval, the NoGo probability, or the stimulus modality) would help optimise understanding of the new data.

Author Contributions: Conceptualisation, R.J.B. and F.M.D.B.; Methodology, R.J.B. and F.M.D.B.; Software, F.M.D.B.; Formal Analysis, F.M.D.B.; Investigation, F.M.D.B., A.T.D. and B.S.M.; Resources, R.J.B.; Data Curation, F.M.D.B.; Writing—Original Draft Preparation, R.J.B. and F.M.D.B.; Writing—Review and Editing, R.J.B., F.M.D.B., A.R.C., A.T.D. and B.S.M.; Visualisation, R.J.B. and F.M.D.B.; Supervision, R.J.B.; Project Administration, R.J.B.; Funding, R.J.B. All authors have read and agreed to the published version of the manuscript.

Funding: This research received no external funding.

Institutional Review Board Statement: This study was conducted in accordance with the Declaration of Helsinki and approved by the joint University of Wollongong/South East Sydney and Illawarra Area Health Service Human Research Ethics Committee: protocol number 2004/105 approved 14 April 2004 (child sample); protocol number 2002/076 approved 10 April 2003 (young adult sample); protocol number 2009/220 approved 30 July 2009 (older adult sample).

Informed Consent Statement: Informed consent was obtained from all participants involved in the study or their parent/guardian (child sample only).

Data Availability Statement: The data presented in this study are available on request from the corresponding author.

Conflicts of Interest: The authors declare no conflicts of interest.

References

1. Barry, R.J.; De Blasio, F.M.; Cave, A.E. Sequential processing in young and older adults in the equiprobable auditory Go/NoGo task. *Clin. Neurophysiol.* **2016**, *127*, 2273–2285. [CrossRef]
2. Barry, R.J.; De Blasio, F.M. Sequential processing in the equiprobable auditory Go/NoGo task: A temporal PCA study. *Int. J. Psychophysiol.* **2013**, *89*, 123–127. [CrossRef] [PubMed]
3. Barry, R.J.; De Blasio, F.M. Performance and ERP components in the equiprobable go/no-go task: Inhibition in children. *Psychophysiology* **2015**, *52*, 1228–1237. [CrossRef]
4. Barry, R.J.; De Blasio, F.M.; Fogarty, J.S. A processing schema for children in the auditory equiprobable Go/NoGo task: ERP components and behaviour. *Int. J. Psychophysiol.* **2018**, *123*, 74–79. [CrossRef] [PubMed]
5. Barry, R.J.; De Blasio, F.M.; Borchard, J.P. Sequential processing in the equiprobable auditory Go/NoGo task: Children vs. adults. *Clin. Neurophysiol.* **2014**, *125*, 1995–2006. [CrossRef] [PubMed]
6. Barry, R.J.; De Blasio, F.M.; Cave, A.E. Development of children's performance and ERP components in the equiprobable Go/NoGo task. *Int. J. Psychophysiol.* **2022**, *171*, 12–19. [CrossRef]
7. Clarke, A.R.; Barry, R.J.; McCarthy, R.; Selikowitz, M. Age and sex effects in the EEG: Development of the normal child. *Clin. Neurophysiol.* **2001**, *112*, 806–814. [CrossRef] [PubMed]
8. Gasser, T.; Verleger, R.; Bächer, P.; Sroka, L. Development of the EEG of school-age children and adolescents. I. Analysis of band power. *Electroencephalogr. Clin. Neurophysiol.* **1988**, *69*, 91–99. [CrossRef]
9. Gasser, T.; Jennen-Steinmetz, C.; Sroka, L.; Verleger, R.; Möcks, J. Development of the EEG of school-age children and adolescents II. Topography. *Electroencephalogr. Clin. Neurophysiol.* **1988**, *69*, 100–109. [CrossRef]
10. John, E.R.; Ahn, H.; Prichep, L.; Trepetin, M.; Brown, D.; Kaye, H. Developmental equations for the electroencephalogram. *Science* **1980**, *210*, 1255–1258. [CrossRef]
11. Mason, L.M.; Barry, R.J.; Clarke, A.R. Age-related changes in the EEG in an eyes-open condition: I. Normal development. *Int. J. Psychophysiol.* **2022**, *172*, 40–45. [CrossRef]

12. Matthis, P.; Scheffner, D.; Benninger, C.; Lipinski, C.; Stolzis, L. Changes in the background activity of the electroencephalogram according to age. *Electroencephalogr. Clin. Neurophysiol.* **1980**, *49*, 626–635. [CrossRef]
13. Anderson, A.J.; Perone, S. Developmental change in the resting state electroencephalogram: Insights into cognition and the brain. *Brain Cogn.* **2018**, *126*, 40–52. [CrossRef] [PubMed]
14. Barry, R.J.; De Blasio, F.M. EEG differences between eyes-closed and eyes-open resting remain in healthy ageing. *Biol. Psychol.* **2017**, *129*, 293–304. [CrossRef]
15. Barry, R.J.; De Blasio, F.M.; Karamacoska, D. Data-driven derivation of natural EEG frequency components: An optimised example assessing resting EEG in healthy ageing. *J. Neurosci. Methods* **2019**, *321*, 1–11. [CrossRef] [PubMed]
16. Sun, H.; Ye, E.; Paixao, L.; Ganglberger, W.; Chu, C.J.; Zhang, C.; Rosand, J.; Mignot, E.; Cash, S.S.; Gozal, D.; et al. The sleep and wake electroencephalogram over the lifespan. *Neurobiol. Aging* **2023**, *124*, 60–70. [CrossRef] [PubMed]
17. Karamacoska, D.; Barry, R.J.; Steiner, G.Z. Electrophysiological underpinnings of response variability in the Go/NoGo task. *Int. J. Psychophysiol.* **2018**, *134*, 159–167. [CrossRef]
18. Karamacoska, D.; Barry, R.J.; Steiner, G.Z.; Coleman, E.P.; Wilson, E.J. Intrinsic EEG and task-related changes in EEG affect Go/NoGo task performance. *Int. J. Psychophysiol.* **2018**, *125*, 17–28. [CrossRef]
19. De Blasio, F.M.; Barry, R.J. Prestimulus delta and theta contributions to equiprobable Go/NoGo processing in healthy ageing. *Int. J. Psychophysiol.* **2018**, *130*, 40–52. [CrossRef]
20. De Blasio, F.M.; Barry, R.J. Prestimulus alpha and beta contributions to equiprobable Go/NoGo processing in healthy ageing. *Int. J. Psychophysiol.* **2020**, *155*, 32–40. [CrossRef]
21. Barry, R.J. Evoked activity and EEG phase resetting in the genesis of auditory Go/NoGo ERPs. *Biol. Psychol.* **2009**, *80*, 292–299. [CrossRef]
22. Barry, R.J.; De Blasio, F.M. EEG frequency PCA in EEG-ERP dynamics. *Psychophysiology* **2018**, *55*, e13042. [CrossRef] [PubMed]
23. Storey, J.E.; Rowland, J.T.J.; Conforti, D.A.; Dickson, H.G. The Rowland universal dementia assessment scale (RUDAS): A multicultural cognitive assessment scale. *Int. Psychogeriatrics* **2004**, *16*, 13–31. [CrossRef]
24. Fulton, S.E.; Lister, J.J.; Bush, A.L.H.; Edwards, J.D.; Andel, R. Mechanisms of the hearing-cognition relationship. *Semin. Hear.* **2015**, *36*, 140–149. [CrossRef] [PubMed]
25. Grassi, M.; Borella, E. The role of auditory abilities in basic mechanisms of cognition in older adults. *Front. Aging Neurosci.* **2013**, *5*, 59. [CrossRef]
26. Delorme, A.; Makeig, S. EEGLAB: An open source toolbox for analysis of single-trial EEG dynamics including independent component analysis. *J. Neurosci. Methods* **2004**, *134*, 9–21. [CrossRef]
27. Dien, J. The ERP PCA Toolkit: An open source program for advanced statistical analysis of event-related potential data. *J. Neurosci. Methods* **2010**, *187*, 138–145. [CrossRef]
28. Gratton, G.; Coles, M.G.H.; Donchin, E. A new method for off-line removal of ocular artifact. *Electroencephalogr. Clin. Neurophysiol.* **1983**, *55*, 468–484. [CrossRef] [PubMed]
29. Kim, H.; Luo, J.; Chu, S.; Cannard, C.; Hoffmann, S.; Miyakoshi, M. ICA's bug: How ghost ICs emerge from effective rank deficiency caused by EEG electrode interpolation and incorrect re-referencing. *Front. Signal Process.* **2023**, *3*, 1064138. [CrossRef]
30. Foti, D.; Hajcak, G.; Dien, J. Differentiating neural responses to emotional pictures: Evidence from temporal-spatial PCA. *Psychophysiology* **2009**, *46*, 521–530. [CrossRef]
31. Barry, R.J.; De Blasio, F.M.; Fogarty, J.S.; Karamacoska, D. ERP Go/NoGo condition effects are better detected with separate PCAs. *Int. J. Psychophysiol.* **2016**, *106*, 50–64. [CrossRef]
32. JASP Team. *JASP*, Version 0.18.3. Computer Software. University of Amsterdam: Amsterdam, The Netherlands, 2024.
33. Bonett, D.G.; Wright, T.A. Sample size requirements for estimating Pearson, Kendall and Spearman correlations. *Psychometrika* **2000**, *65*, 23–28. [CrossRef]
34. Fozard, J.L.; Vercruyssen, M.; Reynolds, S.L.; Hancock, P.A.; Quilter, R.E. Age differences and changes in reaction time: The Baltimore longitudinal study of aging. *J. Gerontol.* **1994**, *49*, P179–P189. [CrossRef] [PubMed]
35. MacDonald, S.W.S.; Li, S.-C.; Bäckman, L. Neural underpinnings of within-person variability in cognitive functioning. *Psychol. Aging* **2009**, *24*, 792–808. [CrossRef]
36. Jenkinson, N.; Brown, P. New insights into the relationship between dopamine, beta oscillations and motor function. *Trends Neurosci.* **2011**, *34*, 611–618. [CrossRef]
37. Westbrook, A.; Braver, T.S. Dopamine does double duty in motivating cognitive effort. *Neuron* **2016**, *89*, 695–710. [CrossRef] [PubMed]

brain sciences

MDPI

Article

Development of Gamma Oscillation during Sentence Processing in Early Adolescence: Insights into the Maturation of Semantic Processing

Mohammad Hossein Behboudi [1,2], Stephanie Castro [3], Prasanth Chalamalasetty [1] and Mandy J. Maguire [1,2,*]

[1] School of Behavioral and Brain Sciences, The University of Texas at Dallas, Richardson, TX 75080, USA; behboudi@utdallas.edu (M.H.B.)

[2] Callier Center for Communication Disorders, The University of Texas at Dallas, Dallas, TX 75235, USA

[3] Department of Human Development and Family Sciences, The University of Texas at Austin, Austin, TX 78705, USA

* Correspondence: mandy.maguire@utdallas.edu

Abstract: Children's ability to retrieve word meanings and incorporate them into sentences, along with the neural structures that support these skills, continues to evolve throughout adolescence. Theta (4–8 Hz) activity that corresponds to word retrieval in children decreases in power and becomes more localized with age. This bottom-up word retrieval is often paired with changes in gamma (31–70 Hz), which are thought to reflect semantic unification in adults. Here, we studied gamma engagement during sentence processing using EEG time–frequency in children (ages 8–15) to unravel the developmental trajectory of the gamma network during sentence processing. Children heavily rely on semantic integration for sentence comprehension, but as they mature, semantic and syntactic processing units become distinct and localized. We observed a similar developmental shift in gamma oscillation around age 11, with younger groups (8–9 and 10–11) exhibiting broadly distributed gamma activity with higher amplitudes, while older groups (12–13 and 14–15) exhibited smaller and more localized gamma activity, especially over the left central and posterior regions. We interpret these findings as support for the argument that younger children rely more heavily on semantic processes for sentence comprehension than older children. And like adults, semantic processing in children is associated with gamma activity.

Keywords: gamma; theta; language development; EEG; semantic

1. Introduction

Efficiently retrieving and integrating word meanings into a coherent sentence is a complex process in which children perform quickly and seemingly effortlessly. Nevertheless, children continue to enhance their language skills through middle childhood and adolescence [1–8]. As children progress through their school years, their ability to retrieve semantic information becomes faster and more efficient [7,9]. Their ability to integrate this information to comprehend a sentence changes with age as well [6,10–12]. Recent research using time–frequency analysis of EEG in adults has shown that changes in neural oscillations, particularly in theta (4–8 Hz) and gamma (31–70 Hz), are associated with retrieving and integrating semantic information, respectively [5,13–18]. Despite these findings, there remains a dearth of research investigating the neural oscillatory activity that supports sentence processing in children. A handful of studies have shown developmental differences in theta response during sentence processing in children [2,19]; however, further research is needed to fully understand the complex neural mechanisms that children use as they retrieve words and integrate them into coherent sentences.

In adults, gamma oscillations are strongly associated with the process of semantic unification during sentence processing [14,18,20,21]. Specifically, semantically correct sentences generate a significantly greater gamma power increase than word lists and sentences with semantic violations, even when the syntax remains intact, and this effect intensifies further when semantic integration becomes more challenging [14,18,20,21]. Furthermore, there is a link between gamma activity and prediction in sentence processing, suggesting that gamma oscillations may be involved in predicting upcoming linguistic input [15,22,23]. Beyond language, gamma has been implicated in other areas of processing including motor preparation [24–26], visual processing [27], memory encoding [28–30], and memory maintenance [31–34]. Here, we will focus on gamma's role in sentence processing.

Related Work

Neurodevelopmental studies offer invaluable insights into the maturation of cognitive functions, revealing that while basic cognitive processes are established by childhood, the sophisticated coordination of these abilities continues to develop throughout adolescence [35–37]. This ongoing development can be attributed to brain maturation and the child's experiences [36–39]. By early childhood, aspects such as cortical folding, brain size, and regional functional specialization of the brain largely achieve a form similar to that of adults. However, significant enhancements, including the expansion of dendritic arborization, synaptic pruning, and increased myelination, continue well into adolescence [37].

Broca's area and Wernicke's area, two key language-related regions in the human brain, in both adults and children are studied in [40]. The findings suggest that the connection between these two regions, which is facilitated by the arcuate fasciculus/superior longitudinal fasciculus (AF/SLF), is less mature in children, potentially due to lower myelination. In children, an alternative pathway to Wernick's area via the ventral pathway is observed [41]. The temporal lobe matures uniquely, with early maturation of the temporal poles and late maturation of the posterior superior temporal gyrus [40,42]. This area integrates functions of other relatively developed association areas, such as memory, audiovisual association, and object recognition [43]. Aside from neural structure development, neural oscillation development enhances our understanding of language maturation.

In children's sentence processing, similar to adults', theta is associated with word retrieval as it appears in response to word presentation [2,44]. However, theta activity in children exhibits a distinct topographic and temporal pattern [5,19,44]. Specifically, in children aged 8–11 years, theta activity is larger in amplitude and is broadly distributed, while in adolescents, it tends to localize to the left central-posterior areas, and the temporal pattern is temporally prolonged compared to adults [2,44,45]. These findings suggest that although theta activity is involved in language processing at an early age, memory systems that involve semantic retrieval and unification may take longer to fully develop.

While our understanding of children's theta oscillations in sentence processing is growing, the role of gamma oscillations is still relatively unexplored. Resting-state or baseline gamma activity is associated with language processing and vocabulary [46–48]. Children with language-learning impairments who take part in intervention exhibit changes in their resting-state gamma activity [49]. Differences in resting-state gamma also mediate the relationship between socioeconomic status and language abilities [47]. However, understanding the dynamics of gamma oscillations during language comprehension tasks in children will offer valuable insights into the involvement of gamma oscillations in language networks.

Behavioral, fMRI, and ERP studies have provided important insights about how children's semantic processing develops. Around age 5, children rely heavily on semantic integration for sentence comprehension, but as they mature, semantic and syntactic processing units become distinct and localized, improving into early adulthood [3,4,6,50]. The early reliance on semantic processing during sentence comprehension is thought to be due to the ventral pathway, which responds to the semantic context, serving as a primary connection within the language network during early language development [42,51–55]. In adults, syntactic processing is mainly observed in the superior temporal gyrus (STG) and pars opercularis [4,6,52,53,56,57]. However, in children, semantics and syntax units overlap and engage additional areas of the left and right Inferior Frontal Gyrus (IFG) [4,6,8,58].

This study aims to investigate the differences in gamma engagement between middle childhood and adolescence during sentence processing. By using time–frequency analysis, we will provide new insights into how gamma activity, which corresponds to semantic integration in adults, is engaged to support sentence comprehension between middle childhood and early adolescence. We use the same data set as in our previous study on theta engagement [2], but expand it to gamma. In the previous work, we found that theta power increases and then decreases quickly as children are exposed to each word in the sentence. We predict the temporal pattern of gamma activity will be different showing a longer period of heightened activity spanning across words to support the process of semantic unification as the sentence unfolds [14,18,59]. Across age, we expect to see this increase in gamma amplitude becomes smaller and more localized [3,4,6,50]. This would be interpreted as evidence that younger children heavily rely on semantic integration for sentence comprehension, whereas as they age, their reliance on this semantic unification decreases and the associated neural networks become more specialized and localized.

2. Materials and Methods

2.1. Participants

Ninety-eight children, spanning ages 8 to 15 years, were included in this analysis. This sample size surpasses those of previous developmental and adult language studies allowing for robust statistical analysis [5,10,14,44]. This group was selected from a larger study of 150 participants, and those with noisy EEG signals were excluded. Participants were grouped by age: 8–9 years (Mean = 9.15, SD = 0.44, N = 20; 6 female), 10–11 years (Mean = 11.04, SD = 0.47, N = 28; 18 female), 12–13 years (Mean = 13.02, SD = 0.61, N = 24; 13 female), 14–15 years (Mean = 14.77, SD = 0.55, N = 26; 14 female). All of the participants included in the analysis were right-handed and, based on self-reports, had no history of significant neurological issues, no language development delay or disorders, and no reading disabilities at the time of the study. Furthermore, all the participants were English-dominant. Following a brief introductory period, the consent documents were completed by the parents, and the assent documents were completed by the children.

2.2. Linguistic Stimuli

Each participant was presented with 100 sets of sentence triplets, wherein each sentence ended with a pseudoword that served as a noun and was preceded by an article or a possessive pronoun (See Table 1). The sentence length varied between 6 and 9 words wherein the words included were selected from the MacArthur-Bates Communicative Inventories, a corpus of words spoken by children between the ages of 4 and 5 years. The sentence structures vary such that the location and frequency of verbs are different across sentences. The verb occurs between words one and seven, but most commonly is the third or fourth (mean = 3.38, See Table 2 for some examples). Each word appeared for 500 milliseconds (ms) and a 300 ms blank screen appeared between each word. The participant was then asked what word best fit all the sentences in the triplet. In our analysis, we focused on the first five words of each sentence to study neural oscillation, ensuring that we examined the same amount of data across all sentences.

Table 1. Example of sentence triplets presented to the participant.

Sentence 1	Go outside and play with the neg	Target word: **Cat**
Sentence 2	I'm covered with hair from the neg	
Sentence 3	All night, I heard the meowing of the neg	

Table 2. Verb position in the stimuli.

He **spilled** food on his wesh.
The dog **slept** on the presh.
To **keep** the sun away, I **wear** a tuch.
Hi favorite toy of all time **is** the therg.
When I **sit** there, I **can** see the pleg.

2.3. EEG Acquisition

To collect EEG data, each participant was asked to sit on a chair 1 m away from the monitor that was used to present stimuli. They were told that they will see three sentences, word by word, and at the end of each triplet, they needed to guess what word fits the last word for all three sentences. EEG data were acquired with a 64-electrode Neuroscan cap, in which the electrodes were placed in the standard 10-10 system. EEG data were recorded continuously using a Neuroscan SynAmps2 amplifier and CURRY software (V7, https://compumedicsneuroscan.com/curry-epilepsy-evaluation/, (accessed on 25 November 2023)) sampled at 1 kHz with impedances typically below 5 kΩ. Data were recorded with the ground at Fz and the reference electrode located near the vertex.

2.4. EEG Pre-Processing

To pre-process EEG signals, data were re-sampled to 250 Hz. Then, a zero-phase FIR filter using a hamming window was used to filter the data with a low-pass filter at 80 Hz, and high-pass filter at 0.1 Hz. The noise line (60 Hz) was removed using a notch filter with a 2 Hz transition bandwidth. This filter, which is zero phase and non-causal, has cutoff frequencies at −6 dB from 59 to 61 Hz. All bad electrodes were removed using the "clean_rawdata" EEGLab plug-in to detect signals beyond the normal amplitude, which were removed by considering +/−75 μv as the threshold [60,61]. To remove non-cortical signals, we used ICA decomposition and the MARA plugin, a machine-learning artifact rejection algorithm [62,63], in the EEGLab toolbox [61]. MARA is a robust, supervised method designed to identify and reject various types of artifacts, including but not limited to eye movement, heartbeats and muscle movement artifacts, thereby offering a comprehensive solution for artifact rejection [62,63]. After eliminating artifacts and bad channels, the data were referenced to the average potential across all the electrodes. Then, the data were epoched from 500 ms before the onset of the sentences to 3700 ms after (end of word 5). Then, the baseline of the data was removed by considering 350 ms before the onset of the first word as the baseline.

2.5. Time–Frequency Analysis

Time–frequency analysis was used in our study to compute event-related spectral perturbations (ERSP). All eligible participants entered the study design and were grouped into four different conditions based on age (8–9 years, 10–11 years, 12–13 years, and 14–15 years). For ERSP, the calculation was based on the Morlet wavelet with the wavelet cycle beginning from 2 cycles and increasing linearly by 0.5 cycles in order to have more cycles for higher frequencies. This resulted in the wavelet width ranging from 2 cycles at 3 Hz to 23.70 cycles at 71 Hz. The EEGLab toolbox was used to calculate the ERSPs between 3 and 71 Hz for each electrode and across each sentence (trial). The average power for each trial across all electrodes and frequencies was calculated and normalized by subtracting the baseline mean for that specific electrode and frequency. This allows us to track changes in brain activity over time and across different frequencies. Then, to obtain the topographic map of gamma (31 to 70 Hz) activity, we average the activity across the time and frequency over the scalp to see the spatial distribution of the activity over the scalp.

2.6. Statistical Analysis

After calculating all the ERSPs for each group, a between-subject analysis of variance (ANOVA) was conducted to compare the neural activity (dependent variable: ERSP in each

Time × Frequency × Electrode points) between different age groups (independent variable: age groups: 8–9 years, 10–11 years, 12–13 years, and 14–15 years). The non-parametric permutation-based ANOVA was calculated using EEGLab with a p-value of 0.05 to find the main effect of age on ERSP. To correct for multiple comparison hypothesis testing, False Discovery Rate (FDR) correction was used. FDR is a technique used in multiple hypothesis testing to account for the anticipated percentage of false positive outcomes among the rejected null hypotheses (Type I Error) [64,65]. In order to keep the expected rate of false discoveries among all made discoveries under control at a specified level (here 5%), it modifies the p-value threshold used to declare statistical significance. This is helpful when several tests are run since it lowers the possibility of creating a lot of false positive findings and improves the specificity of the outcomes. For further analysis, Tukey's HSD post-hoc analysis was conducted. This ANOVA was run on both theta and gamma frequencies to determine if there was a significant difference in the activity pattern of these frequencies across ages. The findings related to the theta frequency band have been previously documented in our earlier publication [2]. In the current manuscript, our primary focus is on the gamma frequency band. However, we make references to the theta analysis where relevant for a comprehensive understanding. Region of interest (ROI) analysis was included, as opposed to reporting findings from each individual electrode, to quantify the ERSP findings better and to simplify and streamline the ANOVA results. Electrodes were categorized into 6 regions of interest and the average ERSP in each ROI was calculated for further statistical analysis. The six ROIs are Left Frontal (FP1, AF3, F7, F5, F3, F1); Left Central (FC5, FC3, C5, C3, C1, CP5, CP3, CP1); Left Posterior (P7, P5, P3, P1, PO7, PO5, PO3, O1); Right Frontal (FP2, AF4, F2, F4, F6, F8); Right Central (FC2, FC4, FC6, C2, C4, C6, CP2, CP4, CP6); Right Posterior (P2, P4, P6, P8, PO4, PO6, PO8, O2) electrodes.

3. Results

3.1. Spatial Analysis

To understand the development of gamma activity during sentence processing, we performed an omnibus one-way ANOVA (Figure 1), which revealed significant differences between age groups in the topography of gamma engagement averaging over the course of sentences (0–3700 ms). Specifically, younger children display higher gamma amplitudes over the scalp, but for older children, this effect is much smaller and occurs only over the left frontocentral and parietal regions. The main effect of age is statistically significant ($F(3, 94) = 4.08$, $p = 0.009$; ${\eta_p}^2 = 0.12$) over the left frontal region as well as the left central region ($F(3, 94) = 3.35$, $p = 0.022$; ${\eta_p}^2 = 0.10$), right frontal ($F(3, 94) = 4.23$, $p = 0.007$; ${\eta_p}^2 = 0.12$), and right central region ($F(3, 94) = 3.91$, $p = 0.011$; ${\eta_p}^2 = 0.11$). The difference between gamma power over the right and left posterior regions is not significantly different among this age group.

The results of the Tukey post-hoc analysis revealed statistically significant differences in various electrode measurements across different age group pairs (Figure 2). Specifically, children aged 8–9 years exhibited significant differences from those aged 12–13 years in the left frontal and right fronto-central electrodes. Additionally, they also differed significantly from the 14–15 years old group in the right fronto-central and central electrodes. On the other hand, children in the 10–11 years age group showed significant differences from the 12–13 years old group in the left frontal, central, and centroparietal electrodes. Furthermore, this age group also exhibited significant differences from the 14–15 years old group in the central electrodes. These findings highlight the distinct variations in electrode measurements across different age groups.

3.2. Spatiotemporal Analysis

The spatial analysis revealed the main effect of age on gamma oscillation across the scalp. By studying changes in gamma activity over the course of the sentence, with a specific focus on when the words are displayed and when they disappear (Figure 3), we can observe how gamma networks act at each stage of sentence processing and as the sentence unfolds, As mentioned in the previous section (Section 3.1). An omnibus one-way ANOVA was conducted at each step of sentence processing to identify the specific stages that contribute to the observed significant differences in gamma oscillation among the different age groups. The results show no obvious differences between the gamma response to a word and to a blank between words, unlike what was observed in theta activity [2]. Instead, this effect changes over the course of the sentence, peaking around words 3 and 4 (1600 to 2900 ms). Our analyses revealed no age group differences at word 1, but after word 2, the gamma response becomes larger for younger age groups, eventually resulting in a significant age group differences (around word 3, 1600 ms after the onset of the first word and word 4, 2400 ms after the onset of the first word). This effect is largest over the right hemisphere for word 3 and becomes more widely distributed by word 4 before decreasing throughout the remainder of the sentence.

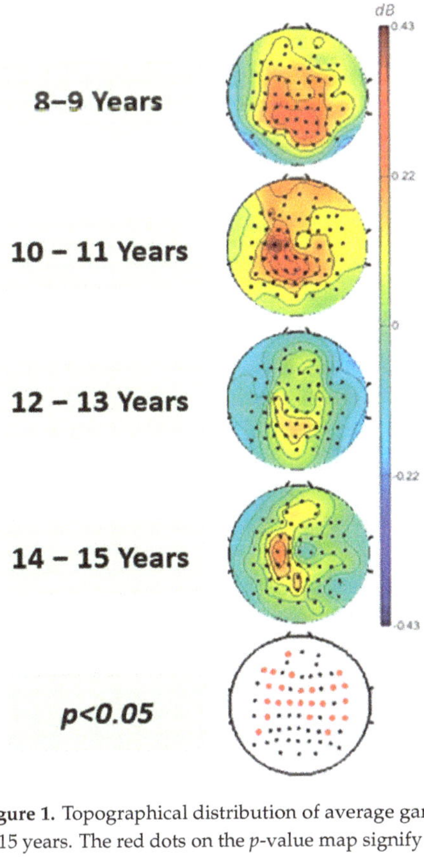

Figure 1. Topographical distribution of average gamma activity across sentences in children aged 8–15 years. The red dots on the p-value map signify electrodes that have yielded statistically significant results in the ANOVA test, thereby indicating the main effect of age.

A detailed examination of the ANOVA was conducted using Tukey's HSD post-hoc analysis, focusing on pairwise comparisons between different age groups as the sentence unfolds. The analysis revealed statistically significant differences in gamma power during sentence processing between younger children (8–11 years old) and older children (12–15 years old), as illustrated in Figure 4. Notably, the most pronounced differences were observed between children aged 10–11 years and those aged 12–13 years, with variations extending across multiple electrodes. Also, the largest amount of significantly different electrodes occurs between word 3 and word 4 (1600–2900 ms).

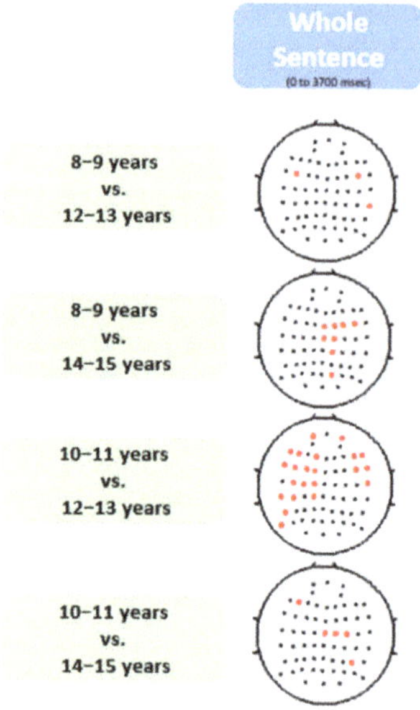

Figure 2. Post-hoc analysis on ANOVA results in Figure 1. The significant electrodes are highlighted in red in each pairwise comparison, with an alpha level set at 0.05. The rows represent p-values for pairwise comparisons between different age groups: the first and second rows compare 8–9-year-olds with 12–13 and 14–15-year-olds, respectively, while the third and fourth rows compare 10–11-year-olds with 12–13 and 14–15-year-olds, respectively. Notably, the pairwise comparisons between 8–9 and 10–11year-olds, and between 12–13 and 14–15-year-olds, were not statistically significant.

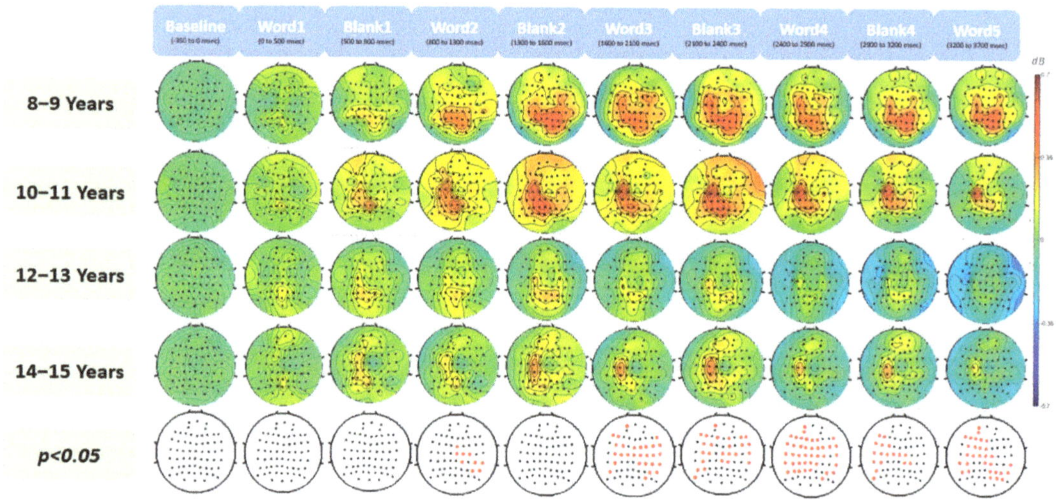

Figure 3. Topographical distribution of gamma band event-related spectral perturbation (ERSP) changes during the course of sentence processing in children aged 8–15 years. The red dots in the p-value map indicate the electrodes that show a significant effect of age in the ANOVA analysis amongst all age groups ($p < 0.05$).

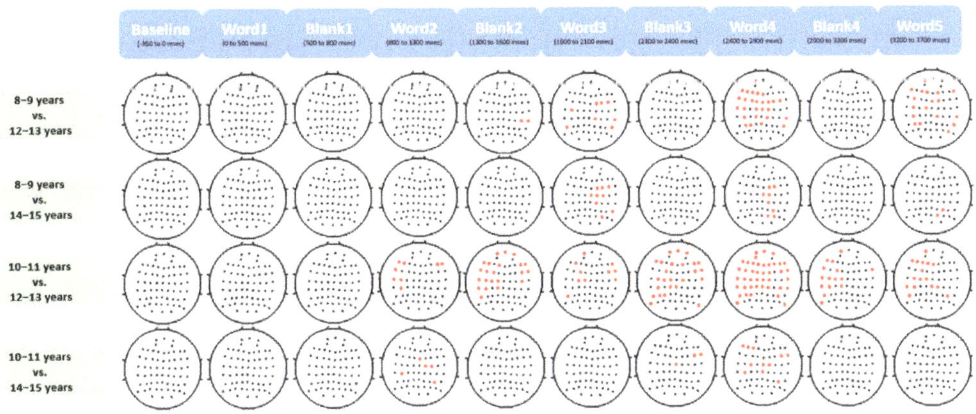

Figure 4. Post-hoc analysis on ANOVA results in Figure 3. The significant electrodes are highlighted in red in each pairwise comparison as sentence unfolds, with an alpha level set at 0.05. The columns show different stages of sentence presentation (i.e., word presentation and blank between words). The rows represent *p*-values for pairwise comparisons between different age groups: the first and second rows compare 8–9-year-olds with 12–13 and 14–15-year-olds, respectively, while the third and fourth rows compare 10–11-year-olds with 12–13 and 14–15-year-olds respectively. Notably, the pairwise comparisons between 8–9 and 10–11-year-olds, and between 12–13 and 14–15-year-olds, were not statistically significant.

3.3. Temporal Analysis

The temporal analysis was conducted to elucidate the temporal pattern of gamma activity during sentence processing as each word is presented and as sentence unfolds. The average ERSP across the gamma frequency band (31–70 Hz) was calculated, focusing on the electrodes that showed significant results in the ANOVA (Figure 1). These selected electrodes included FP1, F7, F5, F1, F8, FC5, FC3, FCz, FC2, FC6, FT8, C5, C3, C1, Cz, C2, C4, T8, CP5, CPz, TP8, CP6, P7, and P6. The results are presented in Figure 5. The overall pattern from the temporal analysis indicates that gamma activity increases with the presentation of each word with a delay relative to the theta activity observed in [2]. This increase in gamma activity is more prolonged in children aged 8–11 years. For instance, the gamma increase for the first word begins at approximately t = 400 ms, and for children aged 8–11 years, it continues even after the presentation of the second word (around t = 1000 ms after the onset of the sentence). However, in children aged 12–15 years, the increase ceases around 800 ms, followed by a decrease in gamma activity. In the 8–11 age group, even after the decrease phase, the gamma activity remains higher than the baseline. This represents the engagement of a neural network in younger children that is even observable during word presentation. Conversely, for ages 12–15, average gamma amplitudes decrease after words 2 and 3 (t = 2400 to 3700 ms), such that by the end of word 5, the gamma amplitude returns to baseline levels. This suggests distinct patterns of neural processing across different age groups during sentence comprehension. The independent sample t-test was conducted on a selection of 20 random participants from each age group, with the aim of identifying statistically significant timepoints across the different age groups. The findings revealed a notable distinction in the temporal pattern of gamma activity between younger children (aged 8–11 years) and older children (aged 12–15 years) with the alpha level of 0.05. Specifically, the results indicated that the gamma activity in younger children (8–11 years old) exhibited a higher amplitude from the onset of the second word (t = 800 ms) in comparison to the older children (12–15 years old). This heightened activity persisted throughout the entire sentence until it unfolded.

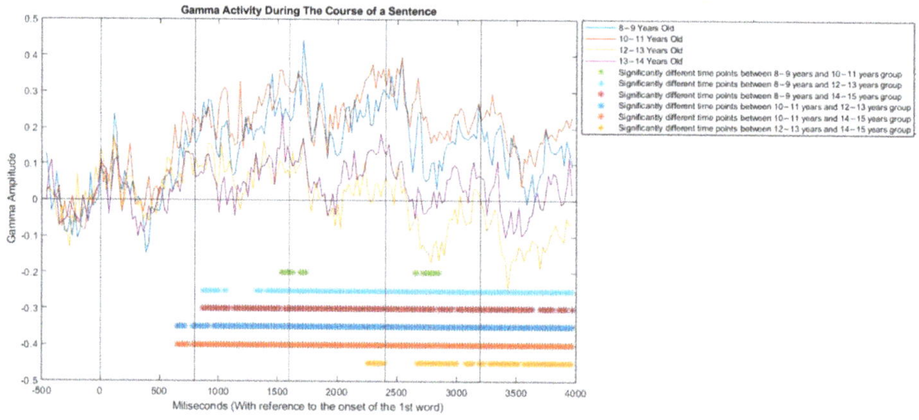

Figure 5. The averages of gamma band power oscillation (in dB) across the electrodes that show significant results in the ANOVA (Figure 1) during the course of sentence processing in children aged 8–15 years. Each vertical line represents the onset of word presentation, and the horizontal line indicates the average of baseline level of gamma power. The starts show represents the time points that are significantly ($p < 0.05$) different in each pairwise comparison.

4. Discussion

This is the first paper to study how gamma power related to sentence processing changes between middle childhood and early adolescence. We observed that gamma activity increased gradually from the beginning of the sentence (slightly later than theta activity) and remained elevated even between words as the sentence unfolds (Figure 2). This is consistent with adult studies that showed gamma is related to semantic unification, or the prediction of the upcoming word in sentence processing and gamma is present along the sentence, rather than a response to each word [14,15,18,20–22,66]. We also observed that gamma shows a developmental shift around age 11, with the younger groups (8–9 and 10–11) exhibiting broadly distributed gamma activity with higher amplitudes and older groups (12–13 and 14–15) exhibiting smaller and more localized gamma activity, especially in the left central and posterior regions (Figure 1). Similar to studies of children's sentence comprehension, this indicates that children's reliance on semantic processing during sentence comprehension decreases during middle childhood and adolescence and becomes more localized [6,8,12,50,58].

Gamma and theta have been associated with semantic aspects of sentence processing, with theta reflecting individual lexical retrieval and gamma reflecting integration. Our data support this distinction because the temporal aspects of gamma in our study are notably different from what has been found in the past associated with theta [2,5,45]. Previous studies [2,5,45] observed theta increase in response to individual words in a sentence, which is broadly distributed in younger children and decreases in amplitude and localizes to over the left centro-parietal regions in older children. Further, these bursts in theta activity occur shortly after the onset of each word and largely attenuate between words. As this study utilized data from the previously mentioned study conducted by Maguire et al. 2022 [2], which specifically investigated the effect of theta on sentence processing, our study did not report detailed results regarding theta. Our contribution lies in the integration of gamma activity along with the well-established frequency band in children's sentence processing, theta.

Through word-level temporal analysis, we observed an increase in gamma activity, after the theta burst, which typically occurs at the onset of individual words [2,44]. The rise in gamma power suggests the activation of a neural network responsible for integrating new words with their preceding context (Figure 5). The peak of this gamma activity occurred around the third and fourth words, followed by a decrease in activity (Figure 2). Older children exhibited a faster and more drastic decrease in gamma activity, returning to baseline levels by the end of the fifth word. Additionally, we observed a gradual increase in gamma activity up to the point that we anticipate to be the maximum semantic unification load followed by a gradual decrease thereafter. We interpret this finding as support for the argument that gamma activity is associated with semantic unification during sentence processing [14,20,21,28]. While our study does not specify the type of unification, previous adult studies associated gamma with semantic unification rather than syntactic unification [14,20,21,28].

In adults, there is also evidence that gamma is related to the congruency between the expected and observed input or prediction [15,66]. However, our observation of a gradual decrease in gamma power after the third or fourth word does not align with these interpretations as words appearing later in a sentence are often more predictable and therefore easier to process [66,67].

The temporal pattern of our gamma results is particularly interesting because it appears to peak around the sentence's main verb, indicating that verbs are strengthening the semantic integration of sentences. While verbs are not the main focus of this study, previous research has shown the significant role of verbs in the semantic structure of sentences [68,69]. Future studies will need to investigate the relationship between verb phrase processing and gamma activity during development.

Although our main goal was not related to where in the brain the effect originates, topographic maps can give us a general idea of how the neural engagement changes with age. Topographically, younger children exhibited a broader distribution of gamma activity, whereas in older children, this activity is mostly localized, specifically to electrodes over the left central and posterior regions (Figure 1). These patterns appear to be consistent with findings that syntactic units develop later than semantic units and, as children get older, semantic and syntactic processing units are segregating gradually and become more specific [4,6,8,42]. This is in line with previous work showing that in adults, syntactic processing primarily engages the superior temporal gyrus (STG) and pars opercularis [6,52,53,57]. On the other hand, children showed overlapping brain activity for both semantics and syntax. This supports previous work that indicates children engage additional areas of the left and right inferior frontal gyrus (IFG) [4,6,58,70]. The reliance on semantic processing during sentence comprehension is thought to be related to the ventral pathway being a primary connection within the language network during early language development [42,52,53].

While our study provides valuable insights into gamma-band oscillations during sentence processing in children, it does have certain limitations. Although we observed effects at the main verb in our gamma analysis, our study was not specifically designed to investigate verb location. This is because the sentences were not aligned in such a way that the verb occupied a specific location. Therefore, while we can suggest a correlation, we cannot definitively confirm this effect. Furthermore, our current focus on gamma-band oscillations could be expanded in future research. For instance, a cross-frequency coupling analysis could be employed to explore the relationship between changes in theta oscillations and changes in gamma oscillations. This approach could offer a more comprehensive understanding of the connection between different cognitive functions in sentence processing. Ultimately, it could lead to a more holistic understanding of the oscillatory dynamics underlying sentence processing in children.

This paper addressed the development of gamma activity during sentence processing tasks from childhood to adolescence. The analysis of gamma frequency during sentence processing in children alongside the previous theta studies advanced our knowledge about the change in children's semantic unification ability. This paper adds to the small but growing body of literature about the neural oscillatory correlates of language processing and how they continue to change throughout middle childhood and adolescence. In addition to expanding our knowledge about developmental cognitive neuroscience, this work will build an essential foundation for studies focused on how processes might differ in children with language delays and disorders.

Author Contributions: Conceptualization, M.H.B. and M.J.M.; Data curation, M.H.B. and M.J.M.; Formal analysis, M.H.B.; Funding acquisition, M.J.M.; Investigation, M.H.B. and M.J.M.; Methodology, M.H.B. and M.J.M.; Project administration, M.H.B.; Resources, M.J.M.; Supervision, M.J.M.; Validation, M.H.B. and M.J.M.; Visualization, M.H.B.; Writing—original draft, M.H.B. and M.J.M.; Writing—review and editing, M.H.B., S.C. and P.C. All authors have read and agreed to the published version of the manuscript.

Funding: This work is funded in part by the US National Science Foundation Division of Behavioral and Cognitive Science (PI: Maguire, Co-PI: Abel; NSF 1551770).

Institutional Review Board Statement: The study was conducted in accordance with the Declaration of Helsinki, and approved by the Institutional Review Board of The University of Texas at Dallas (IRB-24-105).

Informed Consent Statement: Informed consent was obtained from all subjects involved in the study.

Data Availability Statement: The data for this study are available online as part of the Open Science Framework: https://osf.io/z38rb/ (accessed on 25 November 2023).

Conflicts of Interest: The authors declare no conflict of interest.

References

1. Lew-Williams, C.; Fernald, A. Young Children Learning Spanish Make Rapid Use of Grammatical Gender in Spoken Word Recognition. *Psychol. Sci.* **2007**, *18*, 193–198. [CrossRef] [PubMed]
2. Maguire, M.J.; Schneider, J.M.; Melamed, T.C.; Ralph, Y.K.; Poudel, S.; Raval, V.M.; Mikhail, D.; Abel, A.D. Temporal and topographical changes in theta power between middle childhood and adolescence during sentence comprehension. *Dev. Cogn. Neurosci.* **2022**, *53*, 101056. Available online: https://www.sciencedirect.com/science/article/pii/S1878929321001456 (accessed on 15 November 2023). [CrossRef] [PubMed]
3. Nippold, M.A.; Hesketh, L.J.; Duthie, J.K.; Mansfield, T.C. Conversational versus expository discourse: A study of syntactic development in children, adolescents, and adults. *J. Speech Lang. Hear. Res.* **2005**, *48*, 1048–1064. [CrossRef]
4. Nuñez, S.C.; Dapretto, M.; Katzir, T.; Starr, A.; Bramen, J.; Kan, E.; Bookheimer, S.; Sowell, E.R. fMRI of syntactic processing in typically developing children: Structural correlates in the inferior frontal gyrus. *Dev. Cogn. Neurosci.* **2011**, *1*, 313–323. Available online: https://www.sciencedirect.com/science/article/pii/S1878929311000259 (accessed on 15 November 2023). [CrossRef]
5. Schneider, J.M.; Abel, A.D.; Ogiela, D.A.; McCord, C.; Maguire, M.J. Developmental differences in the neural oscillations underlying auditory sentence processing in children and adults. *Brain Lang.* **2018**, *186*, 17–25. Available online: https://www.sciencedirect.com/science/article/pii/S0093934X17301190 (accessed on 15 November 2023). [CrossRef] [PubMed]
6. Skeide, M.A.; Brauer, J.; Friederici, A.D. Syntax gradually segregates from semantics in the developing brain. *NeuroImage* **2014**, *100*, 106–111. Available online: https://www.sciencedirect.com/science/article/pii/S1053811914004650 (accessed on 15 November 2023). [CrossRef]
7. Wang, S.; Allen, R.J.; Lee, J.R.; Hsieh, C.E. Evaluating the developmental trajectory of the episodic buffer component of working memory and its relation to word recognition in children. *J. Exp. Child Psychol.* **2015**, *133*, 16–28. Available online: https://www.sciencedirect.com/science/article/pii/S0022096515000144 (accessed on 15 November 2023). [CrossRef]
8. Wu, C.Y.; Vissiennon, K.; Friederici, A.D.; Brauer, J. Preschoolers' brains rely on semantic cues prior to the mastery of syntax during sentence comprehension. *NeuroImage* **2016**, *126*, 256–266. Available online: https://www.sciencedirect.com/science/article/pii/S1053811915009477 (accessed on 15 November 2023). [CrossRef]
9. Hurks, P.P.M.; Vles, J.S.H.; Hendriksen, J.G.M.; Kalff, A.C.; Feron, F.J.M.; Kroes, M.; van Zeben, T.M.C.B.; Steyaert, J.; Jolles, J. Semantic Category Fluency Versus Initial Letter Fluency Over 60 Seconds as a Measure of Automatic and Controlled Processing in Healthy School-aged Children. *J. Clin. Exp. Neuropsychol.* **2006**, *28*, 684–695. [CrossRef]
10. Mahler, N.A.; Chenery, H.J. A Developmental Perspective on Processing Semantic Context: Preliminary Evidence from Sentential Auditory Word Repetition in School-Aged Children. *J. Psycholinguist. Res.* **2019**, *48*, 81–105. Available online: http://www.scopus.com/inward/record.url?scp=85049695118&partnerID=8YFLogxK (accessed on 15 November 2023). [CrossRef]
11. Nouwens, S.; Groen, M.A.; Kleemans, T.; Verhoeven, L. The role of semantic retrieval in children's reading comprehension development in the upper primary grades. *J. Res. Read.* **2018**, *41*, 597–614. Available online: https://onlinelibrary.wiley.com/doi/abs/10.1111/1467-9817.12128 (accessed on 15 November 2023). [CrossRef]
12. Hahne, A.; Eckstein, K.; Friederici, A.D. Brain signatures of syntactic and semantic processes during children's language development. *J. Cogn. Neurosci.* **2004**, *16*, 1302–1318. [CrossRef] [PubMed]
13. Bastiaansen, M.C.M.; van der Linden, M.; ter Keurs, M.; Dijkstra, T.; Hagoort, P. Theta Responses Are Involved in Lexical—Semantic Retrieval during Language Processing. *J. Cogn. Neurosci.* **2005**, *17*, 530–541. [CrossRef]

14. Hald, L.A.; Bastiaansen, M.C.M.; Hagoort, P. EEG theta and gamma responses to semantic violations in online sentence processing. *Brain Lang.* **2006**, *96*, 90–105. Available online: https://www.sciencedirect.com/science/article/pii/S0093934X05001239 (accessed on 15 November 2023). [CrossRef]
15. Lam, N.H.L.; Schoffelen, J.M.; Uddén, J.; Hultén, A.; Hagoort, P. Neural activity during sentence processing as reflected in theta, alpha, beta, and gamma oscillations. *NeuroImage* **2016**, *142*, 43–54. Available online: https://www.sciencedirect.com/science/article/pii/S1053811916002032 (accessed on 15 November 2023). [CrossRef]
16. Maguire, M.J.; Brier, M.R.; Ferree, T.C. EEG theta and alpha responses reveal qualitative differences in processing taxonomic versus thematic semantic relationships. *Brain Lang.* **2010**, *114*, 16–25. Available online: https://www.sciencedirect.com/science/article/pii/S0093934X10000520 (accessed on 15 November 2023). [CrossRef] [PubMed]
17. Meyer, L. The neural oscillations of speech processing and language comprehension: State of the art and emerging mechanisms. *Eur. J. Neurosci.* **2018**, *48*, 2609–2621. Available online: https://onlinelibrary.wiley.com/doi/abs/10.1111/ejn.13748 (accessed on 15 November 2023). [CrossRef] [PubMed]
18. Penolazzi, B.; Angrilli, A.; Job, R. Gamma EEG activity induced by semantic violation during sentence reading. *Neurosci. Lett.* **2009**, *465*, 74–78. Available online: https://www.sciencedirect.com/science/article/pii/S0304394009011719 (accessed on 15 November 2023). [CrossRef]
19. Schneider, J.M.; Maguire, M.J. Developmental differences in the neural correlates supporting semantics and syntax during sentence processing. *Dev. Sci.* **2019**, *22*, e12782. Available online: https://onlinelibrary.wiley.com/doi/abs/10.1111/desc.12782 (accessed on 15 November 2023). [CrossRef]
20. Bastiaansen, M.; Hagoort, P. Oscillatory neuronal dynamics during language comprehension. *Prog. Brain Res.* **2006**, *159*, 179–196. Available online: https://www.sciencedirect.com/science/article/pii/S0079612306590120 (accessed on 15 November 2023).
21. Bastiaansen, M.; Hagoort, P. Frequency-based Segregation of Syntactic and Semantic Unification during Online Sentence Level Language Comprehension. *J. Cogn. Neurosci.* **2015**, *27*, 2095–2107. [CrossRef] [PubMed]
22. Wang, L.; Zhu, Z.; Bastiaansen, M. Integration or Predictability? A Further Specification of the Functional Role of Gamma Oscillations in Language Comprehension. *Front. Psychol.* **2012**, *3*, 187. Available online: https://www.frontiersin.org/articles/10.3389/fpsyg.2012.00187 (accessed on 15 November 2023). [CrossRef]
23. Prystauka, Y.; Lewis, A.G. The Power of Neural Oscillations to Inform Sentence Comprehension: A Linguistic Perspective. *Lang Linguist. Compass* **2019**, *13*, e12347. Available online: https://www.ncbi.nlm.nih.gov/pmc/articles/PMC7546279/ (accessed on 15 November 2023). [CrossRef] [PubMed]
24. Kaiser, J.; Lutzenberger, W. Human gamma-band activity: A window to cognitive processing. *NeuroReport* **2005**, *16*, 207. Available online: https://journals.lww.com/neuroreport/Fulltext/2005/02280/Combined_EEG_and_MEG_recordings_of_visual_40_Hz.1.aspx?casa_token=e-lSlYwy7ncAAAAA:3IAdC0s2Z-QNQRufwgdGtUKnefB63_R97NGfpSeZ2yw6gfsvosTlMopYaIwSefPe27aDf1KprpjfU-hegz02SC0kFA (accessed on 15 November 2023). [CrossRef] [PubMed]
25. Ulloa, J.L. The Control of Movements via Motor Gamma Oscillations. *Front. Hum. Neurosci.* **2022**, *15*, 787157. Available online: https://www.frontiersin.org/articles/10.3389/fnhum.2021.787157 (accessed on 15 November 2023). [CrossRef]
26. Muthukumaraswamy, S.D. Functional Properties of Human Primary Motor Cortex Gamma Oscillations. *J. Neurophysiol.* **2010**, *104*, 2873–2885. Available online: https://journals.physiology.org/doi/full/10.1152/jn.00607.2010 (accessed on 15 November 2023). [CrossRef]
27. Keil, A.; Gruber, T.; Müller, M.M. Functional correlates of macroscopic high-frequency brain activity in the human visual system. *Neurosci. Biobehav. Rev.* **2001**, *25*, 527–534. Available online: https://www.sciencedirect.com/science/article/pii/S0149763401000318 (accessed on 15 November 2023). [CrossRef]
28. Friese, U.; Köster, M.; Hassler, U.; Martens, U.; Trujillo-Barreto, N.; Gruber, T. Successful memory encoding is associated with increased cross-frequency coupling between frontal theta and posterior gamma oscillations in human scalp-recorded EEG. *NeuroImage* **2013**, *66*, 642–647. Available online: https://www.sciencedirect.com/science/article/pii/S105381191201097X (accessed on 15 November 2023). [CrossRef]
29. Gruber, T.; Tsivilis, D.; Montaldi, D.; Müller, M.M. Induced gamma band responses: An early marker of memory encoding and retrieval. *NeuroReport* **2004**, *15*, 1837. Available online: https://journals.lww.com/neuroreport/fulltext/2004/08060/Induced_gamma_band_responses__an_early_marker_of.30.aspx?casa_token=G5ex5piwVk4AAAAA:0WiZDlFm54-hHMYBi4XQaFtjR8wjJctAyelblKH_QysP3_PQkB3OCnfw7moUyr_xoz-zoE6N9boOGeBkl_z9S32xaQ (accessed on 15 November 2023). [CrossRef]
30. Osipova, D.; Takashima, A.; Oostenveld, R.; Fernández, G.; Maris, E.; Jensen, O. Theta and Gamma Oscillations Predict Encoding and Retrieval of Declarative Memory. *J. Neurosci.* **2006**, *26*, 7523–7531. Available online: https://www.jneurosci.org/content/26/28/7523 (accessed on 15 November 2023). [CrossRef]
31. Honkanen, R.; Rouhinen, S.; Wang, S.H.; Palva, J.M.; Palva, S. Gamma Oscillations Underlie the Maintenance of Feature-Specific Information and the Contents of Visual Working Memory. *Cereb. Cortex* **2015**, *25*, 3788–3801. [CrossRef]
32. Roux, F.; Uhlhaas, P.J. Working memory and neural oscillations: Alpha–gamma versus theta–gamma codes for distinct WM information? *Trends Cogn. Sci.* **2014**, *18*, 16–25. Available online: https://www.cell.com/trends/cognitive-sciences/abstract/S1364-6613(13)00231-3 (accessed on 15 November 2023). [CrossRef]
33. van Vugt, M.K.; Schulze-Bonhage, A.; Litt, B.; Brandt, A.; Kahana, M.J. Hippocampal Gamma Oscillations Increase with Memory Load. *J. Neurosci.* **2010**, *30*, 2694–2699. Available online: https://www.jneurosci.org/content/30/7/2694 (accessed on 15 November 2023). [CrossRef]

34. Chaieb, L.; Leszczynski, M.; Axmacher, N.; Höhne, M.; Elger, C.E.; Fell, J. Theta-gamma phase-phase coupling during working memory maintenance in the human hippocampus. *Cogn. Neurosci.* **2015**, *6*, 149–157. [CrossRef]
35. Lenroot, R.K.; Giedd, J.N. Brain development in children and adolescents: Insights from anatomical magnetic resonance imaging. *Neurosci. Biobehav. Rev.* **2006**, *30*, 718–729. Available online: https://www.sciencedirect.com/science/article/pii/S0149763406000455 (accessed on 15 November 2023). [CrossRef]
36. Diamond, A.; Goldman-Rakic, P.S. Comparison of human infants and rhesus monkeys on Piaget's AB task: Evidence for dependence on dorsolateral prefrontal cortex. *Exp. Brain Res.* **1989**, *74*, 24–40. [CrossRef] [PubMed]
37. Luna, B.; Sweeney, J.A. Studies of brain and cognitive maturation through childhood and adolescence: A strategy for testing neurodevelopmental hypotheses. *Schizophr. Bull.* **2001**, *27*, 443–455. [CrossRef]
38. Poudel, S.; Denicola-Prechtl, K.; Nelson, J.A.; Behboudi, M.H.; Benitez-Barrera, C.; Castro, S.; Maguire, M.J. Rethinking household size and children's language environment. *Dev. Psychol.* **2023**. [CrossRef] [PubMed]
39. Maguire, M.J.; Schneider, J.M. Socioeconomic status related differences in resting state EEG activity correspond to differences in vocabulary and working memory in grade school. *Brain Cogn.* **2019**, *137*, 103619. Available online: https://www.sciencedirect.com/science/article/pii/S0278262618304755 (accessed on 1 November 2023). [CrossRef] [PubMed]
40. Brauer, J.; Anwander, A.; Friederici, A.D. Neuroanatomical Prerequisites for Language Functions in the Maturing Brain. *Cereb. Cortex* **2011**, *21*, 459–466. [CrossRef]
41. Friederici, A.D.; Brauer, J.; Lohmann, G. Maturation of the Language Network: From Inter- to Intrahemispheric Connectivities. *PLoS ONE* **2011**, *6*, e20726. Available online: https://journals.plos.org/plosone/article?id=10.1371/journal.pone.0020726 (accessed on 15 November 2023). [CrossRef]
42. Brauer, J.; Anwander, A.; Perani, D.; Friederici, A.D. Dorsal and ventral pathways in language development. *Brain Lang.* **2013**, *127*, 289–295. Available online: https://www.sciencedirect.com/science/article/pii/S0093934X13000709 (accessed on 15 November 2023). [CrossRef] [PubMed]
43. Gogtay, N.; Giedd, J.N.; Lusk, L.; Hayashi, K.M.; Greenstein, D.; Vaituzis, A.C.; Nygent, T.F., III; Herman, D.H.; Clasen, L.S.; Toga, A.W.; et al. Dynamic mapping of human cortical development during childhood through early adulthood. *Proc. Natl. Acad. Sci. USA* **2004**, *101*, 8174–8179. Available online: https://www.pnas.org/doi/abs/10.1073/pnas.0402680101 (accessed on 15 November 2023). [CrossRef] [PubMed]
44. Krause, C.M.; Salminen, P.A.; Sillanmäki, L.; Holopainen, I.E. Event-related desynchronization and synchronization during a memory task in children. *Clin. Neurophysiol.* **2001**, *112*, 2233–2240. Available online: https://www.sciencedirect.com/science/article/pii/S1388245701006848 (accessed on 15 November 2023). [CrossRef] [PubMed]
45. Spironelli, C.; Angrilli, A. Developmental aspects of language lateralization in delta, theta, alpha and beta EEG bands. *Biol. Psychol.* **2010**, *85*, 258–267. Available online: https://www.sciencedirect.com/science/article/pii/S0301051110001924 (accessed on 15 November 2023). [CrossRef] [PubMed]
46. Benasich, A.A.; Gou, Z.; Choudhury, N.; Harris, K.D. Early cognitive and language skills are linked to resting frontal gamma power across the first 3 years. *Behav. Brain Res.* **2008**, *195*, 215–222. [CrossRef] [PubMed]
47. Cantiani, C.; Piazza, C.; Mornati, G.; Molteni, M.; Riva, V. Oscillatory gamma activity mediates the pathway from socioeconomic status to language acquisition in infancy. *Infant Behav. Dev.* **2019**, *57*, 101384. Available online: https://www.sciencedirect.com/science/article/pii/S0163638319301390 (accessed on 15 November 2023). [CrossRef]
48. Lum, J.A.G.; Clark, G.M.; Bigelow, F.J.; Enticott, P.G. Resting state electroencephalography (EEG) correlates with children's language skills: Evidence from sentence repetition. *Brain Lang.* **2022**, *230*, 105137. Available online: https://www.sciencedirect.com/science/article/pii/S0093934X22000670 (accessed on 15 November 2023). [CrossRef]
49. Heim, S.; Keil, A.; Choudhury, N.; Thomas Friedman, J.; Benasich, A.A. Early gamma oscillations during rapid auditory processing in children with a language-learning impairment: Changes in neural mass activity after training. *Neuropsychologia* **2013**, *51*, 990–1001. Available online: https://www.sciencedirect.com/science/article/pii/S0028393213000171 (accessed on 15 November 2023). [CrossRef]
50. Friederici, A.D. Children's sensitivity to function words during sentence comprehension. *De Gruyter Mouton* **1983**, *21*, 717–740. Available online: https://www.degruyter.com/document/doi/10.1515/ling.1983.21.5.717/html?lang=en (accessed on 15 November 2023). [CrossRef]
51. Duffau, H.; Gatignol, P.; Mandonnet, E.; Peruzzi, P.; Tzourio-Mazoyer, N.; Capelle, L. New insights into the anatomo-functional connectivity of the semantic system: A study using cortico-subcortical electrostimulations. *Brain* **2005**, *128 Pt 4*, 797–810. [CrossRef]
52. Friederici, A.D. Pathways to language: Fiber tracts in the human brain. *Trends Cogn. Sci.* **2009**, *13*, 175–181. [CrossRef]
53. Friederici, A.D. The brain basis of language processing: From structure to function. *Physiol. Rev.* **2011**, *91*, 1357–1392. [CrossRef]
54. Sakai, K.L. Language Acquisition and Brain Development. *Science* **2005**, *310*, 815–819. Available online: https://www.science.org/doi/10.1126/science.1113530 (accessed on 15 November 2023). [CrossRef]
55. Vigneau, M.; Beaucousin, V.; Hervé, P.Y.; Duffau, H.; Crivello, F.; Houdé, O.; Mazoyer, B.; Tzourio-Mazoyer, N. Meta-analyzing left hemisphere language areas: Phonology, semantics, and sentence processing. *NeuroImage* **2006**, *30*, 1414–1432. Available online: https://www.sciencedirect.com/science/article/pii/S1053811905024511 (accessed on 15 November 2023). [CrossRef] [PubMed]
56. Caplan, D. Functional neuroimaging studies of syntactic processing. *J Psycholinguist. Res.* **2001**, *30*, 297–320. [CrossRef] [PubMed]

57. Santi, A.; Grodzinsky, Y. fMRI adaptation dissociates syntactic complexity dimensions. *NeuroImage* **2010**, *51*, 1285–1293. Available online: https://www.sciencedirect.com/science/article/pii/S1053811910003216. (accessed on 15 November 2023). [CrossRef]
58. Brauer, J.; Friederici, A.D. Functional neural networks of semantic and syntactic processes in the developing brain. *J. Cogn. Neurosci.* **2007**, *19*, 1609–1623. [CrossRef] [PubMed]
59. Rommers, J.; Dijkstra, T.; Bastiaansen, M. Context-dependent Semantic Processing in the Human Brain: Evidence from Idiom Comprehension. *J. Cogn. Neurosci.* **2013**, *25*, 762–776. [CrossRef]
60. Giannopoulos, A.E.; Zioga, I.; Kontoangelos, K.; Papageorgiou, P.; Kapsali, F.; Capsalis, C.N.; Papageorgiou, C. Deciding on Optical Illusions: Reduced Alpha Power in Body Dysmorphic Disorder. *Brain Sci.* **2022**, *12*, 293. Available online: https://www.mdpi.com/2076-3425/12/2/293 (accessed on 15 November 2023). [CrossRef]
61. Delorme, A.; Makeig, S. EEGLAB: An open source toolbox for analysis of single-trial EEG dynamics including independent component analysis. *J. Neurosci. Methods* **2004**, *134*, 9–21. Available online: https://www.sciencedirect.com/science/article/pii/S0165027003003479 (accessed on 15 November 2023). [CrossRef]
62. Winkler, I.; Haufe, S.; Tangermann, M. Automatic classification of artifactual ICA-components for artifact removal in EEG signals. *Behav. Brain Funct.* **2011**, *7*, 30. [CrossRef]
63. Winkler, I.; Brandl, S.; Horn, F.; Waldburger, E.; Allefeld, C.; Tangermann, M. Robust artifactual independent component classification for BCI practitioners. *J. Neural Eng.* **2014**, *11*, 035013. [CrossRef] [PubMed]
64. Benjamini, Y.; Hochberg, Y. Controlling the False Discovery Rate: A Practical and Powerful Approach to Multiple Testing. *J. R. Stat. Soc. Ser. B (Methodol.)* **1995**, *57*, 289–300. Available online: https://onlinelibrary.wiley.com/doi/abs/10.1111/j.2517-6161.1995.tb02031.x (accessed on 15 November 2023). [CrossRef]
65. Muller, P.; Parmigiani, G.; Rice, K.; FDR and Bayesian Multiple Comparisons Rules. Johns Hopkins University, Dept of Biostatistics Working Papers 18 July 2006. Available online: https://biostats.bepress.com/jhubiostat/paper115 (accessed on 15 November 2023).
66. Fedorenko, E.; Scott, T.L.; Brunner, P.; Coon, W.G.; Pritchett, B.; Schalk, G.; Kanwisher, N. Neural correlate of the construction of sentence meaning. *Proc. Natl. Acad. Sci. USA* **2016**, *113*, E6256–E6262. Available online: https://pnas.org/doi/full/10.1073/pnas.1612132113 (accessed on 15 November 2023). [CrossRef] [PubMed]
67. Marslen-Wilson, W.; Tyler, L.K. Processing structure of sentence perception. *Nature* **1975**, *257*, 784–786. Available online: https://www.nature.com/articles/257784a0 (accessed on 15 November 2023). [CrossRef] [PubMed]
68. Gentner, D. Verb semantic structures in memory for sentences: Evidence for componential representation. *Cogn. Psychol.* **1981**, *13*, 56–83. Available online: https://www.sciencedirect.com/science/article/pii/0010028581900049 (accessed on 15 November 2023). [CrossRef]
69. Priyastini, N.; Puspani, I.A.M.; Antari, N.M.S. The Semantic Roles of the English Action Verbs. *J. Hum. Fak. Ilmu Budaya Unud* **2016**, *16*, 159–166. Available online: https://ojs.unud.ac.id/index.php/sastra/article/download/33496/20298 (accessed on 15 November 2023).
70. Wang, J.; Wagley, N.; Rice, M.L.; Booth, J.R. Semantic and syntactic specialization during auditory sentence processing in 7–8-year-old children. *Cortex* **2021**, *145*, 169–186. [CrossRef]

brain sciences

MDPI

Article

Changes in Spatiotemporal Dynamics of Default Network Oscillations between 19 and 29 Years of Age

Thorsten Fehr [1,2,*], Sophia Mehrens [1], Marie-Christine Haag [1], Anneke Amelung [1] and Kilian Gloy [1]

[1] Institute for Psychology, University of Bremen, 28357 Bremen, Germany; k.gloy@uni-bremen.de (K.G.)
[2] Center for Advanced Imaging, University of Bremen, 28357 Bremen, Germany
* Correspondence: fehr@uni-bremen.de; Tel.: +49-421-218-68741

Abstract: The exploration of functional resting-state brain developmental parameters and measures can help to improve scientific, psychological, and medical applications. The present work focussed on both traditional approaches, such as topographical power analyses at the signal space level, and advanced approaches, such as the exploration of age-related dynamics of source space data. The results confirmed the expectation that the third life decade would show a kind of stability in oscillatory signal and source-space-related parameters. However, from a source dynamics perspective, different frequency ranges appear to develop quite differently, as reflected in age-related sequential network communication profiles. Among other discoveries, the left anterior cingulate source location could be shown to reduce bi-directional network communication in the lower alpha band, whereas it differentiated its uni- and bidirectional communication dynamics to sub-cortical and posterior brain locations. Higher alpha oscillations enhanced communication dynamics between the thalamus and particularly frontal areas. In conclusion, resting-state data appear to be, at least in part, functionally reorganized in the default mode network, while quantitative measures, such as topographical power and regional source activity, did not correlate with age in the third life decade. In line with other authors, we suggest the further development of a multi-perspective approach in biosignal analyses.

Keywords: EEG; resting state; default mode network; neural oscillations; source model; brain dynamics

1. Introduction

1.1. About the Functional Neuroanatomy of the Human Default Mode Network (DMN) and Its Developmental Changes

One of the problems with which the neuroscientist is confronted while examining brain physiological changes over time is their non-linear developmental nature [1–6]. Furthermore, individual factors, as well as contextual and learning histories, were discovered largely to modulate the elaboration of complex perception–action networks, particularly in the heteromodal association cortices [7]. Anatomical properties such as symmetrically closed (anisotropic) and asymmetrically open (isotropic) field cytoarchitectonic [8] and related physiological parameters, such as, for example, local field potentials and brain oscillations in different frequency ranges and neural sub-systems [5,9], were discovered to substantially modulate brain function and oscillations via dynamic neural network interaction [10]. There is evidence that oscillatory brain activity is modulated by white [11] and grey [12] matter plasticity/maturation, and/or vice versa, which would also, at least in part, explain the substantial changes in (resting-state) brain oscillations during healthy and pathological development.

Electroencephalography (EEG) and magnetoencephalography (MEG) provide powerful scientific and non-invasive tools for the examination of fast, dynamic brain oscillatory fluctuations and their changes during brain development. Popov et al. [13] reported good and very good retest reliability for several resting-state EEG parameters, such as power estimates on both scalp and source levels, individual alpha peak power and frequency, microstate measures, and connectivity parameters. Whereas at the signal-space level (i.e., topographical signals at electrode positions from standardized scalp locations), data mainly provide information of high temporal resolution, source space analyses provide less spatially smeared but also highly temporal information [14]. However, EEG source space analyses confront us with the so-called inverse problem [15]. As discrete source analyses are based on approximation algorithms, there is an infinite number of solutions that can explain a particular scalp topography. In the present work, we suggest, therefore, a hypothesis-driven ad hoc model [14,16–18] to be seeded into source space for further spatiotemporal analyses of band-pass filtered data in different pre-defined frequency ranges.

There is a variety of seminal work on changes in oscillatory brain activity over the life span, and these changes appear to be of a rather non-linear/ordinal nature [13,19–21], accompanied by more or less dramatic reorganization and maturation of the neural network architecture, such as prominent phases of dendritic proliferation (e.g., during childhood), pruning (e.g., during adolescence), and myelinization (i.e., till the end of the third life decade), progressing predominantly from primary to heteromodal association cortices [3,4,6,22]. Most prominently, resting-state alpha oscillations showed a peak around ten Hz at posterior electrode sites in most individuals (e.g., [23–25]). Whereas alpha peak activation has been reported to be intra-individually stable on the short-term scale [26], it appears to increase from childhood to adolescence [19,27,28], shows a plateau during early adulthood, and decreases again in later adulthood. Multi-channel MEG data were reported to reveal larger general absolute power values in adolescents as compared to adults in delta and theta, as well as lower alpha power (1.5–10.5 Hz), lower beta frequency band power (13.0–21.5 Hz), and larger higher alpha power (10.5–13.0 Hz) at frontal and frontotemporal sensor sites [20]. Adolescents showed relatively larger relative power values in delta and theta bands and lower relative power values in alpha, beta, and gamma frequency ranges. Cross-sectional correlations within groups showed a regional decrease in absolute delta and theta in both adolescents and adults and a general decrease in absolute lower alpha power in adolescents. Absolute beta and lower gamma band power partially decreased mainly in adults, and less prominently in adolescents. Higher absolute gamma power (65–100 Hz) increased over central sites prominently in adults. Relative power values showed a general decrease in delta and an increase in beta frequency ranges in adults and a decrease in relative lower alpha band power in adolescents. Summarizing, it appears that brain oscillations across childhood, adolescence, and early and late adulthood change in a non-linear—or to say non-uniform—manner, whereas the third life decade (i.e., during the twentieth) was discovered to show a kind of plateau in oscillatory brain development [13]. In the present work, we addressed this question from different exploratory perspectives to stimulate further prospective hypotheses building and testing.

Brain areas discovered to be involved in the DMN were discovered to become especially active during the mental resting state, which was associated with different kinds of internal mental states, such as introspection and mind wandering [29,30], but also self-referential thought [31,32]. It also appears that DMN activity decreases during tasks with external stimulation and demands (e.g., [33]), which would make the DMN a complementary counterpart for perceptually and executively driven mental processes. Functional neuroimaging approaches revealed differences in neural network dynamics (e.g., in connectivity) across age during adulthood [34–36], especially involving communication between anterior and posterior parts of the DMN, with medial prefrontal areas serving as an important neural node [37–39].

Chow et al. [40] speculated that because most work on age-related neural network dynamics in the DMN came from functional neuroimaging studies (i.e., fMRI and PET), the discussion might predominantly be directed towards slower network activation (i.e., as functional neuroimaging approaches mostly relied on slow haemodynamic brain responses), while higher frequency dynamics might not be considered as much as they should [41,42]. Approaches with higher temporal resolution, such as EEG and MEG, should come more into play for the examination of age-related changes in higher-frequency-related network dynamics (e.g., [40]). However, as Chow and colleagues [40] critically mentioned, most EEG work on age-related resting-state network dynamics was based on signal space data (i.e., at electrode sensor level and, respectively, smeared surface topographies), leading to valuable but quite heterogeneous findings (e.g., [43–46]) on the basis of different methodological approaches for the determination of respective neural network dynamics in theta, alpha, and beta frequency ranges (e.g., [43–48]). Chow et al. [40] used source space data to explore age-related differences in EEG phase synchronization of the DMN in resting-state (between one and 50 Hz), particularly between medial prefrontal areas and other brain regions of the DMN. The reported data reflected a decrease in higher alpha and an increase in theta and beta frequency band connectivity between medial prefrontal and other brain areas in the DMN, indicating, according to the author's conclusions, a non-uniform frequency depending on the nature of the neural DMN dynamics [40].

1.2. Which Brain Areas Shall Be Included in an Ad Hoc Model for the Exploration of Respective Neural Generator Dynamics in the Default Mode Network (DMN)?

1.2.1. Neurophysiological Mapping of the Default Mode Network DMN

One has to seriously differentiate between structural and functional brain networks. Brain development changes, especially during the first three life decades, while structural sprouting, pruning, and myelinization follow non-linear trajectories with potential modulatory influence on functional development [3,4,6,7,22,49]. Together with individually experienced context factors and respective learning histories, this appears to massively shape mental perceptual and conceptual—or differential psychological—processing concepts at different complexity levels. This has been discussed in relation to an elaboration of neural perception–action cycle networks representing stereotypes of individual cognitive style and mental strategy [7,22,50,51]. Both (micro-)structural and (psycho-)functional development might be discussed in relation to a kind of default network readiness of the brain across individual development. Involved neural structures in the DMN have been identified by examining their functional involvement, particularly during mental resting condition. Association cortices and networks with particular structural connectivity have been discussed as important neural instances [31,52]. It has been shown that regions within the DMN are part of a group of nodes that have stronger connections with each other than with other regions of the cortex [53]. For example, the precuneus was identified as one such potentially important hub [52], which is in line with the discussion of the precuneus as part of the so-labeled heteromodal association cortices (HACs), which become more and more involved during individual learning history for the neural elaboration of complex cognitive and emotional mental processing [7,22,51,54]. The precuneus is strongly and reciprocally connected to different association areas, such as the superior parietal lobule, temporal polymodal cortices, and prefrontal lobe structures. These brain areas were discovered to be important for both the functional organization of the brain and the DMN, as they coordinate multiple streams of information also related to other heteromodal and paralimbic association cortices that have been discovered to provide relevant neural instances for the integration of multiple streams of cognitive and emotional information [52].

Additional sub-network units have been discovered to be involved in the DMN, such as, for example, the frontoparietal network (FPN) and the dorsal attention network (DAN) [55]. FPN, DAN, and HAC networks were discovered to be interactively involved in state-dependent changes in younger and older adults. Functional connectivity between these networks at rest and during cognitive demands (i.e., during the multi-source interference task, MSIT) showed differences in the interactions in the FPN and DAN and between these sub-networks and the DMN. In particular, the FPN was more strongly coupled with the DMN at resting, while the DAN was stronger recruited during MSIT processing in both age groups [56]. The increased interaction between the FPN and the DMN during resting-states suggests that frontal brain regions modulate the activity of the DMN, especially when the brain switches to an internal processing mode. In the following sub-sections, a network of particular brain structures is discussed as a series of potentially important neural instances for the regulation of the DMN.

1.2.2. Frontal Brain Structures and the Default Mode Network (DMN)

Brain structures discovered to be involved in the DMN have also been discovered to be recruited during autobiographical, social, and emotional task processing [57]. Connectivity between right inferior frontal gyrus and several DMN areas in the left hemisphere were related to emotional interference processing, regulating, among other things, the interplay between internal and externally driven network activities [58]. It has also been discovered that the medial superior frontal gyrus and the middle frontal gyrus potentially provide important nodes in the DMN [57,59], and the ventromedial prefrontal cortex (vmPFC) has been discovered to be involved in ongoing and non-externally driven thought patterns, particularly those with episodic or social content [57,60]. The vmPFC appears to be also involved in the suppression of stimulus-related tasks associated with the DMN [59,61]. The anterior ventrolateral inferior frontal cortex [61] and the anterior cingulate cortex (ACC) has also been discussed as parts of the DMN [59,62–66]. Koch and colleagues [66] used fMRI data to investigate the strength of the interconnectivity of DMN regions and the extent of DMN coactivation. A conjunction analysis across twelve fMRI studies further illustrates the role of the ACC in task-related deactivation [66].

1.2.3. Parietal and Occipital Brain Structures and the Default Mode Network (DMN)

The posterior cingulate cortex (PCC) has repeatedly been shown to be connected to other areas of the DMN [61,65,67–69]. It appears as one of the nodes that shows the highest functional coactivation with other regions of the DMN, such as the ACC, the superior frontal gyrus, and the hippocampus [70]. A significant reduction in this coactivation in the PCC was observed in late adulthood [65], which could analogously be found for the right precuneus [59,67,70,71]. Davey et al. [72] discussed the inferior parietal lobule (IPL) as one of the key structures of the DMN that was related to activation patterns in the ventral posterior cingulate (PCC) and the medial prefrontal cortex (medPFC) during rest compared to situations requiring external attention. The IPL could also be related to processes of self-referencing [71]. The intraparietal sulcus (IPS) was discussed as a major instance in spatial orientation and visual attention [73], but it was also discussed as an important part of the DMN [57]. The caudal IPS has further been reported to show functional connectivity to several cortical areas, including the lateral and medial prefrontal cortex, the posterior cingulate cortex, the anterior middle frontal cortex, the inferior temporal cortex, and the parahippocampal gyrus [74,75].

Wu et al. [76] observed that better task performance was associated with increased suppression of irrelevant information in the PCC, in occipitotemporal cortex areas, and the extrastriate cortex, which has been concluded to be consistent with the generally known assumption about the function of the DMN [76].

1.2.4. Temporal Brain Structures and the Default Mode Network (DMN)

Both lateral and medial temporal brain areas were related to the DMN. A reduction of cognitive load during decision making has been related to an increase in activation in medial temporal areas but also in particular parietal and frontal brain regions [77]. Based on resting-state connectivity analyses, Briggs and colleagues [78] discussed the middle temporal gyrus from posterior parts to the anterior pole as an integrative instance of the DMN modulating semantic processing (i.e., learning and, in particular, verbal memory) and language production. Furthermore, medial middle temporal gyrus was discovered to be involved in the modulation of semantic memory and control networks [79]. These and other findings emphasize the assumption that the MTG plays a central role in integrating automatic retrieval in the default mode network with cognitively demanding, goal-directed thinking tasks and serves as a central hub in the DMN [57,70].

1.2.5. Subcortical Brain Structures and the Default Mode Network (DMN)

The regions that are part of the default mode network (DMN) are reported inconsistently across studies. One of the regions that is not always associated with the DMN is the thalamus, with its manyfold specific and unspecific nuclei and connections to numerous cortical and sub-cortical brain networks. The thalamus acts as a central hub in the brain that receives sensory information from various sensory organs and relays it to the appropriate regions of the brain, including areas associated with the DMN. Furthermore, the thalamus also sends feedback signals to sensory regions, emphasizing its role in modulating cortical activity and integrating information [80]. Additionally, the thalamus has been shown to connect with other key regions of the DMN, such as the posterior cingulate cortex. These anatomical and functional connections between the thalamus and core areas of the DMN suggest that the thalamus plays an essential role in coordinating the activity of the DMN and should therefore be considered in a respective DMN model [61,80–83]. Furthermore, thalamus nuclei were associated with episodic memory processing and spatial navigation, which were also associated with activations of the DMN [57,84]. The hippocampus is an additional structure that should be considered in a DMN model as functional connectivity studies provided evidence for its functional coupling to DMN-structures such as the precuneus and both the dorsal and the ventral mPFC [81].

1.2.6. Can Particular Brain Areas Be Excluded from the Default Mode Network (DMN)?

From an equipotentiality and holistically driven standpoint, it appears inappropriate to exclude particular brain areas from any kind of more or less complex mental process, but from a modularity point of view, it would be appropriate [22,85]. Thus, Fehr [22] suggested a hybrid strategy when discussing local neural network involvement during the processing of any kind of information processing, but particularly the organization of complex mental processing, which was discovered to be reflected in terms of HOW information was processed, not WHAT concrete information [5,9,22,50,51]).

Smallwood et al. [77] suggested a model for the DMN that predominantly includes those brain areas which show the highest distance to the so-called topographical, primary cortices, such as primary and retinotopically organized visual, tonotopically organized auditory, and body landmark (i.e., homunculus)-oriented somatotopic, and primary motor cortices. Later, primary cortices were also discovered to host so-called phyletic memories, from which all other memories emerge to secondary and heteromodal cortices during individual learning histories [50,51]. Analogously, Fehr [22] further suggested developmental starting nodes in the middle and superior, medial frontal, and intra-parietal areas, which he discussed to provide important neural instances, from which complex mental processing performance (i.e., executive, working memories, integrative information processing, canonical, spatial coordination, etc.) emerges and plastically recruits more and more heteromodal, adjacent brain areas during individual development. Patterns of activation during complex mental processing were discovered to reflect individual mental strategies (i.e., the HOW) applied during task processing. Furthermore, many studies using too-restrictive statistical alpha-error correction might predominantly report a kind of by-product activation pattern, relying on these developmental starting points and not the relevant adjacent and more individually recruited task-related neural resources [86]. Latter developmental neural starting points might also provide good candidates to be considered for a DMN model. In Section 2.3.2, respective anatomical center coordinates are defined and listed in detail.

1.3. The Present Work, Exploratory Questions, and Several Working Hypotheses

Data reported so far imply that there is a lot more to do in terms of the exploratory qualification of the nature of age-related neural network dynamics in the DMN. The present work addresses changes in brain oscillations and the follow-up dynamics of respective regional source equivalents during the third life decade via cross-sectional correlation analyses. Here, we followed both a traditional and a data-driven exploratory approach. Both topographical signal and ad hoc modeled source space network data were related to age and several state- and trait-related covariates such as ratings of current psychological and physiological well-being, acute stress level, and several personality scales. The following main hypotheses and exploratory questions were addressed:

(1) It was expected that both relative and absolute power spectra peak at around 10 Hz with topographical dominance over posterior brain areas.

(2) Both absolute and relative topographically distributed power spectra were expected to substantially corelate with age over the third life decade.

(3) The carefully deduced default mode network source model (DNMSM, see Section 2.3.2 for technical details) will produce less than 10% residual variance when applied to band-pass-filtered single epoch signal space data in different traditionally defined frequency ranges between one and 35 Hz (i.e., delta, theta, alpha low and high, beta low and high, and gamma).

(4) Follow-up communication dynamics in the DNMSM were explored, and it was expected that different frequency band oscillations might produce different profiles of functional changes across the third life decade. The present data shall provide a hypothetical basis for subsequent studies considering different mental functions, age-cohorts, and pathological conditions.

2. Materials and Methods

For the present work, both traditional and data-driven exploratory approaches were considered. In a first step, resting-state EEG data from two laboratories were analyzed via fast Fourier transform and topographical correlation analyses with age and several state and trait variables. In a second step, a default-mode network-oriented source model was introduced and applied to explore potential age-related changes in neurophysiological dynamics in resting-state EEG data across the third life decade.

2.1. Sample characteristics

Resting-state EEG data and several test-scale values of 54 participants from two different studies on similar social decision making tasks (not published so far) were obtained and analyzed. The used experimental setups were designed according to the Code of Ethics of the World Medical Association (Declaration of Helsinki, published in the British Medical Journal, 18 July 1964), and stimulus materials were approved by the local ethics committee. The participant group consists of 39 male and 15 female individuals (range: 19 to 29; mean age: 23.7+/−2.8 years, with four 19-, four 20-, four 21-, nine 22-, five 23-, ten 24-, two 25-, three 26-, six 27-, six 28-, and one 29-year-old participants). All of them were right-handed, as determined by the Edinburgh Inventory [87], and all participants gave written informed consent with respect to their participation and did not report any chronic psychological or physiological disorders. Before the examination, participants were asked for psychological, neurological, and other severe and/or chronic diseases to exclude individuals with the respective pathological backgrounds from the sample. Thus, in the present work, only healthy young adults (all of them university students) in their third life decade (i.e., between 19 and 29 years, see also above) of brain development were included as participants.

2.2. Experimental Procedures

Different variable categories were considered in the present work: AGE (in years) was included as the most important independent variable. The state variables PSYCH (acute psychological well-being), PHYS (acute physiological well-being), and STRESS (acute stress level) were assessed by ten-leveled self-rating Likert scales (1–10). The so-called "big five" personality trait dimensions—Neuroticism, Extraversion, Openness to Experience, Agreeableness, and Conscientiousness—were assessed with the 60-item NEO-FFI-inventory, introduced by Costa and McRae [88] (German version by [89]).

Resting EEG data included in the present work were measured after participants performed a series of quasi-naturalistic social decision tasks [90]. During rest, participants were asked to watch a smoothed dot in the center of a black screen in front and wipe all thoughts that come into their mind beside. All participants reported that they could easily do that and that this relaxing phase was a welcome break after the previous experimental runs.

In both laboratories, participants were familiarized with the measurement environment, and recordings were performed in a dimly lit and comfortably heated room without any noise contamination.

2.3. Multi-Channel-Electroencephalography

Data from two different laboratories were included, both using a comparable 64-electrode measurement setup. In the first laboratory, at Bremen University, EEG recording was conducted with 64 Ag/AgCl scalp electrodes placed according to the extended international standardized 10-10 system (data were measured average referenced). These electrodes included FP1, FP2, AFz, AF3, AF4, AF7, AF8, Fz, F1, F2, F3, F4, F5, F6, F7, F8, FCz, FC1, FC2, FC3, FC4, FC5, FC6, FC7, FC8, Cz, C1, C2, C3, C4, C5, C6, T7, T8, CPz, CP1, CP2, CP3, CP4, CP5, CP6, TP7, TP8, TP9, TP10, Pz, P1, P2, P3, P4, P5, P6, P7, P8, POz, PO3, PO4, PO7, PO8, PO9, PO10, Oz, O1, and O2 positions. In both participating laboratories, an electro-oculogram (EOG) was recorded with four additional Ag/AgCl electrodes that were attached to the left and right canthi (horizontal EOG) and below and above the right eye (vertical EOG). The ground electrode was placed on the left masseter muscle near the chin. The EEG data were recorded with EEmagine software (version 3.3). The signal was amplified (REFA 136 multi-channel system; TMS international) and digitized with a sampling rate of 512 Hz. The impedance of each channel was kept below 10 kOhm.

In the second laboratory, at the University of Cologne (Germany), EEG data were recorded with 64 Ag/AgCl scalp electrodes distributed over the scalp according to the 10-10 system (Cz, FC2, CP2, CP1, FC1, F4, C4, P4, Pz, P3, C3, F3, Fz, FC6, CP6, CP5, FC5, FP2, F8, T8, P8, O2, Oz, O1, P7, T7, F7, FP1, TP10, PO10, PO9, TP9, FCz, C2, CPz, C1, F2, FC4, CP4, P2, P1, CP3, FC3, F1, AFz, AF4, F6, C6, P6, PO4, POz, PO3, P5, C5, F5, AF3, AF8, FT8, TP8, PO8, PO7, TP7, FT7, AF7) and using a machine by BioSemi; the model was ActiveTwo AD-box ADC-12. Data were digitized with a sampling rate of 512 Hz. The default reference electrodes were CSM and DLR. The recorded data are always referenced to these electrodes; during the processing of the data, however, the reference was changed to average reference. EEG data were recorded with the software ActiView 720 LoRes. The ground electrode was placed on the left cheek near the chin.

2.3.1. Signal Space: Topographical Fast Fourier Analyses (Absolute and Relative Power Spectra)

FFT was calculated for continuous data recorded during resting condition, and values were obtained by averaging power spectra revealed by the application of FFT procedures on moving data time segments of 2000 ms (28.8+/−2.7 trials). Each data segment overlapped

50% with the next segment and was multiplied by a cosine squared (cos2) window. This combination of overlap and windowing ensures that each time point contributes equally to the mean spectrum (since cos2(x) + sin2(x) = 1) (see also BESA® -Software, version 5.2.2).

FFT data were divided into seven frequency bands: delta [1.0–4.0 Hz], theta [4.0–8.0 Hz], alpha-low [8.0–10.5 Hz], alpha-high [10.5–13.0 Hz], beta-low [13.0–21.5 Hz], beta-high [21.5–30.0 Hz], and one gamma band [30.0–35.0 Hz]. In addition to the absolute power band analyses, relative spectra were calculated (band-related power values divided by absolute power in the whole spectrum).

After careful visual inspection of the averaged referenced raw data, several channels were interpolated (no more than five electrode positions per data set), and artefact contaminated trials were excluded from further analyses on the basis of individual visual trial-by-trial inspection. About 30 prototypic blinks were averaged for each individual to produce topographical templates for their respective artefact correction implemented in BESA®-software (version 5.2.2). The respective procedure follows an adaptive correction strategy. Brain activity is estimated by a defined data epoch (here, an averaged blink epoch as mentioned above). Data epochs were considered to represent brain activity when the correlation between data and artifact topography did not exceed a certain threshold and when the signal amplitudes were below a particular threshold. The remaining data segments served as a basis for a principal component analysis (PCA); identifying all PCA components that explained more than a specified variance level provided the topographical basis for the artefact correction procedure. Through this procedure, the data were decomposed using all respective topographies into a linear combination of brain and artifact activities. Finally, the estimated artifact signals are much less overlapped with brain activity and can be subtracted from the original signals without substantial functional signal loss [91].

For topographical, correlative signal space analyses, FFT data were, respectively, averaged over the following electrode clusters: midline frontal (f: AF3, AFz, AF4, F1, Fz, F2), central (c: FC1, FCz, FC2, C1, Cz, C2), parietal (p: CP1, CPz, CP2, P1, Pz, P2), occipital (o: PO3, POz, PO4, O1, Oz, O2), lateral frontotemporal left (ftl: F7, F5, F3, FT7, FC5, FC3), frontotemporal right (ftr: F4, F6, F8, FC4, FC6, FT8), parietotemporal left (ptl: T7, C5, C3, TP7, CP5, CP3), parietotemporal right (ptr: C4, C6, T8, CP4, CP6, TP8), occipitotemporal left (otl: P7 P5 P3 P9 PO7), and occipitotemporal right (otr: P4 P6 P8 PO8 P10).

2.3.2. Source Space: Default Network Model for Ad Hoc Seeding Procedures

For the present work, we introduced a Default Mode Network Source Model (DMNSM, see Figure 1A, middle panel, and Table 1 for anatomical labels and shortcuts), and respective Talairach center coordinates [92] to be seeded ad hoc [15] and used to obtain source wave forms (in nAm units) from band-pass-filtered data (see Section 2.3.1 for respective bands and frequency borders). This procedure was chosen to circumvent the inverse problem during free source fitting procedures usually based on maximum likelihood algorithms [14,15,18,93–95].

The here-proposed Default Mode Network Source Model (DMNSM) was based on prior knowledge derived from functional neuroimaging studies and further elaborate knowledge about discussed issues pertaining to the functional neuroanatomy of the so-labeled Default Mode Network (DMN; [96]). In the following sections, the importance of interdependent neural networks in the DMN addressing particular principal brain structures is elaborated and considered in relation to the above-mentioned DMNSM.

Based on the previous sections and the introduction, the here-introduced Default Mode Network Source Model (DMNSM) comprises 28 approximate center coordinates (according to the Talairach taxonomy [92]) of 14 anatomical structures, in parallel, for the left and right hemispheres (see Table 1 for details and Figure 1A for illustration). This ad hoc model serves as a basis for seeding procedures into band-pass-filtered signal space data. The resulting source wave forms (nAm) were further analyzed for dynamic properties and correlative relationships with AGE and several state and trait variables (see Sections 2.3.3 and 2.3.4 for more details).

Table 1. Number of regional sources (RSs), anatomical-label, short-label, and approximated Talairach center coordinates for their respective RSs to be seeded for further analyses (i.e., Default Mode Network Source Model/DMNSM model).

RS	Anatomical Label	Short Label	TAL-x	TAL-y	TAL-z
1	Ventromedial Prefrontal left	lvmP	−8	30	−24
2	Ventromedial Prefrontal right	rvmP	8	30	−24
3	Anterior Ventrolateral Inferior Frontal left	lantVIF	−28	25	−24
4	Anterior Ventrolateral Inferior Frontal right	rantVIF	28	25	−24
5	Medial Superior Frontal Gyrus left	lmSFG	−10	19	52
6	Medial Superior Frontal Gyrus right	rmSFG	10	19	52
7	Middle Frontal Gyrus left	lMFG	−38	40	20
8	Middle Frontal Gyrus right	rMFG	38	40	20
9	Anterior Cingulate Cortex left	lACC	−10	45	17
10	Anterior Cingulate Cortex right	rACC	10	45	17
11	Intraparietal Sulcus left	lIPS	−35	−60	50
12	Intraparietal Sulcus right	rIPS	35	−60	50
13	Inferior Parietal Lobule left	lIPL	−56	−36	28
14	Inferior Parietal Lobule right	rIPL	56	−36	28
15	Precuneus left	lPC	−15	−58	36
16	Precuneus right	rPC	15	−58	36
17	Middle Temporal Gyrus left	lMTG	−44	−66	17
18	Middle Temporal Gyrus right	rMTG	44	−66	17
19	Medial Temporal Cortex left	lMTC	−55	14	10
20	Medial Temporal Cortex right	rMTC	55	14	10
21	Hippocampus left	lHIP	−31	−25	−8
22	Hippocampus right	rHIP	31	−25	−8
23	Posterior Cingulate Cortex left	lPCC	−10	−58	11
24	Posterior Cingulate right	rPCC	10	−58	11
25	Thalamus left	lTHA	−13	−17	8
26	Thalamus right	rTHA	13	−17	8
27	Occipitotemporal junction left	lOTJ	−44	−71	2
28	Occipitotemporal junction right	rOTJ	44	−71	2

2.3.3. Source Space: Quantification of Dynamic Follow-Up (FU) Generator Activities

For the FU analyses, a 30 s resting-state data period was extracted and divided in 1 s trial epochs. In additional to the eye–artefact correction procedures described in Section 2.3.1, all 1 s trial epochs that contained time bins with more than two standard deviations above the mean nAm values were rejected from further analyses. In the end, there were 26.3 (+/−5.9) 1 s epochs left.

The source model (DMNSM), introduced in Section 2.3.2, was, respectively, applied to band-pass-filtered signal space data (Figure 1A). Afterwards, the resulting source wave forms (SWFs, in nAm units) for each regional source (RS) of the DMNSM were written for further analyses. For the present approach, SWF data (Figure 1B, panel 1) were down-sampled to the double of the upper border (minimum criterion according to the Nyquist theorem) of the respective, pre-defined frequency band (see Section 2.3.1). For each new down-sampled time bin (delta: 8.0 Hz; theta: 16 Hz; alpha low: 22.0 Hz; alpha high: 26.0 Hz; beta low: 43 Hz; beta high: 60 Hz; gamma: 70 Hz), nAm values were summed up accordingly (Figure 1B, panel 2).

As both the amount of neural generator activities and their contribution to surface topographical mapping depend on depth, local cytoarchitecture (electrical open/closed field constellation) [8], and concepts like sparse coding in neural network communication [97], in the present approach, data were source-wise normalized to the maximum nAm value of each included one-second epoch. Thus, the resulting data further represented percentage values representing dynamic fluctuations without providing absolute amplitude information for a direct comparison between RSs (Figure 1B, panel 3).

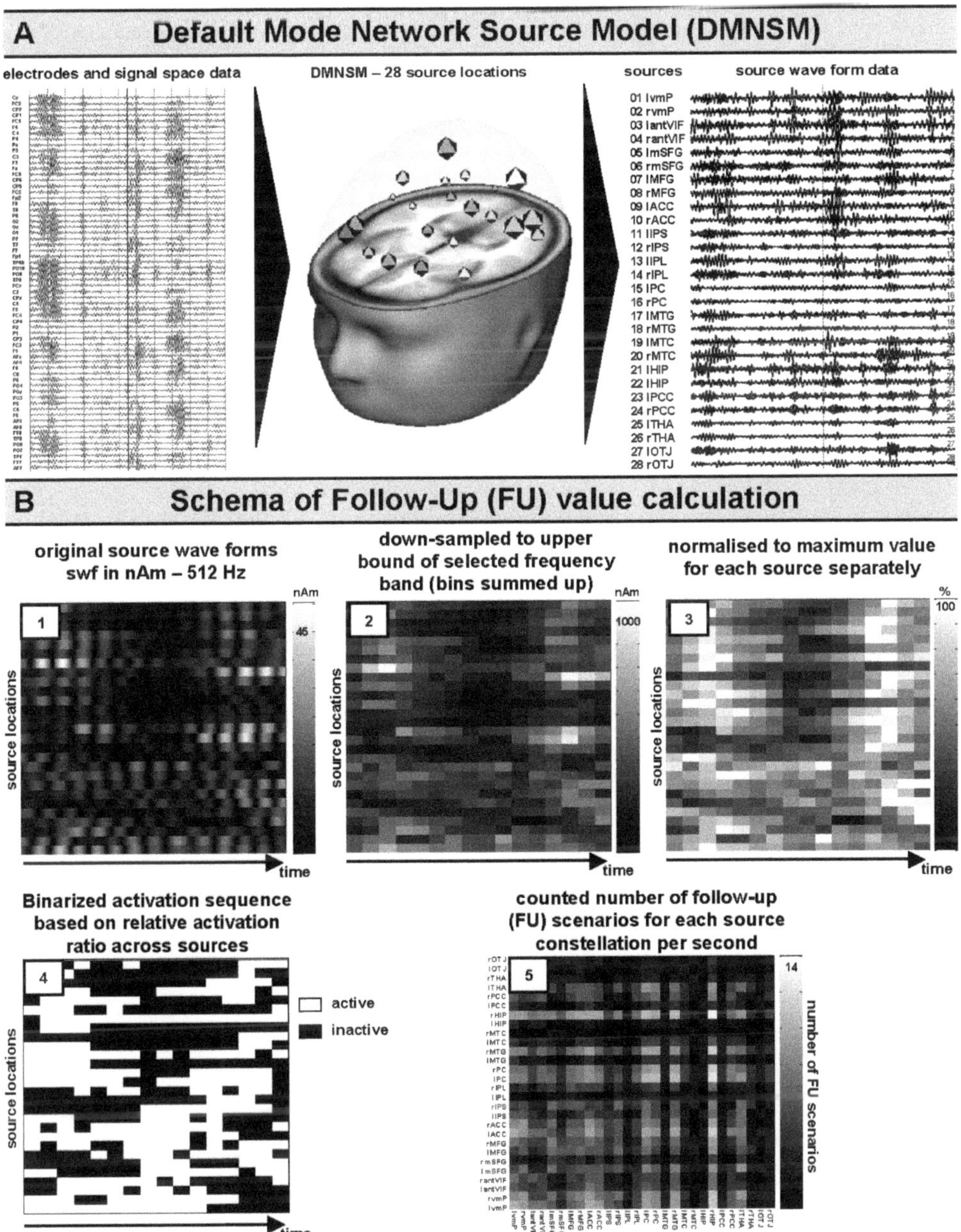

Figure 1. Panel (**A**): illustration of the appliance of the default network mode source model (DNMSM); panel (**B**): illustration of analysis steps for the determination of source follow-up (FU) frequencies. Please see text for details to processing steps in panels 1–5.

In the next step, activation dynamics between sources were determined by estimating the ratio of percentage values of each RS for each time bin separately [98,99]. In cases where the percentage value of a RS represented more than the nth part (n = number of RS = 28 in the present DMNSM) of the sum of all RSs, it was labeled as 1 = active (otherwise, as 0 = not active). This procedure resulted in a binary matrix of dynamic activations and deactivations (see Figure 1B, panel 4, for illustration).

Finally, the resulting binary matrices were analyzed with a strictly non-linear and descriptive approach, which was further labeled as the follow-up (FU) analyses. Each active RS (indicated by 1 in the binary matrix) was examined for other active sources on the consecutive following time bin. If so, the respective follow-up constellation between two sources, A and B, was incremented by one, resulting in follow-up matrices (i.e., FU frequencies for 28 by 28 RSs; see Figure 1B, panel 5, for illustration [single trial example], and Figure 5A,B, first column, for further frequency-band-related mean FU plots) that were averaged for each individual over the respective one-second epoch trials.

The here-suggested approach tried to account for the non-linear and periodic nature of oscillatory bursts. In this approach, concrete phase and amplitude information became lost, but it was possible to quantify the average relative amount of follow-up (FU) activations per second over numerous one-second epoch trials. This information was used to explore and illustrate cross-sectional AGE-dependent changes in frequency-band-related communication dynamics (i.e., bi- and uni-directional) over the third life decade (see Figure 5A,B, third column, for the respective causal network diagrams, CNDs).

2.3.4. Correlation Analyses

Absolute (see Figure 3) and relative (see Figure 4) band-pass-related regional FFT power spectrum values (for details see Section 2.3.1) were correlated (via non-parametric Spearman rank correlation) with AGE in years and several state- and trait-related covariates (see Section 2.2 for an introduction to covariates).

Average regional source (RS) moment values (mean nAm) derived from different frequency bands were correlated (via Pearson correlation) with AGE in years and several state- and trait-related covariates (see also above). Furthermore, frequency-band-related mean follow-up (FU) frequency values, as introduced in Section 2.3.3, were correlated with AGE in years. Pearson correlation coefficients, significances (second column), and causal network diagrams (CND) were illustrated in Figure 5A,B.

3. Results

Mean FFT-values and standard deviations were illustrated for absolute and relative spectra in Figure 2A,B, respectively. As expected, both absolute and relative power spectra showed a peak of around 10 Hz with regional prominence over posterior (i.e., parietal and occipital) brain areas. In the following sub-sections, correlation analyses, including non-physiological and physiological parameters, will be reported, respectively.

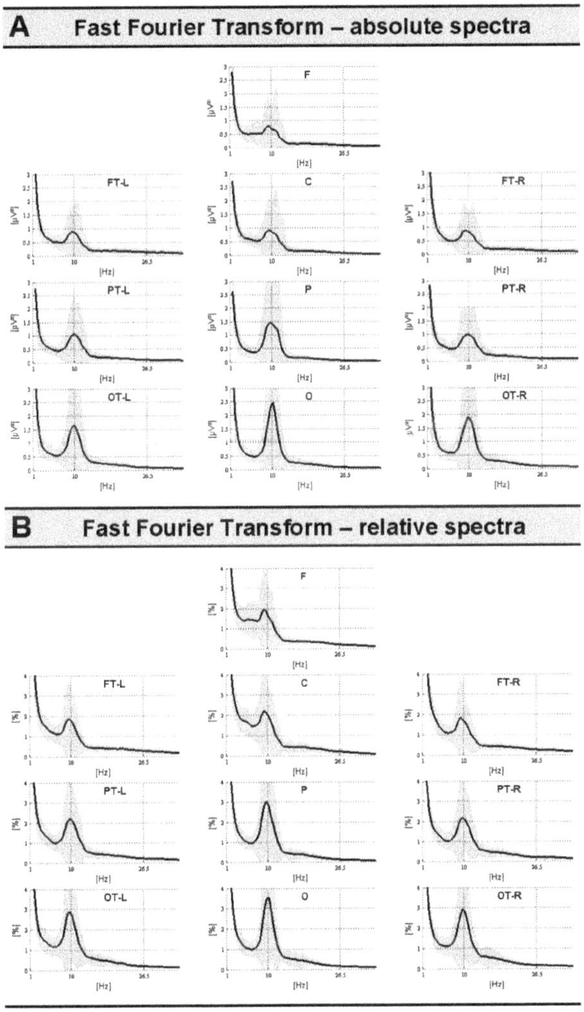

Figure 2. Mean regional power spectra (FFT) for absolute (panel **A**) and relative (panel **B**) spectra. Shaded areas indicate standard deviations (SDs).

3.1. Correlation Analyses

Neither state (i.e., PSYCH [7.6+/−1.5], PHYS [7.7+/−1.4], and STRESS [5.4+/−2.1]) nor trait (i.e., Neuroticism [1.6+/−0.6], Extraversion [2.6+/−0.6], Openness to Experience [2.8+/−0.5], Agreeableness [2.6+/−0.5], and Conscientiousness [2.6+/−0.6]) ratings revealed any significant correlative (i.e., Spearman rank correlation) relationship with AGE. Self-rated physiological (PHYS) and psychological (PSYCH) well-being was negatively correlated (PHYS: R = −0.36, $p < 0.01$; PSYCH: R = −0.41. $p < 0.01$), and stress level was positively correlated (STRESS: R = 0.28, $p < 0.05$) with AGE.

3.1.1. Signal Space Analyses: Regional Absolute and Relative Power Spectra

Exploratory correlation analyses revealed numerous relationships between regional resting-state-related brain oscillations in different frequency bands and several state- and trait-related variables (see Figures 3 and 4 for illustrations). AGE did not show any covariations with regional brain oscillatory activities in signal space. Absolute and relative regional FFT values mostly showed a different correlation profile with their respective covariates. Trait variable Conscientiousness (C) showed a comparable profile over absolute and relative regional power values, reflecting lower alpha and higher high-frequency brain activity. The self-rated stress-level appears to be related to higher beta activity over frontal brain areas.

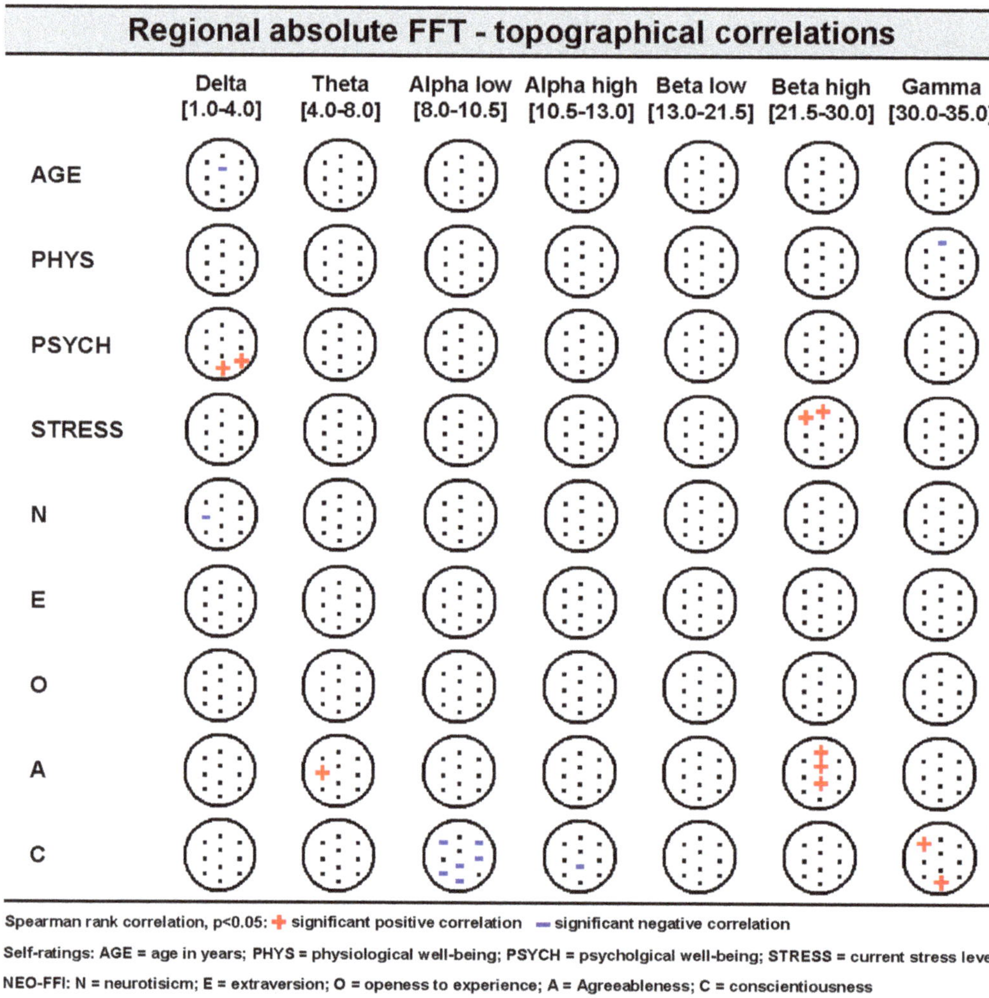

Figure 3. Non-parametric (Spearman) topographical rank correlation between seven absolute power spectrum frequency bands (delta, theta, alpha low, alpha high, beta low, beta high, and gamma) and nine covariates (AGE, PSYCH, PHYS, STRESS, N, E, O, V, G) (see Section 2.3.1 and legend for more details).

3.1.2. Source Space Analyses

Source space analyses were performed on source moment wave forms (SWFs) derived from the appliance of the source model (DMNSM) introduced in Section 2.3.2, comprising 28 regional sources (RSs, see Figure 1A, and Table 1 for details). The appliance of the DMNSM on the selected band-bass-filtered resting-state data epochs (in average 26.3 +/−5.9 one second epochs) revealed the following variance explanations: delta: 7.3+/−3.6%; theta: 3.2+/−2.2%; alpha low: 2.4+/−2.0%; alpha high: 2.7+/−3.8; beta low: 3.8+/−1.9%; beta high: 5.3+/−2.0%; gamma: 6.5+/−2.2%. One-sample t-test showed that all model fits ranked statistically below 10% residual variances (all $p < 0.001$). Theta and both alpha bands ranked below 5% residual variances (all $p < 0.001$).

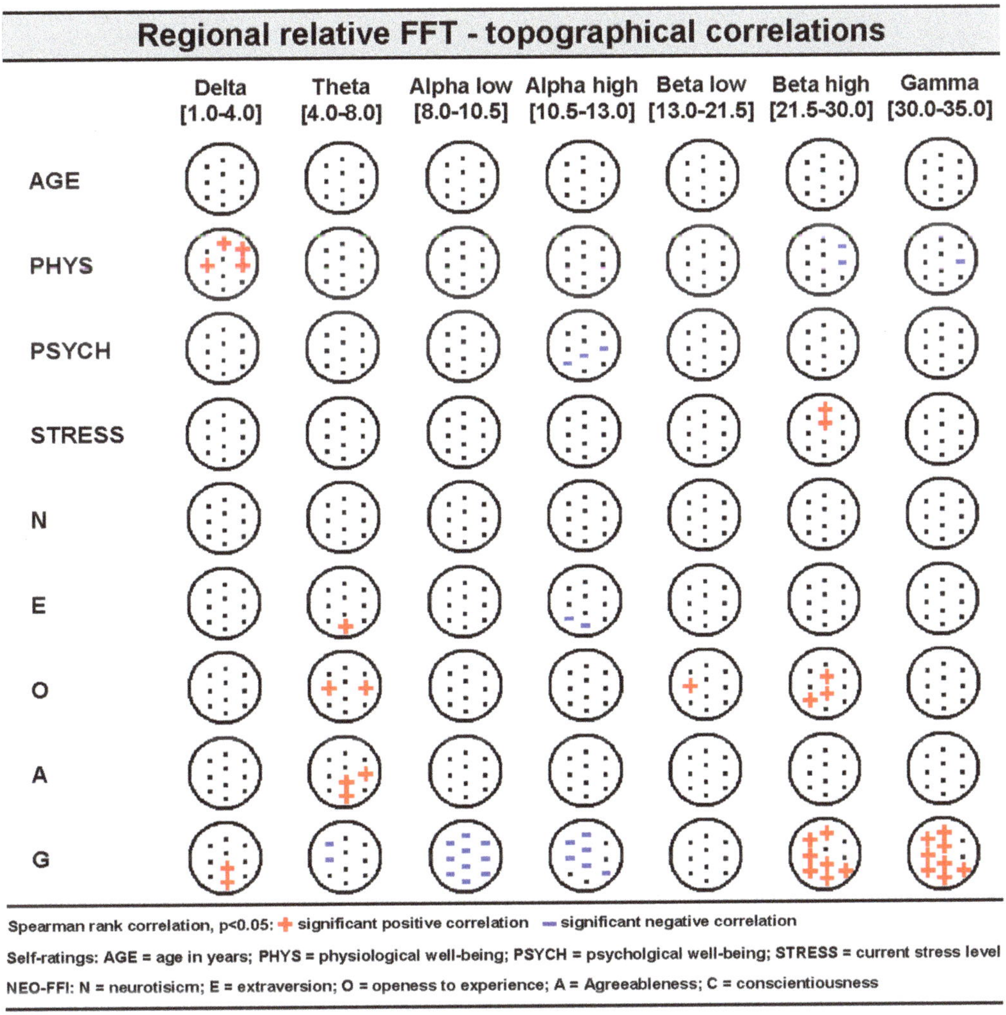

Figure 4. Non-parametric (Spearman) topographical rank-correlation between seven relative power spectrum frequency bands (delta, theta, alpha low, alpha high, beta low, beta high, and gamma) and nine covariates (AGE, PSYCH, PHYS, STRESS, N, E, O, V, and G) (see Section 2.3.1 and legend for more details).

Correlations between Mean Regional Source (RS) Moments and AGE

RS values did not show any significant cross-sectional correlations with AGE over the third life decade. This highly corresponded to the signal space data. There was only a trend towards significance for a negative correlation between AGE and mean delta source activity in the left medial superior frontal gyrus (lmSFG, $p = 0.06$) and left occipitotemporal junction (lOTJ, $p = 0.07$).

Spatiotemporal Dynamics—Follow-Up (FU) Activations and AGE-Related Causal Network Plots (CND)

Spatiotemporal brain dynamics (as quantified by follow-up frequency calculations (see Section 2.3.3 for details) showed differential cross-sectional correlation patterns with AGE as covariate in different oscillatory frequency bands (see Figure 5A,B, second column). Follow-up frequency plots (see Figure 5A,B, first column) suggest complex patterns of regional and inter-regional source dynamics, while steady-state activations appear dominant, as indicated by relatively large values along the respective plot diagonal.

Causal network diagrams (CNDs, see Figure 5A,B, third column) indicate numerous significant (Pearson correlation, *t*-test, $p < 0.05$) positive and negative correlations, indicating both uni-directional and bi-directional neural generator dynamics cross-sectionally related to developmental changes across the third life decade. Changes in the delta oscillations suggest an age-related enhancement of uni-directional left posterior input coming from subcortical and prefrontal brain areas. The thalamus appeared to decrease its input to the left intra parietal sulcus and right posterior cingulate (see Figure 5A, third column, first CND for more details).

The CND for theta oscillations indicated a complex picture of age-related changes of positive and negative, uni- and bi-directional dynamics across the whole brain work. Numerous inter-hemispheric, uni- and bi-directional communication frequencies involving, for example, the right occipitotemporal junction, right middle temporal cortex, left middle temporal gyrus, right intra-parietal sulcus, left anterior cingulate, and left anterior ventrolateral inferior frontal areas appear to decrease their resting communication dynamics over the third life decad; whereas others, such as left medial temporal cortex, right middle temporal gyrus, right inferior parietal lobule, and left and right superior frontal gyri appear to increase their resting-state communication frequencies during the third life decade (see Figure 5A, second column, first CND for more details).

Age-related changes in communication frequencies in the lower alpha band were mainly characterized by a decrease in resting brain dynamics between the left anterior cingulate and numerous other brain areas, such as, for example, the right medial temporal cortex, left thalamus, left hippocampus, left middle frontal gyrus, and bilateral superior frontal gyri (see Figure 5A, third column, third CND for more details). Resting brain dynamics in the upper alpha band appear to predominantly increase between the left and right thalamus and several brain areas, such as, for example, the right ventromedial prefrontal areas, left and right medial superior frontal areas, right inferior parietal lobule, and left medio-temporal cortex. The left ventromedial prefrontal and left anterior cingulate areas showed a partial decrease in communication dynamics (see Figure 5B, third column, first CND for more details).

Lower beta band oscillations showed a partial age-related decrease in communication between the left ventromedial prefrontal cortex and posterior and temporal brain areas. In all, the lower beta band appears quite stable in resting-state brain dynamics during the third life decade (see Figure 5B, third column, second CND for more details). Higher beta band brain dynamics appear to increase during the third life decade predominantly within the frontal brain areas and between frontal and parietal locations mainly bi-directionally (see Figure 5B, third column, third CND for more details). Similarly, gamma band brain dynamics between 30 and 35 Hz showed an age-related increase in predominant bi-directional dynamics between frontal brain areas and in frontoparietal networks (see Figure 5B, third column, forth CND for more details).

Summarizing, the cross-sectional increases and decreases in brain dynamics over the third life decade showed a differential pattern for the seven considered oscillatory frequency bands.

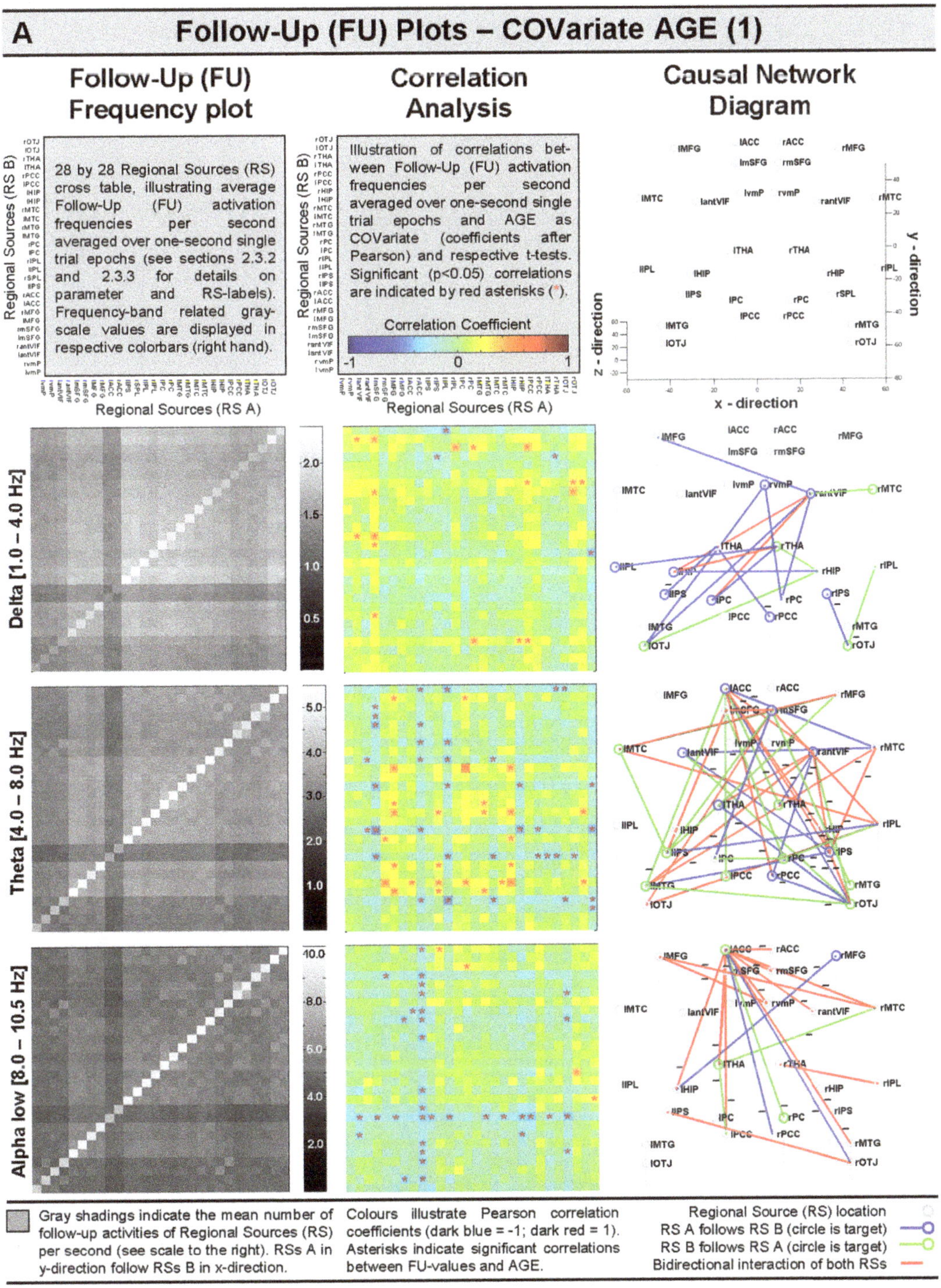

Figure 5. *Cont.*

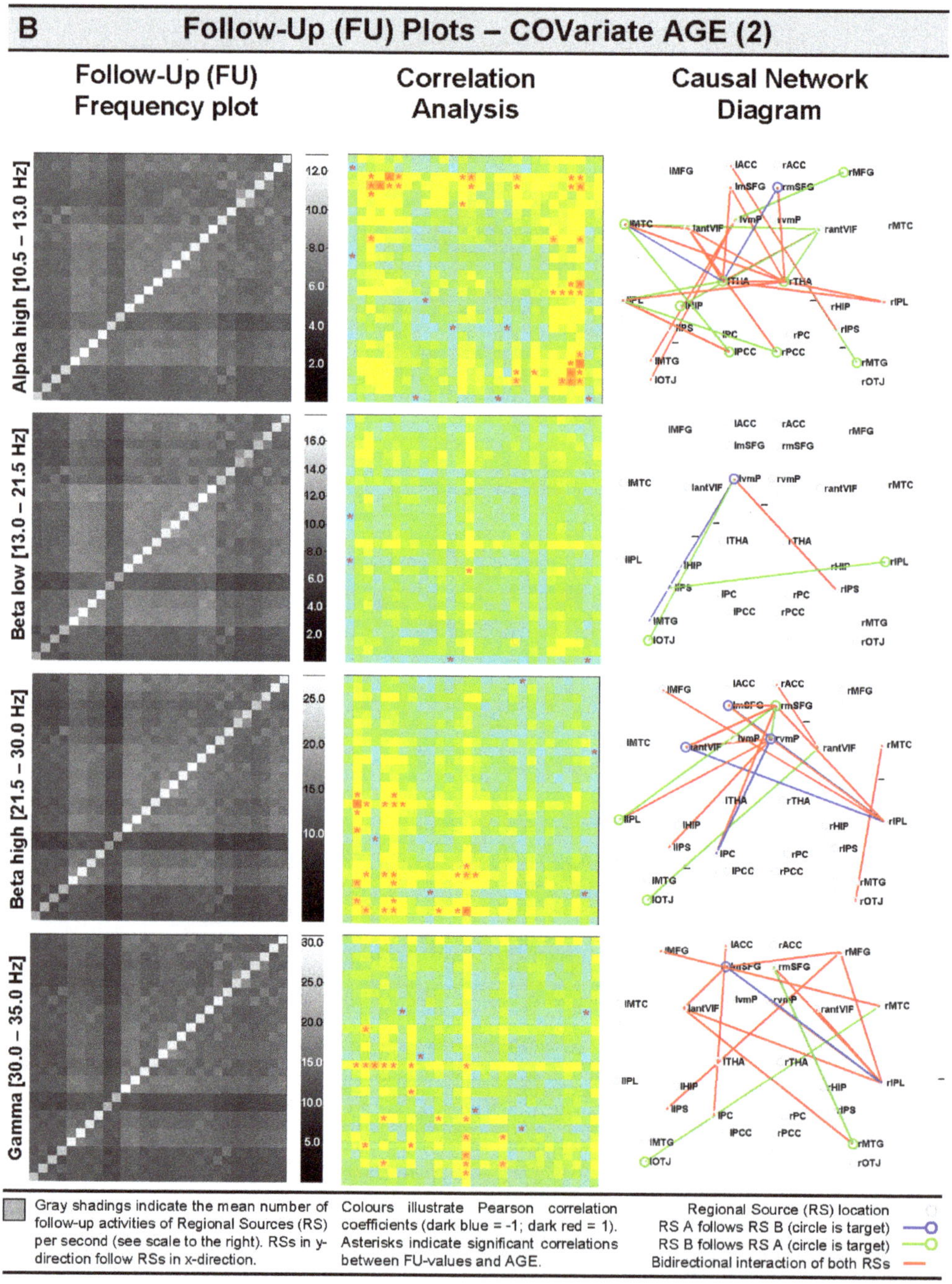

Figure 5. (**A**,**B**): Follow-up (FU) frequency plots (left column); heat maps for the illustration of correlations between frequency-band-related FU dynamics and AGE (middle column); and causal network diagrams (CNDs) illustrating the directionality of age-related changes between source locations of the DMNSM (right column). For more detailed information, see first panel line of (**A**), legend at the bottom, and text.

4. Discussion

The human brain is perhaps the most complicated thing that we know of besides the universe itself. The here-presented data appear to substantiate this idea. Functional neuroimaging made it popular to examine data in a three dimensional—or cartesian-way; whereas combined with biosignal analyses (e.g., EEG and MEG), it is possible to examine spatiotemporal brain dynamics with high temporal resolution [5,13,22]. The present work relies on knowledge from functional neuroanatomy and functional neuroimaging to suggest a default network (DMN) model for the exploration of follow-up brain dynamics during the resting state in different brain oscillatory frequency bands.

First, the present signal space data confirmed the expectation that both absolute and relative power spectra should peak around 10 Hz, with dominant amplitudes over posterior sites, and that the respective resting alpha power values do not vary with age during adulthood [13,20,23–25,100]. Furthermore, it could be seen that personality scores did not correlate with age across the third life decade, which confirms a traditionally discussed trait-stability in personality [101,102]. The exploration of topographical brain oscillation data, considering several state and trait variables, revealed correlations between self-ratings of psychological well-being and absolute slow wave oscillations at posterior electrode sites. Ratings of stress correlated with both absolute and relative power in higher beta band over frontal areas and with age. Ratings of physical well-being correlated with relatively slow wave oscillations over the frontal and frontotemporal areas, and both self-rated psychological and physical well-being correlated negatively with age. Summarizing, signal-space data confirmed the stability of brain oscillatory activity across early adulthood [20,27,28], but it appeared to counteract MEG data that reflected a decrease in slow-wave and an increase in higher-frequency oscillations [20]. However, the sample of individuals included by Fehr et al. [20] ranged from early adulthood to elderly participants in their sixties.

The here-used regional default mode network source model (DMNSM) revealed good-to-very-good variance explanations, especially for theta and alpha frequency ranges. Mean source wave form resting-state data revealed only a statistical trend for an age-related decrease in slow wave activity, which was, however, in line with data reported by Fehr [20], who also reported a decrease in delta source activity (Minimum-Norm, L2) but also reported an increase in higher-frequency bands. It appears that age-related changes in oscillatory resting-state activation follow quite-subtle rules and cannot sufficiently be examined through just a few methodological approaches. More functional [13]—but also external—variables (e.g., individual learning history) and individual parametrization of brain physiological data [86,98,99] appears to be necessary for a sufficient explanation of age-related changes in resting-state brain activity, particularly during the third life decade.

Here, resting-state source wave form data were also examined for dynamic, oscillatory properties. The parameter that was used quantified the frequency of mean relative follow-up (FU) activations per second. FU frequency plots showed most stable activations within their respective source locations, suggesting prolonged activations within each principal region, which holds for all considered frequency bands between 1 and 35 Hz. This finding might reflect a certain relationship to microstates, discussed as a reliable measure for the occurrence of transiently stable brain activations that can, among other things, be related to temporal brain dynamics, brain developmental issues, and modalities of thinking [13,103–105]. In the present work, we point to the idea of counting how often certain source locations in the DMNSM follow up others and correlate this information with age (i.e., brain dynamics). The resulting partially positive and partially negative correlations showed quite different profiles, indicating age-related changes in uni-directional and bi-directional neural source sequencing for all considered frequency bands (see causal network diagrams, CNDs, Figure 5A,B). These findings suggest that there is a differential involvement of oscillatory neural networks during the development of resting-state brain activity in the third life decade. Chow and colleagues [40] also reported a non-uniform frequency depending on the involvement of different frequency oscillator dynamics as reflected in an age-related change in connectivity between the medial prefrontal and other brain areas associated with the DMN. Higher alpha connectivity decreased and both theta and beta connectivity increased with age. The here-presented CNDs confirmed growing network dynamics between the medial superior frontal and other (e.g., posterior and temporal) brain areas in higher beta and gamma bands (21.5–35 Hz) but decreased partial network dynamics in lower beta bands between the left medial prefrontal and posterior and temporal areas. Lower beta band showed rather few age-related changes in neural network dynamics, reflecting an age-related decrease in dynamics between the left ventromedial prefrontal cortex and posterior and temporal areas.

Delta oscillations showed a mainly uni-directional, age-related increase in particularly left thalamocortical network dynamics. Right anterior ventrolateral inferior location showed increasing uni-directional and bi-directional dynamics to several posterior and temporal locations. Dynamics in theta oscillation provided the most complex picture of all considered frequency bands. There were a lot of positive and negative, uni- and bi-directional dynamics across the whole DMNSM in relation to age. The predominantly involved locations were the left anterior cingulate cortex, right occipitotemporal junction, left and right intra-parietal sulci, and right anterior ventrolateral inferior frontal cortex, suggesting the involvement of executive, spatial, and object-related nodes of mental processing, potentially involved in the differential development (i.e., reorganization of related functional networks) of, for example, resting-state-related introspection and mind wandering [29,30].

Age-related changes in lower alpha dynamics appeared to be very much dominated by a decrease in follow-up communication between the left anterior cingulate cortex and numerous cortical and sub-cortical locations associated with the DMN. Thus, theta and lower alpha oscillations appear to reduce dynamic resting-state communication over the third life decade, whereas theta oscillations appear to compensate for the reduction via the establishment of new functional sub-networks. Higher alpha dynamics were dominated by an age-related increase in thalamocortical interaction, involving numerous frontal, posterior, and left temporal locations. It appears convenient to differentiate between lower and higher alpha brain dynamics.

Several limitations of the present work shall be mentioned here. Though widely accepted, the definition of the frequency band ranges used for band-pass-filtering might be revised by more fine-graded oscillatory analyses in future studies. Furthermore, longitudinal approaches provide more valid data than cross-sectional approaches. Thus, the present work shall be understood as a proof of concept that has to be approved in subsequent respective repeated examination designs. Furthermore, higher gamma band ranges above 35 Hz should also be considered.

5. Conclusions

Different methodological approaches often lead to apparently divergent outcomes. The present work includes traditional (i.e., topographical FFT power analyses), advanced (i.e., seeded source network analyses), and innovative approaches, exploring age-related follow-up properties in a seeded default network source model. FFT power analyses revealed the expected alpha frequency peak at about ten Hz. Topographical power analyses did not show any age-related effects, which was not completely unexpected [13]. Chow and colleagues [40] also mentioned the heterogeneity of findings on age-related resting-state examinations (e.g., [43–46]). It appears that both reliability proofs [13] and new innovative explorations are necessary for substantial and multi-perspective progress in this field. The present findings might be interpreted in such a way that quantity (i.e., power or source strength) in oscillatory resting-state processing does not reflect the most important age-related effect, but the HOW resting-state is processed [22]. According to the idea of sparse neural coding [97], quality (the reorganization of the functional HOW) might change without a change in quantity (e.g., power).

The here-introduced causal network diagrams (CNDs) shall be improved, particularly when network communications become more complicated. Video animations of respective network dynamics can help us to acquire better insights into the underlying neural processes. Furthermore, dynamic non-linear auto-correlation procedures might help not only to identify microstates [94,103,105] but also transient attractor circuits of different durations. Alongside correlative approaches, as reported in the present work, future studies should include different developmental age cohorts with different educational backgrounds, considering different experimental modalities (i.e., children, adolescents, and elderly individuals, e.g., [106,107]) and patient groups (e.g., [108]) to identify complex systemic and multi-dimensional, structural, and functional biomarkers for developmental and/or neurodegenerative diseases (e.g., [109]).

Author Contributions: Conceptualization, T.F., S.M., M.-C.H. and A.A.; methodology, T.F. and K.G.; software, T.F. and K.G.; validation, T.F., S.M., M.-C.H., A.A. and K.G.; formal analysis, T.F., S.M., M.-C.H. and A.A.; investigation, T.F. and K.G.; resources, T.F.; data curation, T.F.; writing—original draft preparation, T.F., S.M., M.-C.H. and A.A.; writing—review and editing, S.M., M.-C.H., A.A. and K.G.; visualization, T.F., S.M. and K.G.; supervision, T.F.; project administration, T.F.; All authors have read and agreed to the published version of the manuscript.

Funding: This research received no external funding.

Institutional Review Board Statement: The used experimental setups of the main studies were designed according to the Code of Ethics of the World Medical Association (Declaration of Helsinki, published in the British Medical Journal, 18 July 1964) and the used stimulus materials (videos) were approved by the local ethics committee at Bremen University (no protocol code available, year of approval was 2009). All participants gave written informed consent to their participation.

Informed Consent Statement: Informed consent was obtained from all study participants involved in the study.

Data Availability Statement: As physiological data, such as multi-channel EEG recordings, have been shown to be relatable to individuals like a fingerprint and therefore bear the risk of (medical, economical) misuse, we decided to make the respective data not available online. Participants gave their written informed consent among others based on this warranty declaration. However, scientists are welcome to visit us at Bremen University and perform analyses offline in cooperation.

Conflicts of Interest: The authors declare no conflicts of interest.

References

1. La, C.; Mossahebi, P.; Nair, V.A.; Bendlin, B.B.; Birn, R.; Meyerand, M.E.; Prabhakaran, V. Age-Related Changes in Inter-Network Connectivity by Component Analysis. *Front. Aging Neurosci.* **2015**, *7*, 237. [CrossRef]
2. Thatcher, R.W. Cyclic cortical reorganization: Origins of human cognitive development. In *Human Behavior and the Developing Brain*; Dawson, G., Fischer, K.W., Eds.; The Guilford Press: New York, NY, USA, 1994; pp. 232–266.
3. Giedd, J.N.; Blumenthal, J.; Jeffries, N.O.; Castellanos, F.X.; Liu, H.; Zijdenbos, A.; Paus, T.; Evans, A.C.; Rapoport, J.L. Brain development during childhood and adolescence: A longitudinal MRI study. *Nat. Neurosci.* **1999**, *2*, 861–863. [CrossRef]
4. Sowell, E.R.; Thompson, P.M.; Tessner, K.D.; Toga, A.W. Mapping continued brain growth and gray matter density reduction in dorsal frontal cortex: Inverse relationships during postadolescent brain maturation. *J. Neurosci.* **2001**, *21*, 8819–8829. [CrossRef]
5. Basar, E. *Brain-Body-Mind in the Nebulous Cartesian System: A Holistic Approach*; Springer: Berlin/Heidelberg, Germany, 2011.
6. Fehr, T. Chancen und Grenzen von Methoden der kognitiven Neurowissenschaften—Funktionelle Magnetresonanztomographie und Biosignalanalyse im Kontext der Entwicklungsneurophysiologie. *Z. Gestalt.* **2009**, *20*, 29–43.
7. Fehr, T. The Neural Architecture of Violence-Related Socialization—Evidence from Functional Neuroimaging. In *Handbook of Anger, Aggression, and Violence*; Martin, C.R., Preedy, V.R., Patel, V.B., Eds.; Springer: Cham, Swtizerland, 2023.
8. Lorente de No, R. *A Study of Nerve Physiology: Studies from the Rockefeller Institute of Medical Research*, 1st ed.; The Rockefeller Institute of Medical Research: New York, NY, USA, 1947.
9. Basar, E. The theory of the whole-brain-work. *Int. J. Psychophysiol.* **2006**, *60*, 133–138. [CrossRef] [PubMed]
10. Weiss, A.; Faber, D.S. Field effects in the CNS play functional roles. *Front. Neural Circuits* **2010**, *4*, 1541. [CrossRef] [PubMed]
11. Hindriks, R.; Woolrich, M.; Luckhoo, H.; Joensson, M.; Mohseni, H.; Kringelbach, M.L.; Deco, G. Role of white-matter pathways in coordinating alpha oscillations in resting visual cortex. *Neuroimage* **2015**, *106*, 328–339. [CrossRef]
12. Babiloni, C.; Del Percio, C.; Boccardi, M.; Lizio, R.; Lopez, S.; Carducci, F.; Marzano, N.; Soricelli, A.; Ferri, R.; Triggiani, A.I.; et al. Occipital sources of resting-state alpha rhythms are related to local gray matter density in subjects with amnesic mild cognitive impairment and Alzheimer's disease. *Neurobiol. Aging* **2015**, *36*, 556–570. [CrossRef]

13. Popov, T.; Tröndle, M.; Baranczuk-Turska, Z.; Pfeiffer, C.; Haufe, S.; Langer, N. Test–retest reliability of resting-state EEG in young and older adults. *Psychophysiology* **2023**, *60*, e14268. [CrossRef]
14. Scherg, M.; Berg, P.; Nakasato, N.; Beniczky, S. Taking the EEG back into the brain: The power of multiple discrete sources. *Front. Neurol.* **2019**, *10*, 855. [CrossRef]
15. Scherg, M. Functional imaging and localization of electromagnetic brain activity. *Brain Topogr.* **1992**, *5*, 103–111. [CrossRef] [PubMed]
16. Scherg, M.; Berg, P. New concepts of brain source imaging and localization. *Electroencephalogr. Clin. Neurophysiol. Suppl.* **1996**, *46*, 127–137. [PubMed]
17. Scherg, M.; Berg, P. Use of prior knowledge in brain electromagnetic source analysis. *Brain Topogr.* **1991**, *4*, 143–150. [CrossRef] [PubMed]
18. Scherg, M. BESA Source Analysis Combining EEG and fMRI. E-Book. 2004. Available online: www.besa.de (accessed on 19 June 2024).
19. Somsen, R.J.; van't Klooster, B.J.; van der Molen, M.W.; van Leeuwen, H.M.; Licht, R. Growth spurts in brain maturation during middle childhood as indexed by EEG power spectra. *Biol. Psychol.* **1997**, *44*, 187–209. [CrossRef] [PubMed]
20. Fehr, T.; Bott, C.; Haeberle, A.; Rockstroh, B. MEG power spectrum and age: Differences between adolescents and adults. In *BIOMAG 2002—13th International Conference on Biomagnetism, Jena, Germany, 10–14 August 2002*; Nowak, H., Haueisen, J., Gießler, F., Huonker, R., Eds.; VDE Verlag GmbH: Berlin, Germany; Offenbach, Germany, 2002; pp. 320–322.
21. Finley, A.J.; Angus, D.J.; van Reekum, C.M.; Davidson, R.J.; Schaefer, S.M. Periodic and aperiodic contributions to theta-beta ratios across adulthood. *Psychophysiology* **2022**, *59*, e14113. [CrossRef] [PubMed]
22. Fehr, T. A hybrid model for the neural representation of complex mental processing in the human brain. *Cogn. Neurodyn.* **2013**, *7*, 89–103. [CrossRef] [PubMed]
23. Basar, E. A review of alpha activity in integrative brain function: Fundamental physiology, sensory coding, cognition and pathology. *Int. J. Psychophysiol.* **2012**, *86*, 1–24. [CrossRef] [PubMed]
24. Adrian, E.D.; Matthews, B.H.C. The Berger rhythm: Potential changes from the occipital lobes in man. *Brain* **1934**, *57*, 355–385. [CrossRef]
25. Compston, A. The Berger rhythm: Potential changes from the occipital lobes in man. *Brain* **2010**, *133*, 3–6. [CrossRef]
26. Grandy, T.H.; Werkle-Bergner, M.; Chicherio, C.; Schmiedek, F.; Lovden, M.; Lindenberger, U. Peak individual alpha frequency qualifies as a stable neurophysiological trait marker in healthy younger and older adults. *Psychophysiology* **2013**, *50*, 570–582. [CrossRef]
27. Cragg, L.; Kovacevic, N.; McIntosh, A.R.; Poulsen, C.; Martinu, K.; Leonard, G.; Paus, T. Maturation of EEG power spectra in early adolescence: A longitudinal study. *Dev. Sci.* **2011**, *14*, 935–943. [CrossRef]
28. Miskovic, V.; Ma, X.; Chou, C.A.; Fan, M.; Owens, M.; Sayama, H.; Gibb, B.E. Developmental changes in spontaneous electrocortical activity and network organization from early to late childhood. *Neuroimage* **2015**, *118*, 237–247. [CrossRef]
29. Mason, M.F.; Norton, M.I.; Van Horn, J.D.; Wegner, D.M.; Grafton, S.T.; Macrae, C.N. Wandering minds: The default network and stimulus-independent thought. *Science* **2007**, *315*, 393–395. [CrossRef]
30. Andrews-Hanna, J.R. The Brain's default network and its adaptive role in internal mentation. *Neuroscientist* **2012**, *18*, 251–270. [CrossRef]
31. Buckner, R.L.; Andrews-Hanna, J.R.; Schacter, D.L. The brain's default network: Anatomy, function, and relevance to disease. *Ann. N. Y. Acad. Sci.* **2008**, *1124*, 1–38. [CrossRef]
32. Wen, T.; Mitchell, D.J.; Duncan, J. The functional convergence and heterogeneity of social, episodic, and self-referential thought in the default mode network. *Cereb. Cortex* **2020**, *30*, 5915–5929. [CrossRef]
33. Anticevic, A.; Cole, M.W.; Murray, J.D.; Corlett, P.R.; Wang, X.-J.; Krystal, J.H. The role of default network deactivation in cognition and disease. *Trends Cogn. Sci.* **2012**, *16*, 584–592. [CrossRef]
34. Beckmann, C.F.; Arigita, E.J.S.; Barkhof, F.; Scheltens, P.; Stam, C.J.; Smith, S.M.; Rombouts, S.A.R.B. Reduced resting-state brain activity in the "default network" in normal aging. *Cereb. Cortex* **2008**, *18*, 1856–1864. [CrossRef]
35. Hafkemeijer, A.; van der Grond, J.; Rombouts, S.A.R.B. Imaging the default mode network in aging and dementia. *Biochim. Biophys. Acta Mol. Basis Dis.* **2012**, *1822*, 431–441. [CrossRef]
36. Campbell, K.; Grigg, O.; Saverino, C.; Churchill, N.; Grady, C. Age differences in the intrinsic functional connectivity of default network subsystems. *Front. Aging Neurosci.* **2013**, *5*, 73. [CrossRef]
37. Andrews-Hanna, J.R.; Snyder, A.Z.; Vincent, J.L.; Lustig, C.; Head, D.; Raichle, M.E.; Buckner, R.L. Disruption of large-scale brain systems in advanced aging. *Neuron* **2007**, *56*, 924–935. [CrossRef]
38. Persson, J.; Lustig, C.; Nelson, J.K.; Reuter-Lorenz, P.A. Age differences in deactivation: A link to cognitive control? *J. Cogn. Neurosci.* **2007**, *19*, 1021–1032. [CrossRef]
39. Grady, C.L.; Protzner, A.B.; Kovacevic, N.; Strother, S.C.; Afshin-Pour, B.; Wojtowicz, M.; Anderson, J.A.E.; Churchill, N.; McIntosh, A.R. A Multivariate analysis of age-related differences in default mode and task-positive networks across multiple cognitive domains. *Cereb. Cortex* **2010**, *20*, 1432–1447. [CrossRef]
40. Chow, R.; Rabi, R.; Paracha, S.; Hasher, L.; Anderson, N.D.; Alain, C. Default Mode Network and Neural Phase Synchronization in Healthy Aging: A Resting State EEG Study. *Neuroscience* **2022**, *485*, 116–128. [CrossRef]

41. Cabral, J.; Kringelbach, M.L.; Deco, G. Exploring the network dynamics underlying brain activity during rest. *Prog. Neurobiol.* **2014**, *114*, 102–131. [CrossRef]
42. Ghuman, A.S.; McDaniel, J.R.; Martin, A. A wavelet-based method for measuring the oscillatory dynamics of resting-state functional connectivity in MEG. *Neuroimage* **2011**, *56*, 69–77. [CrossRef]
43. Vysata, O.; Kukal, J.; Prochazka, A.; Pazdera, L.; Simko, J.; Valis, M. Age-related changes in EEG coherence. *Neurol. Neurochir. Pol.* **2014**, *48*, 35–38. [CrossRef]
44. Scally, B.; Burke, M.R.; Bunce, D.; Delvenne, J.-F. Resting-state EEG power and connectivity are associated with alpha peak frequency slowing in healthy aging. *Neurobiol. Aging* **2018**, *71*, 149–155. [CrossRef]
45. Smit, D.J.A.; Boersma, M.; Schnack, H.G.; Micheloyannis, S.; Boomsma, D.I.; Hulshoff Pol, H.E.; Stam, C.J.; de Geus, E.J.C.; Valdes-Sosa, P.A. The brain matures with stronger functional connectivity and decreased randomness of its network. *PLoS ONE* **2012**, *7*, e36896. [CrossRef]
46. Moezzi, B.; Pratti, L.M.; Hordacre, B.; Graetz, L.; Berryman, C.; Lavrencic, L.M.; Ridding, M.C.; Keage, H.A.D.; McDonnell, M.D.; Goldsworthy, M.R. Characterization of young and old adult brains: An EEG functional connectivity analysis. *Neuroscience* **2019**, *422*, 230–239. [CrossRef]
47. Kikuchi, M.; Wada, Y.; Koshino, Y.; Nanbu, Y.; Hashimoto, T. Effect of Normal Aging upon Interhemispheric EEG Coherence: Analysis during Rest and Photic Stimulation. *Clin. Electroencephalogr.* **2000**, *31*, 170–174. [CrossRef]
48. Vecchio, F.; Miraglia, F.; Bramanti, P.; Rossini, P.M. Human brain networks in physiological aging: A graph theoretical analysis of cortical connectivity from EEG data. *J. Alzheimer's Dis.* **2014**, *41*, 1239–1249. [CrossRef]
49. Von Bonin, G.V. *Essay on the Cerebral Cortex*, Charles C. Thomas: Springfield, MA, USA, 1950.
50. Fuster, J.M. The cognit: A network model of cortical representation. *Int. J. Psychophysiol.* **2006**, *60*, 125–132. [CrossRef]
51. Fuster, J.M. Cortex and Memory: Emergence of a New Paradigm. *J. Cogn. Neurosci.* **2009**, *21*, 2047–2072. [CrossRef]
52. van Oort, E.S.; Van Walsum, A.V.C.; Norris, D.G. An investigation into the functional and structural connectivity of the Default Mode Network. *Neuroimage* **2014**, *90*, 381–389. [CrossRef]
53. Van Den Heuvel, M.P.; Sporns, O. Rich-club organization of the human connectome. *J. Neurosci.* **2011**, *31*, 15775–15786. [CrossRef]
54. Achard, S.; Salvador, R.; Whitcher, B.; Suckling, J.; Bullmore, E. A resilient, low-frequency, small-world human brain functional network with highly connected association cortical hubs. *J. Neurosci.* **2006**, *26*, 63–72. [CrossRef]
55. Hellyer, P.J.; Shanahan, M.; Scott, G.; Wise, R.J.; Sharp, D.J.; Leech, R. The control of global brain dynamics: Opposing actions of frontoparietal control and default mode networks on attention. *J. Neurosci.* **2014**, *34*, 451–461. [CrossRef]
56. Avelar-Pereira, B.; Bäckman, L.; Wåhlin, A.; Nyberg, L.; Salami, A. Age-related differences in dynamic interactions among default mode, frontoparietal control, and dorsal attention networks during resting-state and interference resolution. *Front. Aging Neurosci.* **2017**, *9*, 152. [CrossRef]
57. Menon, V. 20 years of the default mode network: A review and synthesis. *Neuron* **2023**, *111*, 2469–2487. [CrossRef]
58. Di Plinio, S.; Ferri, F.; Marzet, L.; Romani, G.L.; Northoff, G.; Pizzella, V. Functional connections between activated and deactivated brain regions mediate emotional interference during externally directed cognition. *Hum. Brain Mapp.* **2018**, *39*, 3597–3610. [CrossRef]
59. Uddin, L.Q.; Kelly, A.M.; Biswal, B.B.; Castellanos, F.X.; Milham, M.P. Functional connectivity of default mode network components: Correlation, anticorrelation, and causality. *Hum. Brain Mapp.* **2009**, *30*, 625–637. [CrossRef]
60. Konu, D.; Turnbull, A.; Karapanagiotidis, T.; Wang, H.-T.; Brown, L.R.; Jefferies, E.; Smallwood, J. A role for the ventromedial prefrontal cortex in self-generated episodic social cognition. *Neuroimage* **2020**, *218*, 116977. [CrossRef]
61. Alves, P.N.; Foulon, C.; Karolis, V.; Bzdok, D.; Margulies, D.S.; Volle, E.; de Schotten, M.T. An improved neuroanatomical model of the default-mode network reconciles previous neuroimaging and neuropathological findings. *Commun. Biol* **2019**, *2*, 370. [CrossRef]
62. Binder, J.R.; Frost, J.A.; Hammeke, T.A.; Bellgowan, P.; Rao, S.M.; Cox, R.W. Conceptual processing during the conscious resting state: A functional MRI study. *J. Cogn. Neurosci.* **1999**, *11*, 80–93. [CrossRef]
63. Damoiseaux, J.S.; Rombouts, S.A.R.B.; Barkhof, F.; Scheltens, P.; Stam, C.J.; Smith, S.M.; Beckmann, C.F. Consistent resting-state networks across healthy subjects. *Proc. Natl. Acad. Sci. USA* **2006**, *103*, 13848–13853. [CrossRef]
64. Greicius, M.D.; Krasnow, B.; Reiss, A.L.; Menon, V. Functional connectivity in the resting brain: A network analysis of the default mode hypothesis. *Proc. Natl. Acad. Sci. USA* **2003**, *100*, 253–258. [CrossRef]
65. Koch, W.; Teipel, S.; Mueller, S.; Buerger, K.; Bokde, A.; Hampel, H.; Coates, U.; Reiser, M.; Meindl, T. Effects of aging on default mode network activity in resting state fMRI: Does the method of analysis matter? *Neuroimage* **2010**, *51*, 280–287. [CrossRef]
66. Schilbach, L.; Eickhoff, S.B.; Rotarska-Jagiela, A.; Fink, G.R.; Vogeley, K. Minds at rest? Social cognition as the default mode of cognizing and its putative relationship to the "default system" of the brain. *Conscious. Cogn.* **2008**, *17*, 457–467. [CrossRef]
67. Bluhm, R.L.; Osuch, E.A.; Lanius, R.A.; Boksman, K.; Neufeld, R.W.; Théberge, J.; Williamson, P. Default mode network connectivity: Effects of age, sex, and analytic approach. *Neuroreport* **2008**, *19*, 887–891. [CrossRef]
68. Das, A.; de Los Angeles, C.; Menon, V. Electrophysiological foundations of the human default-mode network revealed by intracranial-EEG recordings during resting state and cognition. *Neuroimage* **2022**, *250*, 118927. [CrossRef]
69. Dixon, M.L.; Moodie, C.A.; Goldin, P.R.; Farb, N.; Heimberg, R.G.; Zhang, J.; Gross, J.J. Frontoparietal and default mode network contributions to self-referential processing in social anxiety disorder. *Cogn. Affect. Behav. Neurosci.* **2021**, *22*, 187–198. [CrossRef]

70. Laird, A.R.; Eickhoff, S.B.; Li, K.; Robin, D.A.; Glahn, D.C.; Fox, P.T. Investigating the functional heterogeneity of the default mode network using coordinate-based meta-analytic modeling. *J. Neurosci.* **2009**, *29*, 14496–14505. [CrossRef]
71. DeSerisy, M.; Ramphal, B.; Pagliaccio, D.; Raffanello, E.; Tau, G.; Marsh, R.; Posner, J.; Margolis, A.E. Frontoparietal and default mode network connectivity varies with age and intelligence. *Dev. Cogn. Neurosci.* **2021**, *48*, 100928. [CrossRef]
72. Davey, C.G.; Pujol, J.; Harrison, B.J. Mapping the self in the brain's default mode network. *Neuroimage* **2016**, *132*, 390–397. [CrossRef]
73. Binkofski, F.C.; Klann, J.; Caspers, S. Chapter 4—On the Neuroanatomy and Functional Role of the Inferior Parietal Lobule and Intraparietal Sulcus. In *Neurobiology of Language*; Academic Press: Cambridge, MA, USA, 2016; pp. 35–47. [CrossRef]
74. Mars, R.B.; Jbabdi, S.; Sallet, J.; O'Reilly, J.X.; Croxson, P.L.; Olivier, E.; Noonan, M.P.; Bergmann, C.; Mitchell, A.S.; Baxter, M.G.; et al. Diffusionweighted imaging tractography-based parcellation of the human parietal cortex and comparison with human and macaque resting-state functional connectivity. *J. Neurosci.* **2011**, *31*, 4087–4100. [CrossRef]
75. Thomas Yeo, B.T.; Krienen, F.M.; Sepulcre, J.; Sabuncu, M.R.; Lashkari, D.; Hollinshead, M.; Roffman, J.L.; Smoller, J.W.; Zöllei, L.; Polimeni, J.R.; et al. The organization of the human cerebral cortex estimated by intrinsic functional connectivity. *J. Neurophysiol.* **2011**, *106*, 1125–1165. [CrossRef]
76. Wu, S.-C.J.; Jenkins, L.M.; Apple, A.C.; Petersen, J.; Xiao, F.; Wang, L.; Yang, F.-P.G. Longitudinal fMRI task reveals neural plasticity in default mode network with disrupted executive-default coupling and selective attention after traumatic brain injury. *Brain Imaging Behav.* **2020**, *14*, 1638–1650. [CrossRef]
77. Smallwood, J.; Bernhardt, B.C.; Leech, R.; Bzdok, D.; Jefferies, E.; Margulies, D.S. The default mode network in cognition: A topographical perspective. *Nat. Rev. Neurosci.* **2021**, *22*, 503–513. [CrossRef]
78. Briggs, R.G.; Tanglay, O.; Dadario, N.B.; Young, I.M.; Fonseka, R.D.; Hormovas, J.; Dhanaraj, V.; Lin, Y.-H.; Kim, S.J.; Bouvette, A.; et al. The unique fiber anatomy of middle temporal gyrus default mode connectivity. *Oper. Neurosurg.* **2021**, *21*, E8–E14. [CrossRef]
79. Xu, J.; Wang, J.; Fan, L.; Li, H.; Zhang, W.; Hu, Q.; Jiang, T. Tractography-based parcellation of the human middle temporal gyrus. *Sci. Rep.* **2015**, *5*, 18883. [CrossRef] [PubMed]
80. Cunningham, S.I.; Tomasi, D.; Volkow, N.D. Structural and functional connectivity of the precuneus and thalamus to the default mode network. *Hum. Brain Mapp.* **2017**, *38*, 938–956. [CrossRef] [PubMed]
81. Lee, T.W.; Xue, S.W. Functional connectivity maps based on hippocampal and thalamic dynamics may account for the default-mode network. *Eur. J. Neurosci.* **2018**, *47*, 388–398. [CrossRef] [PubMed]
82. Sontheimer, A.; Pontier, B.; Claise, B.; Chassain, C.; Coste, J.; Lemaire, J.-J. Disrupted pallido-thalamo-cortical functional connectivity in chronic disorders of consciousness. *Brain Sci.* **2021**, *11*, 356. [CrossRef] [PubMed]
83. Yuan, R.; Di, X.; Taylor, P.A.; Gohel, S.; Tsai, Y.-H.; Biswal, B.B. Functional topography of the thalamocortical system in human. *Brain Struct. Funct.* **2016**, *221*, 1971–1984. [CrossRef] [PubMed]
84. Trepel, M. *Neuroanatomie: Struktur und Funktion*, 8th ed.; Elsevier Health Sciences: Amsterdam, The Netherlands, 2021.
85. Fodor, J. *The Modularity of Mind*; MIT Press: Cambridge, MA, USA, 1983.
86. Fehr, T.; Milz, P. The individuality index—A measure to quantify the degree of inter-individual variability in intra-cerebral brain electric and metabolic activity. *Cogn. Neurodyn.* **2019**, *13*, 429–436. [CrossRef] [PubMed]
87. Oldfield, R.C. The assessment and analysis of handedness: The Edinburgh inventory. *Neuropsychologia* **1997**, *9*, 97–113. [CrossRef] [PubMed]
88. Costa, P.T., Jr.; McCrae, R.R. *Revised NEO Personality Inventory and NEO Five-Factor Inventory*; Professional Manual; Psychological Assessment Resources: Odessa, FL, USA, 1992.
89. Borkenau, P.; Ostendorf, F. *NEO-Fünf-Faktoren Inventar (NEO-FFI): Nach Costa und McCrae*; Hogrefe: Göttingen, Germany, 1993.
90. Fehr, T.; Achtziger, A. Contextual modulation of binary decisions in dyadic social interactions. *Front. Behav. Neurosci.* **2021**, *15*, 715030. [CrossRef] [PubMed]
91. Ille, N.; Berg, P.; Scherg, M. Artifact correction of the ongoing EEG using spatial filters based on artifact and brain signal topographies. *J. Clin. Neurophysiol.* **2002**, *19*, 113–124. [CrossRef]
92. Talairach, J.; Tournoux, P. *Co-Planar Stereotaxic Atlas of the Human Brain*, 1st ed.; Thieme: New York, NY, USA, 1988.
93. Hopfinger, J.; Khoe, W.; Song, A. Combining Electrophysiology with structural and functional Neuroimaging: ERP's, PET, MRI, fMRI. In *Event-Related Potential: A Methods Handbook*; Handy, T.C., Ed.; The MIT Press: Cambridge, MA, USA, 2005; pp. 345–379.
94. Michel, C.M.; Koenig, T. EEG microstates as a tool for studying the temporal dynamics of whole-brain neuronal networks: A review. *NeuroImage* **2018**, *180*, 577–593. [CrossRef]
95. Bledowski, C.; Prvulovic, D.; Hoechstetter, K.; Scherg, M.; Wibral, M.; Goebel, R.; Linden, D.E. Localizing P300 generators in visual target and distractor processing: A combined event-related potential and functional magnetic resonance imaging study. *J. Neurosci.* **2004**, *24*, 9353–9360. [CrossRef] [PubMed]
96. Buckner, R.L.; Di Nicola, L.M. The brain's default network: Updated anatomy, physiology and evolving insights. *Nat. Rev. Neurosci.* **2019**, *20*, 593–608. [CrossRef] [PubMed]
97. Palm, G. Neural associative memories and sparse coding. *Neural Netw.* **2013**, *37*, 165–171. [CrossRef] [PubMed]
98. Fehr, T.; Achtziger, A.; Hinrichs, H.; Herrmann, M. Interindividual differences in oscillatory brain activity in higher cognitive functions—Methodological approaches in analyzing continuous MEG data. In *The Cognitive Neuroscience of Individual Differences*; Reinvang, I., Greenlee, M.W., Herrmann, M., Eds.; T. Bis Publishers: Oldenburg, Germany, 2003; pp. 101–120.

99. Achtziger, A.; Fehr, T.; Oettingen, G.; Gollwitzer, P.; Rockstroh, B. Strategies of Intention Formation are Reflected in Continuous MEG Activity. *Soc. Neurosci.* **2008**, *4*, 11–27. [CrossRef] [PubMed]
100. Fehr, T. *Lokalisation Langsamer Hirnaktivität bei Schizophrenen Patienten Mittels Magnetenzephalografischer Untersuchungen und Exploration von Zusammenhängen Zwischen Langsamwelliger Hirnaktivität und Symptomatik*; Shaker: Aachen, Germany, 2002.
101. Hampson, S.E.; Goldberg, L.R. Personality Stability and Change over Time. In *The Wiley Encyclopedia of Personality and Individual Differences: Models and Theories*, 1st ed.; Carducci, B.J., Nave, C.S., Eds.; John Wiley & Sons Ltd.: Portland, OR, USA, 2021; Volume 1, pp. 317–321.
102. Zhi, S.; Zhao, W.; Wang, R.; Li, Y.; Wang, X.; Liu, S.; Li, J.; Xu, Y. Stability of specific personality network features corresponding to openness trait across different adult age periods: A machine learning analysis. *Biochem. Biophys. Res. Commun.* **2023**, *672*, 137–144. [CrossRef] [PubMed]
103. Koenig, T.; Prichep, L.; Lehmann, D.; Sosa, P.V.; Braeker, E.; Kleinlogel, H.; Isenhart, R.; John, E.R. Millisecond by millisecond, year by year: Normative EEG microstates and developmental stages. *NeuroImage* **2002**, *16*, 41–48. [CrossRef] [PubMed]
104. Michel, C.M.; Murray, M.M.; Lantz, G.; Gonzalez, S.; Spinelli, L.; de Peralta, R.G. EEG source imaging. *Clin. Neurophysiol.* **2004**, *115*, 2195–2222. [CrossRef] [PubMed]
105. Milz, P.; Faber, P.L.; Lehmann, D.; Koenig, T.; Kochi, K.; Pascual-Marqui, R.D. The functional significance of EEG microstates--associations with modalities of thinking. *NeuroImage* **2016**, *125*, 643–656. [CrossRef] [PubMed]
106. Stoffel, T.; Vaque-Alcazar, L.; Bartres-Faz, D.; Pero-Cebollero, M.; Canete-Masse, C.; Guardia-Olmos, J. Reduced default mode network effective connectivity in healthy aging is modulated by years of education. *NeuroImage* **2024**, *288*, 120532. [CrossRef]
107. Giannopoulos, A.E.; Zioga, I.; Papageorgiou, P.; Pervanidou, P.; Makris, G.; Chrousos, G.P.; Stachtea, X.; Capsalis, C.; Papageorgiou, C. Evaluating the Modulation of the Acoustic Startle Reflex in Children and Adolescents via Vertical EOG and EEG: Sex, Age, and Behavioral Effects. *Front. Psychol.* **2022**, *16*, 798667. [CrossRef]
108. Giustiniani, A.; Danesin, L.; Bozzetto, B.; Macina, A.; Benavides-Varela, S.; Burgio, F. Functional changes in brain oscillations in dementia: A review. *Rev. Neurosci.* **2023**, *34*, 25–47. [CrossRef] [PubMed]
109. Moreno-De-Luca, A.; Myers, S.M.; Challman, T.D.; Moreno-De-Luca, D.; Evans, D.W.; Ledbetter, D.H. Developmental brain dysfunction: Revival and expansion of old concepts based on new genetic evidence. *Lancet Neurol.* **2013**, *12*, 406–414. [CrossRef] [PubMed]

brain sciences

MDPI

Review

A Review of Childhood Developmental Changes in Attention as Indexed in the Electrical Activity of the Brain

Sirel Karakaş

Psychology Department, Doğuş University, İstanbul 34775, Turkey; sirel@neurometrika-tech.com

Abstract: This review aims to present age-related changes in the neuroelectric responses of typically developing children (TDC) who are presumed to meet developmental stages appropriately. The review is based on findings from the frequently used neuropsychological tasks of active attention, where attention is deliberately focused versus passive attention where attention is drawn to a stimulus, facilitatory attention, which enhances the processing of a stimulus versus inhibitory attention, which suppresses the processing of a stimulus. The review discusses the early and late stages of attentional selectivity that correspond to early and late information processing. Age-related changes in early attentional selectivity were quantitatively represented in latencies of the event-related potential (ERP) components. Age-related changes in late attentional selectivity are also qualitatively represented by structural and functional reorganization of attentional processing and the brain areas involved. The purely bottom-up or top-down processing is challenged with age-related findings on difficult tasks that ensure a high cognitive load. TDC findings on brain oscillatory activity are enriched by findings from attention deficit hyperactivity disorder (ADHD). The transition from the low to fast oscillations in TDC and ADHD confirmed the maturational lag hypothesis. The deviant topographical localization of the oscillations confirmed the maturational deviance model. The gamma-based match and utilization model integrates all levels of attentional processing. According to these findings and theoretical formulations, brain oscillations can potentially display the human brain's wholistic–integrative functions.

Keywords: typically developing children; development of attention; clinical model of attention; connectivity; oscillatory dynamics; event-related brain activity; resting electroencephalogram

1. Introduction

This review focuses on the typical development of attention due to its impact on almost all other cognitive functions within the human information processing system. Among numerous external or internal sources, selecting those the information processing system will treat is a function of attention. Complex modulation of the selectivity and/or the intensity of cognitive functions (for example, the strength of learning and the amount of information retained) at different stages of the system is another function of attention [1–3]. However, this extensive impact has also made attention an influential confounding variable for studies on other cognitive functions. Attention has a deterministic (as an independent variable) or, if not duly controlled, a confounding effect, irrespective of whether a behavioral approach or a cognitive neuroscience approach is employed. Consequently, any discourse on life span or age-related changes in brain activity should consider attention, even when the study is on other cognitive functions [4,5].

The way age influences attention is a crucial point that needs to be addressed before treating the issue of age-related changes in attentional processing. Ridderinkhof and van der Stelt [5] used between-channel selection and channel interference tasks to study the effect of age on event-related potentials (ERPs), which are associated with attention (N2 and P300). Two of the research hypotheses were overruled. One of the hypotheses assumed a global and undifferentiated age effect on all attention-related components, and the other assumed an age effect on the earliest component, which then propagated to the later one. The third hypothesis, which assumed an independent and differential effect of age on parameters of attention-related components, was confirmed. This finding justifies the present review on age-related changes in attentional processing.

2. Stages of Attentional Processing

The primary function of attention is selection. Attention helps to select environmental (external or internal) stimuli with specific properties, focus on the relevant stimulus, and sustain this focus. Attentional selectivity occurs at an early and late stage of information processing. Table 1 provides a summary, and Sections 2.1 and 2.2 provide information on stages of selectivity, the tasks by which these stages are experimentally produced, and the associated cognitive processes.

Table 1. Experimental Tasks, Types of Interference and Attention According to Stages of Attentional Selectivity and Information Processing.

Stage of Attentional Selectivity	Type of Information Processing	Relevant Experimental Tasks	Interference	Type of Attention
Early Attentional Selectivity	Filtering	Between-channel tasks	Perceptual conflict	Active/facilitatory attention
		Channel interference tasks	Perceptual conflict, response competition	Active/facilitatory and inhibitory attention
Late Attentional Selectivity	Selective-Set	Active deviance detection tasks	Target detection with response competition	Active/facilitatory attention
		Passive deviance detection tasks		Passive/facilitatory attention
		Target search tasks		Active/facilitatory attention
		Attention priming tasks		Active/facilitatory attention
		Attention shifting tasks		Active/facilitatory and inhibitory attention

2.1. *Early Attentional Selectivity, Filter Tasks and Cognitive Correlates*

Early attentional selectivity originates from the currently obsolete model by Broadbent [6]. According to the model, the stimuli in both the external and the internal environments are analyzed by the information processing system for their physical attributes (e.g., novelty, intensity, abrupt occurrence). Attention then acts as a filter and selects stimuli based on these physical attributes.

According to the current concept of early attentional selectivity (Table 1), stimuli are analyzed through an iterative interaction between the sensory register and long-term memory; this stage involves automatic processes of the pre- or unconscious [7]. In this early stage, attentional selectivity is perceptual; perceptual conflicts are resolved at the perceptual level, and processes of attentional selectivity are quantitively represented. In event-related potentials (ERPs), attentional selectivity is represented, for example, by component amplitudes and latencies. In the brain's oscillatory activity, attentional selectivity is represented, for instance, by component power, phase locking, or frequency locking in the resting electroencephalogram (EEG) and event-related oscillations (EROs).

Early attentional selectivity is associated with filtering and is studied using filter tasks [5,8]. One group of filter tasks requires between-channel selection (e.g., spatial selection task). Here, participants are to select and attend to a stimulus in a given channel while they ignore the stimulus at the other channel. Between-channel selection tasks are used when the focus of interest is perceptual conflict. A second group of filter tasks produces channel interference (e.g., Stroop task, dichotic listening task). Here, participants are to perform a discriminative response to one stimulus channel. However, this response is interfered with the stimulus in a competing channel. In this case, channel interference is a perceptual conflict that is represented by the prolonged latency of the earlier perceptual components. Interference may also be a response competition; in this case, prolonged latencies are displayed in motor components and overt behavioral responses (e.g., reaction time: RT).

2.2. Late Attentional Selectivity, Selective-Set Tasks, and Cognitive Correlates

The concept of late attentional selectivity originates from the model developed by Deutsch and Deutsch [9]. According to this model, external and internal stimuli are analyzed for physical and semantic attributes. In contrast to early selectivity, late selectivity involves response selection. Attention acts as a filter to select a functional response from various possible alternatives.

According to the current concept of late attentional selectivity (Table 1), the previously perceived stimuli are processed via integrative associational processing within relevant brain networks [7,10]. In contrast to early attentional selectivity, late selectivity involves consciously monitored controlled processing. The selectivity in task-relevant responding involves target detection and response selection. Any conflict is thus due to response competition that is also resolved at the response level. Late attentional selectivity processes are quantitively and qualitatively (structurally) represented, for example, in a changing topography of the components and the specific processes responsible for attentional selectivity [11–13].

Late attentional selectivity involves the formation of selective sets [5,8]. Among the most frequently used selective-set tasks are deviance tasks. Active deviance tasks (e.g., active oddball task) ask the participant to select a rarely occurring stimulus (target) and to respond to it (task-relevant response). Passive deviance tasks (e.g., passive oddball task) use the same experimental procedure, but the task is to perform an irrelevant task. Active deviance is related to active, facilitatory attention in the form of focused and selective attention. Passive deviance is related to passive inhibitory attention. It includes an involuntary orientation reaction [14] and preconscious, passive attention [15]. Other tasks involve priming tasks (e.g., auditory priming), which produce focused attention (active facilitatory attention) via a priming stimulus, and shifting tasks, where attention (both facilitatory and inhibitory active attention) has to be moved between targets [8].

3. Age-Related Childhood Changes in ERP Components and Attentional Processing

The time domain methodology and analysis of brain responses were first recorded by Davis [16] as evoked potentials (EPs). The summation technique strengthened the methodology (Dawson [17]) because it made the very low amplitude brain responses, which occur within the context of an irregular and large amplitude background activity, discernable. The first event-related cognitive component (P300) was recorded by Sutton and colleagues [18,19]. Early studies on cognitive processes and the development of attention used the ERP methodology, and this trend still endures [20,21]. As the following sections show, the ERP methodology is also used in most of the studies on brain responses of TDC [5,8,21,22].

3.1. Early Stage of Attentional Processing: Filtering

According to a meta-analysis by Verhaeghen and De Meersman [23], the delay that channel interference produces (the Stroop effect) is a typical cognitive event that stays invariant throughout adulthood [24]. In TDC, there is a strong Stroop effect. In neuropsychological tasks, this effect is represented in longer response durations [25]. However, the ERP components represent the Stroop effect in large amplitudes [26]. These amplitudes decrease throughout childhood (5–12 years) and stabilize at the adult level around 12 years of age [24].

When attention is exogenously cued to relevant aspects of the stimulus, children's performance reaches that of adults [27]. This finding shows that channel interference, which component amplitudes represent, is due to perceptual conflict, and the filter operates at the perceptual stage [28]. Inefficient suppression of inappropriate responses is a characteristic of young children. The Stroop effect is demonstrated by a long latency of the readiness potential (RP) and, in behavioral studies, a long RT [29]. In this case, the locus of the filter is post-perceptual; the filter is operating at the response stage.

When five-year-old children perform between-channel tasks, they respond with the attention-related N2b and P3b components. This is regardless of whether the stimulus is task relevant (to be attended) or task irrelevant (to be ignored [30]). This finding shows that TDC can detect stimuli but cannot filter out and selectively respond to the relevant stimulus. Efficient channel selection becomes observable in 7–9-year-old children [31], becomes possible around 12 years of age, and continues development until 17 years of age [32]. Several ERP components serve as markers of the development of efficient channel selection and different types of attention [33–35]. Two markers show latency variations. Latency of the frontal selection positivity (FSP) represents an early selection between relevant and irrelevant channels. Further processing of the attended channel is represented in the latency of N2b. A third marker shows amplitude variations in the early part of the processing negativity (PN) [36]. Age-related amplitude variations in early PN represent the development of stimulus comparison and selection.

3.2. Late Stage of Attentional Processing: Formation of Selective Sets

The most widely used selective-set tasks of late attentional selectivity are the active and passive oddball (OB) tasks. In adults, the prominent attention-related component of the active OB (OB-a) task is P3b [18,37,38]; that of the passive OB (OB-p) task is mismatch negativity (MMN; [39]).

3.2.1. Findings on Active Deviance Detection

Attention-related ERPs to active deviance detection start in adults with N2, an early selection component. Development is predominantly represented by decreasing latencies between 5–6 and 22 years of age [40,41]. Another developmental representation is the topographical spread of N2. The diffuse representation up to 10–14 years of age evolves to a frontal distribution starting at 17 [41]. According to these N2 findings, the speed by which stimuli are discriminated and compared increases with age, and the cortical spread of the component becomes focalized to the frontal sites.

Another attention-related ERP component in adults is P300 [42,43]. P300 appears as P3a or P3b. In infancy (4–7 months of age), precursors of P3b are two late components. The first is a central–frontal negative component (Nc; onset 140–160 ms, duration 700 ms). Age-related increases in Nc amplitude represent the detection of and attention to deviant stimuli. The second precursor is a distributed late positive wave (Pc/PSW; onset 200–1000 ms, duration 1300 ms). Age-related increases in Pc amplitude represent the extent of memory encoding [44,45].

The third late component is the precursor of the frontally obtained adult P3a [43]. In adults, P3a is an orientation response to novel events and is responsible for novelty detection. It is thus obtained in response to novel, intense, or unexpected events or events that are significant to the species or the individual [37,46,47]. The P3a precursor, the negative slow wave (NSW) becomes apparent at around two years of age [48]. Its amplitude decreases from childhood, and its latency decreases up to seven years of age [11,48,49]. Nevertheless, P3a (of the N2-P3a complex) continues to display age-related changes within the 8–28 age range [50]. Increasing P3a amplitudes and decreasing latencies indicate faster processing of novel events and improved cognitive control.

The relation between P3b latency and information processing speed is a consistent finding [47]. From a young age, latencies progressively decrease until they reach adult levels at 22–25. Findings on P3b amplitude, however, do not make a general conclusion feasible. There are reports about amplitude increases until 21 [47] or 15 years of age [48] and increasing amplitudes in the frontal lobes but not in the parietal lobes, where amplitudes are similar to those of adults [51].

3.2.2. Findings on Passive Deviance Detection

The composition of the stimuli (frequent standards and infrequent deviants that are serially presented in random order) in the OB-p task is identical to that in OB-a; the difference is in the task the participant is asked to perform. In OB-a, the task is to respond to a relevant stimulus (e.g., press a button, count). In OB-p, the task is to perform an irrelevant task throughout the stimulus presentation. The OB-p task produces the mismatch negativity (MMN) component in adults. The component is a frontally distributed negative component (100–300 ms), and it is obtained mainly in response to auditory stimulation (e.g., frequency and tone deviances). MMN is produced by repetitive stimulation; frequent repetition of the standard stimulus makes the sensory representation of the standard stimulus precise, as a result of which detection of even a slight difference in the deviant stimuli becomes possible. This deviance or mismatch detection is automatically performed in the preconscious [39].

However, many of the findings on the age-related changes in MMN latency and amplitude are inconsistent. Reported findings on age-related latency changes include 240 ms in newborns, 207 ms in 8-year-old children, and 140 ms in adults [52], as well as a latency decrease only between 3 and 14 years of age [53] and a stable latency throughout childhood and adulthood [36,54]. Age-related amplitude changes include almost all of the possible findings: large amplitudes throughout childhood [55], decreasing amplitudes from younger to older children [36], and a stable amplitude throughout childhood and adulthood [56].

4. A Bidirectional Serial Processing: Bottom-Up with Top-Down Processing

According to the previous findings, the onset of the development of attention-related components might be as early as 4–7 months of age/infancy (e.g., precursors of P300) or two years of age (e.g., P3a), but the more common initiation is around 5–7 years of age/middle childhood (e.g., N2b of between-channel selection). Depending on the type of ERP component, latencies may attain adult levels as early as seven years of age/middle childhood (e.g., for P3a), at 17 years of age/adolescence (e.g., for N2b and P3b of between-channel selection), and at 22 years of age/early adulthood (e.g., for N2 of active deviance detection) (Sections 3.1 and 3.2).

Broadbent's [6] theory assumes that attentional selectivity (or the filter) operates upon physically analyzed stimuli, and this early selectivity is demonstrated with filter tasks (e.g., between-channel and channel interference tasks). According to its current extensions, this is a bottom-up process. According to the theory of Deutsch and Deutsch [9], attentional selectivity (filter) operates upon semantically analyzed stimuli, and according to its current extension, the process is a top-down one. This early selectivity is demonstrated with the selective-set tasks (e.g., deviance detection tasks and target search tasks) [8].

According to the review article by Karakaş [8], age-related changes in TDC occur in both the early (findings of studies that use the various filter tasks; Section 3.1) and the late stage (findings of studies that use the selective-set tasks; Section 3.2) of attentional processing. The consistently obtained findings are quantitative and occur mainly upon component latencies. Decreasing latencies show that, as children age, they become faster and more efficient at perceptual analysis and filtering; early attentional selectivity that involves bottom-up processing is enhanced. Age-related changes in late attentional selectivity involve top-down processing, represented in the speed and efficiency of target detection, classification, and response selection.

Amplitude represents power and the associated energy; in information processing, amplitude is the biological marker of the intensity of resource allocation. Available resources and power vary according to individuals' transitory and stable characteristics. Therefore, age-related amplitude variations are usually less consistent in TDC than in latency. However, amplitude becomes the critical index under sub-optimal stimulus conditions constraining developmental trends. In adults, task difficulty is represented in P3b amplitude. In TDC, difficulty and cognitive load are represented in the amplitude of entirely different ERP components. In children, the responsible components are the late, slow potentials, such as processing negativity (PN) [36]. In infants, detection is represented by two components unique to infancy: Nc and Pc (Section 3.2.1).

Results of studies that manipulate task difficulty/cognitive load draw attention to the interplay between early and late attentional processing stages. In adults, attentional selectivity is ordinarily perceptual (early stage). In TDC, inefficient perceptual filtering [5] due to task difficulty and cognitive load produces a shift to the later stage of information processing. Age-related changes are then represented not in perceptual but in response-related indices, such as the lateralized readiness potential (LRP), data from electromyography (EMG), behavioral RT, and response accuracy [5].

These ERP findings show that the brain does not serially operate in either a bottom-up or a top-down direction at a given time and under given conditions. Variables related to the task and/or the participant may require a shift between the mode of processing. The transition between perceptual selectivity and response selectivity, and from perceptual to response components, represents a reorganization of the cognitive processes, the ERP components, and the respective brain areas. In contrast to the age-related latency variations, which are quantitative, the reorganization of the cognitive processes and brain neuroelectric activity under task difficulty represent qualitative changes [12,13,57].

5. Age-Related Childhood Changes in EEG and ERO Components of Attentional Processing

As Section 2 shows, age-related changes in attentional processing have mainly used the ERP methodology, which analyzes neuroelectric activity in the time domain. The theoretical framework behind most ERP studies is the current extensions of the classical theories of attention [6,9]. These theories are based on the assumption that human information processing is a serial process that proceeds in the bottom-up or the top-down direction. Accompanying information processing is from the periphery to the center or vice versa; and from sensory/perceptual processes to successively higher cognitive functions or vice versa [7,58].

5.1. Oscillatory Dynamics, Principles, and Theories

An alternative noninvasive methodology analyzes resting electroencephalography (EEG) and event-related activity in the frequency or the more recently formulated time and frequency domains [59]. This approach dates back to Berger [60], who recorded human EEG for the first time and demonstrated the alpha and beta oscillations in EEG. The approach gained impact with Adrian [61], who discovered the event-related oscillation (ERO) in the gamma band.

Studies in 1980–1990 led Mountcastle [62] to pronounce a paradigm change to the oscillatory activity (EEG and EROs) for understanding brain function. This decade witnessed studies on the oscillatory dynamics of humans and many nonhuman species; developmental stages; sensation and cognition; and spontaneous, evoked, emitted, and induced brain activity [63,64]. These findings were replicated and confirmed; accumulated information led to formulating principles of oscillatory dynamics [19]. Erol Başar, who is a pioneer in the brain's oscillatory dynamics, outlines these principles as follows [63–65]. The waveforms in resting EEG, EP, and ERP result from the superposition of the oscillatory components of different frequencies (the principle of superposition); each oscillation has multiple functions, and multiple oscillations represent a given cognitive function (the principle of multiplicity of function); cognitive functions are represented by oscillations that are temporally and spatially integrated within the whole brain (the principle of whole brain work). A demonstration of these principles can be found in two review papers [19,66], one on the historical background of oscillatory dynamics and the other on the theta oscillation of the brain.

Contemporary brain theories are mainly based on the principles of oscillatory dynamics [67,68]. A pioneering theory is that of Erol Başar [69]. According to the theory of whole brain work, (1) bottom-up and top-down processes occur in parallel within selectively distributed neural pathways. (2) This structural organization produces complex and dynamically changing connectivity patterns between temporally and spatially associated brain oscillations. (3) The connectivity patterns are the correlates of cognitive-affective processes of the human information processing system.

5.2. Attentional Processing and Resting-State EEG Components in TDC and a Clinical Model

Time domain ERP findings and formulations dominate the development of attentional processing [21,70]. However, the few studies on the oscillatory dynamics of attention provided clinically useful criteria, formed the basis of a model of attentional processing involving a wholistically functioning brain and had heuristic value for future research on the typical development of attention.

A well-established developmental finding on TPD pertains to the components decomposed from the resting EEG using frequency domain methodology [19]. At the beginning of life, EEG is dominated by spontaneously occurring slow wave activity (e.g., delta). In early childhood, theta and alpha oscillations (the faster activity) become posteriorly visible, and in time, they propagate to the anterior regions. Another fast activity, beta, becomes centrally visible, and in time, it propagates to the anterior (frontal) and the posterior regions (parietal and occipital) [71–74]. By 7.5 years of age, slow oscillations are replaced by high-frequency oscillations [20].

The spontaneously occurring beta oscillation's relevance to conscious attentive wakefulness [60,72,75], alpha oscillation's relevance to vigilance and selective attention [76,77], and the theta oscillation's relevance to the differentiation of states of consciousness [78] are well-established. Some serve as criteria for the standard classification of sleep–wakefulness stages [79].

In DSM-5 [80], the basic symptom of attention deficit hyperactive disorder (ADHD) is inattention. As such, ADHD is a useful clinical model of attentional processing. Following that, a significant part of the studies on the oscillatory dynamics of attention are on children with ADHD rather than on TDC.

In ADHD, maturation of resting EEG is structurally and functionally equivalent to that in TDC, but changes occur later [20]. For example, in TDC, slow-wave ADHD dominance ceases, and high frequencies emerge at around 7.5 years of age [8]. In ADHD, this change is delayed to 10 years of age. Delayed development is the prediction of the maturational lag model of Kinsbourne [81]. This model is supported not only by oscillatory components of EEG but also by neuroanatomical and neuropsychological findings. Neuroanatomically, the most considerable cortical thickness is attained in TDC at 7.5, but in ADHD at 10.5 years of age [82,83]. When different forms of attention are behaviorally assessed using relevant neuropsychological tests, children with ADHD reach the level of age-matched TDC around 12 years of age [25].

However, the maturational lag model provides only a partial explanation; in addition to delayed maturation, there are maturational deviances in ADHD [84,85]. According to the maturational deviance model [86], the deviances are not typical to any age but are relatively permanent. These oscillatory deviances [8] consist of low relative alpha in the parietotemporal region [20,87,88]; low absolute and relative beta in the parietotemporal and frontal areas [20,89]; low absolute EEG power in the theta band [90]; and elevated absolute and relative theta in the frontal regions and the frontal midline area [20,91]. The ratio between the elevated theta and the reduced beta, the theta/beta ratio (TBR), is specific to ADHD, and it thus serves as a biomarker of ADHD and as an auxiliary criterion for ADHD diagnosis [92,93]. Deviant patterns are also revealed in neuropsychological test scores. Using a battery of tests on different aspects of attentional processing, Erdoğan Bakar and Karakaş [25] provided evidence for both the maturational delay and the maturational deviance model that persisted throughout the studied age range (6–12 years). The test scores represent atypical information processing characterized by non-attentional factors. Similarly, the response strategies were based on impulsive and/or hyperactive behavior rather than attentional processes.

Additively, structural and functional resting-state deviancy is coupled in ADHD with deviant or reduced connectivity patterns. Per the whole brain work theory [69], high global integrity typically characterizes the brain nodes [94]. In ADHD, integrity is impaired both structurally and functionally [95]. This impairment is due to a specific redistribution of the regional nodes over the resting-state network, resulting in high local clustering and low global integrity [96,97].

Reduced differentiation and specialization in cortico-cortical circuits are associated with clusters of deviant oscillatory activity [72,75,98]. The cortical hypoarousal cluster of ADHD is characterized by elevated frontal theta, reduced delta, and reduced and deficient beta (hence the elevated TBR). The cortical hyperarousal cluster is characterized by elevated frontal beta, deficient delta, and alpha. The maturational lag cluster is characterized by elevated delta and theta and reduced and deficient beta. The last cluster is characterized by increased frontal beta. These clusters are not only neuroelectric entities; they are accompanied by specific behavior patterns typical to children with ADHD [99].

5.3. Attentional Processing and EROs in TDC and ADHD

The typical development of attentional processing is correlated with a quantitative change involving a decremental trend in the latencies of ERP components. This trend indicates that TDC gradually assumes the adult levels of early and late attentional selectivity. In ADHD, however, attentional processing is represented in component amplitudes. This is a quantitative change, but unlike the consistently decremental trend in component latencies of TDC, the amplitude variations are atypical. They apply to early selection processes, such as perceptual analysis, filtering, and automatic processing; and they represent involuntary attention, orienting disorder, and impairment in selective, focused, and sustained attention [66]. Atypical amplitude variations also apply to late selection processes, such as target detection, controlled processing, and response-related processes, and they represent filtering deficits, selectivity that is principally guided by stimulus salience, and deficits in attentional resource allocation [8].

The power and energy of the waveform amplitudes represent resource availability in information processing [8]. A heuristic from the preceding sections on the oscillatory dynamics of ADHD versus TDC may be the following. The experimental tasks are difficult for children with ADHD, so they invest whatever resources they possess. Despite this effort, they continue displaying the impairments of the disorder and perform poorly on attention-related tasks.

5.3.1. Event-Related Gamma Oscillation: Findings and Theory

The gamma band response (GBR) was discovered later than the other frequency bands, and unlike them, it was found as an event-related component [61]. The GBR was considered among the brain's basic operating rhythms [62,63]. Gamma also became the most frequently studied oscillation in ADHD because gamma band oscillation represents the facilitatory and inhibitory attentional processes that are clinically impaired in ADHD [80] and it modulates dopamine polymorphisms that affect the heritability of ADHD [100].

In line with the multiplicity of function principle, the GBR represents different processes [101]. In the early time window (0–150 ms), a phase-locked GBR takes place in bottom-up processing in sleep and wakefulness [78,102]. This early gamma is associated with sensory processing [103]. However, early gamma response is also associated with higher cognitive functions, such as gender specificity [104] and neuropsychological test scores [105]. A nonphase-locked GBR of the late time window (130–400 ms) is also associated with higher cognitive functions [101].

A third group of early studies on adults asserted that the GBR is related to attentional processing [106–108]. In line with the 'principle of superposition,' event-related gamma is superposed on event-related theta. During bottom-up processing, the gamma/theta superposition represents the allocation of attention. During top-down processing, the gamma/theta superposition represents the activation of memory stores [105,109–113].

To our knowledge, Yordanova and colleagues conducted the only research on TDC's GBR [114–116]. During a between-channel task, TDC (9–12 years of age) responds to the attended stimuli with large, synchronized, and phase-locked (120 ms post-stimulus) GBR. Gamma is observed in the left frontocentral area. Under the same experimental condition, the GBR of older TDC (3–16 years of age) was ipsilateral to the side of stimulation.

Hermann and colleagues [111] integrated previously mentioned processes (bottom up and top down) and functions (attention, sensation, perception, higher cognitive functions) that operate via the GBR within the context of the match and utilization model (MUM). According to the model, the first phase is a bottom-up process. Evoked and phase-locked GBR in the first part of the early time window (0–150 ms) represents the allocation of resources to rapid processing and classification of input. Stimulus features are analyzed and integrated; sensation is transformed into perception. In the latter part of the early time window, processing is top-down. Perceived content is compared with existing memory representations. Attention modulates matching, and the GBR is enhanced with a match or the anticipation of a match [110,112,117].

An induced, nonphase-locked GBR characterizes the second phase in a later time window (approximately 200 ms following stimulus onset). The induced GBR represents the utilization of the information. In this phase, information is stored, and attention is redirected. The induced GBR of the second phase is recorded from highly distributed networks, indicating connectivity between selected areas of the brain [110,112].

Findings on ADHD support the assumptions of the MUM. In the first part of the early phase, episodes of inattention are accompanied by reduced activity, showing impairment of early automatic processing, which includes extraction, analysis, and integration of sensory information [118–120]. A stronger phase locking and synchronization shows alteration in early auditory processing [116].

In the second part of phase one, stimuli are encoded for later stimulus recognition and represented in task-related GBR enhancement. However, in ADHD, the later recognition performance does not accompany this enhancement, indicating that GBR enhancement is associated with a general activation of processing resources not specific to the relevant stimuli [121]. The GBR to known stimuli is not enhanced. According to the MUM, this represents an inability to allocate processing resources or a failure to access and match stimuli with the relevant memory representation [122]. Atypical or impaired processing leads to impaired stimulus filtering. Then, attention cannot be shifted from the irrelevant input or focused on the relevant input [123,124]. The mainly quantitative problems in the early stage lead to inattention and distractibility, leading to atypical involuntary and orienting attention and impaired active attention consisting of selective, focused, sustained, and inhibitory attention [123,124].

Impairments in the first phase influence the processes in the second phase, and children with ADHD make compensatory attempts to accomplish the early-phase processes [21]. The amplitudes of components representing late processing (e.g., P3b, FSP SN) are reduced [122,125–128]. The qualitative problems in the late stage of attentional processing mainly lead to difficulties in target detection, response-related processes, and controlled processing [8].

5.3.2. Event-Related Theta Oscillation: Findings and Theory

Theta was discovered by Jung and Kornmüller [129] in the rabbit hippocampus. Later research found that theta oscillation plays a role in many cognitive processes such as sensation, perception, learning, memory, navigation, and motor processes [14,130–138].

As such, theta is an excellent demonstration of the 'multiplicity of function principle' of oscillatory dynamics [139–142].

The most recent review [66] systematizes the currently known functions and structures of theta under the cortico-hippocampal interplay model. The network where this interplay occurs is highly interconnected. Consequently, information travels over a densely interconnected, highly synchronized system via the theta oscillation. This architecture bestows theta with the many cognitive-affective correlates, one of which is attention. In ADHD, where connectivity is impaired [95], the corticocortical theta circuits are characterized by high clustering and low global integrity, resulting in reduced cortical differentiation and specialization [143].

The primary oscillation that composes the novelty N2-P3a waveform is theta [144]. Wienke and colleagues [50] studied the connectivity strengths through inter-site coherences between 48 pairs of electrodes in participants 8–28 years of age. Age and theta phase coupling were linearly related, showing that theta is a maturation marker from late childhood to young adulthood. Enhanced connectivity strengths were displayed in the short-range frontocentral and the long-range anteroposterior connections. These findings support those of Barry and colleagues [143] and indicate that typical maturation includes increased integration of frontal brain activity (seed electrodes were Fz, FCz, and Pz) within a widely distributed neural network (the long-range anteroposterior connections).

Research [43,76,77,109,141,145–148] indicates that spontaneous theta in the resting EEG is responsible for arousal. Event-related theta or the superposition of theta with oscillations in other frequency bands (e.g., theta superposed on the delta at P3b latency, theta superposed with alpha at 100 ms latency) are responsible for all sorts of attention that filter and selective-set tasks produce (Section 2). These include active and passive attention that the active deviance and passive deviance tasks, respectively, produce; facilitatory attention (selective, focused, and sustained attention) and inhibitory attention (as in the Stroop task); early attentional selectivity at approximately 150 ms (as in the filter task); and late attentional selectivity at approximately 300 ms (as in the selective-set tasks).

During active and passive attention, adults display an enhanced and strongly phase locked early theta response (0–300 ms) and a nonphase-locked late theta response (300–600 ms). Early theta is associated with a N200 component; late theta is associated with a P3b-like P400–700 component [115]. The high-amplitude early theta response of TDC displays weak phase locking. As children age, phase locking gets more robust, and the latency of the theta response decreases [114].

However, despite its relevance to almost all aspects of adult attentional processing, conclusive research was not devoted to developing attention in TDC. Theta may be a possible marker of ADHD or may serve as an auxiliary device for diagnosis. This issue did not capture the attention of the scientific community.

6. Conclusions

The present paper reviews the development of attention via resting state EEG, ERPs, and EROs in children (6–12 years of age). Findings on the earlier stage of attentional processing were studied using filter tasks (e.g., between-channel and channel interference tasks). Those in the later stage of attentional processing were studied using selective-set tasks (e.g., deviance detection, priming). Findings were discussed within serial (bottom up, top down), serial–bidirectional, and holistic–integrative information processing. Qualitative changes that involve a transition from early to late attentional processing were demonstrated in experimental designs that use difficult tasks with high cognitive load. Another approach involved using ADHD as a clinical model of attentional processing.

The review highlights the following issues:

- The task that the experiment uses determines the type and stage of attentional processing. Early attentional selectivity is studied with the filter tasks (between-channel selection, channel selectivity) and late attentional selectivity with the selective-set

tasks (active deviance detection, passive deviance detection, target search, attention priming, shifting).
- Attention has an independent and differential effect on all other aspects of information processing. Thus, it deserves scientific interest as a cognitive phenomenon that must be explained and a confounding variable that must be controlled.
- Typical development of attention is quantitatively represented in latency variations of ERP components. However, attention is qualitatively represented when a task is difficult. Late attentional processing becomes operational, late ERP components appear, and components vary in amplitudes.
- The resting EEG quantitatively represents the typical development of attention, where a transition from a dominance of the slower oscillations to the faster ones occurs.
- ADHD is a useful clinical model for studying attention since the basic and most frequent symptom of the disorder is inattention.
- In ADHD, the resting EEG development follows the TDC trend, albeit at later ages. Such a delay also applies to neuroanatomical development and age-related changes in neuropsychological test scores. These findings support the maturational lag model.
- Some of the ADHD findings indicate atypical and relatively permanent organization of attentional processes and the respective brain areas. These findings support the maturational deviance model.
- ADHD is also a result of deviant connectivity patterns between brain areas. The structural and functional impairment is associated with a nodal redistribution of the default network that involves high local clustering and low global integrity. Findings indicate four types of EEG clusters and the accompanying behavioral patterns resembling the inattention, impulsivity, and hyperactivity symptoms of ADHD.
- According to the MUM, attentional processing is represented by variations in the gamma oscillation within the context of a wholistically functioning brain.
- Studies on clinical models (e.g., ADHD) may produce groundbreaking findings and formulations on the nature of attention.

Time domain ERP findings dominate the literature on typical attentional development. This approach largely ignores the holistic–integrative brain function and implicitly assumes a serial processing of information in either the bottom-up or the top-down direction. Although less abundant, the frequency domain approach involving the brain's oscillatory activity could provide biomarkers and auxiliary criteria for ADHD diagnosis. The match and utilization model demonstrated attentional processing within a selectively distributed parallel processing brain system that used gamma oscillation. The multifunctional theta oscillation has the potential to explain attentional processing and its different components. Future studies focusing on the typical development of attention may invest greater interest in oscillatory dynamics and benefit from its methodology and theoretical background.

Funding: This research received no external funding.

Conflicts of Interest: The author declares no conflicts of interest.

References

1. Banich, M.T.; Compton, R.J. *Cognitive Neuroscience*, 3rd ed.; Wadsworth Cengage Learning: Belmont, CA, USA, 2011.
2. Goldstein, B. *Cognitive Psychology*, 4th ed.; Cengage Learning: Stamford, CT, USA, 2011; pp. 84–117.
3. Wickens, C.D. Processing resources of attention. In *Varieties in Attention*; Parasuraman, R., Davies, R., Eds.; Academic Press: New York, NY, USA, 1984; pp. 120–142.
4. Plude, D.; Enns, J.T.; Brodeur, D. The development of selective attention: A lifespan overview. *Acta Psychol.* **1994**, *86*, 227–272. [CrossRef]
5. Ridderinkhof, K.R.; van der Stelt, O. Attention and selection in the growing child: Views derived from developmental psychophysiology. *Biol. Psychol.* **2000**, *54*, 55–106. [CrossRef] [PubMed]
6. Broadbent, D.E. Perception and Communication. Pergamon: London, UK, 1958.
7. Karakas, S. A descriptive framework for information processing: An integrative approach. *Int. J. Psychophysiol.* **1997**, *26*, 353–368. [CrossRef]

8. Karakas, S. A comparative review of the psychophysiology of attention in children with and without attention deficit hyperactivity disorder. *Int. J. Psychophysiol.* **2022**, *177*, 43–60. [CrossRef] [PubMed]
9. Deutsch, J.A.; Deutsch, D. Attention: Some theoretical considerations. *Psychol. Rev.* **1963**, *70*, 80–90. [CrossRef] [PubMed]
10. Karakas, S.; Basar, E. Models and theories of brain function within a framework of behavioral cognitive psychology. *Int. J. Psychophysiol.* **2006**, *60*, 186–193. [CrossRef] [PubMed]
11. Courchesne, E. Neurophysiological correlates of cognitive development: Changes in long-latency event-related potentials from childhood to adulthood. *Electroencephalogr. Clin. Neurophysiol.* **1978**, *45*, 468–482. [CrossRef] [PubMed]
12. Thomas, K.M.; Nelson, C.A. Age-related changes in the electrophysiological response to visual stimulus novelty: A topographical approach. *Electroencephalogr. Clin. Neurophysiol.* **1996**, *98*, 294–308. [CrossRef] [PubMed]
13. Ridderinkhof, K.R.; Bashore, T.R.; van der Molen, M.W.; Band, G.P.H. The speed of information processing: Qualitative versus quantitative changes with age. In *Cognitive Changes Due to Aging and Fatigue as Revealed in the Electric Brain Activity*; Falkenstein, M., Hohnsbein, J., Ullsperger, P., Eds.; Bundesanstalt fuer Arbeitsschutz und Arbeitsmedizin: Dortmund, Germany, 1999; pp. 172–182.
14. Sokolov, E.N. Higher nervous functions: The orienting reflex. *Annu. Rev. Physiol.* **1963**, *25*, 545–580. [CrossRef]
15. Naatanen, R.; Paavilainen, P.; Rinne, T.; Alho, K. The mismatch negativity (MMN) in basic research of central auditory processing: A review. *Clin. Neurophysiol.* **2008**, *118*, 2544–2590. [CrossRef]
16. Davis, P.A. Effects of acoustic stimuli on the waking human brain. *J. Neurophysiol.* **1939**, *2*, 494–499. [CrossRef]
17. Dawson, G.D. A summation technique for detecting small signals in a large irregular background. *J. Physiol. Lond.* **1951**, *115*, 2–3.
18. Sutton, S.; Braren, M.; Zubin, J.; John, E.R. Evoked-potential correlates of stimulus uncertainty. *Science* **1965**, *150*, 1187–1188. [CrossRef] [PubMed]
19. Karakas, S.; Barry, R. A brief historical perspective on the advent of brain oscillations in the biological and psychological disciplines. *Neurosci. Biobehav. Rev.* **2017**, *73*, 335–347. [CrossRef] [PubMed]
20. Barry, R.; Clarke, R.L.; Johnstone, S.J. A review of electrophysiology in attention-deficit/hyperactivity disorder: I. qualitative and quantitative electroencephalography. *Clin. Neurophysiol.* **2003**, *114*, 171–183. [CrossRef] [PubMed]
21. Barry, R.; Johnstone, S.J.; Clarke, A.R. A review of electrophysiology in attention-deficit/hyperactivity disorder II: Event-related potentials. *Clin. Neurophysiol.* **2003**, *114*, 184–198. [CrossRef] [PubMed]
22. Naatanen, R. The role of attention in auditory information processing as revealed by event-related potentials and other brain measures of cognitive function. *Behav. Brain Sci.* **1990**, *13*, 201–288. [CrossRef]
23. Verhaeghen, P.; De Meersman, L. Aging and the Stroop effect: A meta-analysis. *Psychol. Aging* **1998**, *13*, 120–126. [CrossRef] [PubMed]
24. Comalli, P.E.; Wapner, S.; Werner, H. Interference of Stroop color-word test in childhood, adulthood, and aging. *J. Genet. Psychol.* **1962**, *100*, 47–53. [CrossRef]
25. Erdogan Bakar, E.; Karakas, S. Spontaneous age-related changes of attention in unmedicated boys with attention deficit hyperactivity disorder. *Clin. Neuropsychol.* **2020**, *20*, 664–698. [CrossRef]
26. Shepp, B.E.; Barrett, S.E. The development of perceived structure and attention: Evidence from divided and selective attention tasks. *J. Exp. Child Psychol.* **1991**, *51*, 434–458. [CrossRef]
27. Akhtar, N.; Enns, J.T. Relations between covert orienting and filtering in the development of visual attention. *J. Exp. Child Psychol.* **1989**, *48*, 315–334. [CrossRef] [PubMed]
28. Ridderinkhof, K.R.; van der Molen, M.W.; Band, G.P.H.; Bashore, T.R. Sources of interference from irrelevant information: A developmental study. *J. Exp. Child Psychol.* **1997**, *65*, 315–341. [CrossRef] [PubMed]
29. Ridderinkhof, K.R.; van der Molen, M.W. A psychophysiological analysis of developmental differences in the ability to resist interference. *Child Dev.* **1995**, *66*, 1040–1056. [CrossRef]
30. Van der Molen, M.W. Developmental changes in inhibitory ability evidence from psychophysiological measures. *Biol. Psychol.* **2000**, *54*, 207–239. [CrossRef] [PubMed]
31. Wijker, W. ERP Ontogenesis in Childhood. In *Unpublished Doctoral Dissertation*; University of Amsterdam: Amsterdam, The Netherlands, 1991.
32. Berman, S.; Friedman, D. The development of selective attention as reflected by event-related potentials. xp. *Child Psychol.* **1995**, *59*, 1–31. [CrossRef] [PubMed]
33. Courchesne, E. Event-related brain potentials: Comparison between children and adults. *Science* **1977**, *197*, 589–592. [CrossRef]
34. Hillyard, S.A.; Picton, T.W. Event-related brain potentials and selective information processing in man. In *Progress in Clinical Neurophysiology, Cognitive Components in Cerebral Event-related Potentials and Selective Attention*; Desmedt, J.E., Ed.; Karger: Basle, Switzerland, 1979; Volume 6, pp. 1–52.
35. Naatanen, R. *Attention and Brain Function*; Lawrence Erlbaum Assoc: London, UK, 1992.
36. Van der Stelt, O.; Kok, A.; Smulders, F.T.Y.; Snel, J.; Gunning, W.B. Cerebral event-related.potentials associated with selective attention to color: Developmental changes from childhood to adulthood. *Psychophysiology* **1998**, *35*, 227–239. [CrossRef]
37. Polich, J. Updating P300: An integrative theory of P3a and P3b. *Clin. Neurophysiol.* **2007**, *118*, 128–2148. [CrossRef]
38. Pritchard, W.S. Psychophysiology of P300. *Psychol. Bull.* **1981**, *89*, 306–540. [CrossRef]
39. Naatanen, R.; Schröger, E.; Karakas, S.; Tervaniemi, M.; Paavilainen, P. Development of a memory trace for a complex sound in the human brain. *Neuroreport* **1993**, *4*, 503–506. [CrossRef] [PubMed]

40. Kok, A.; Rooijakkers, J.A.J. Comparison of event-related potentials of young children and adults in a visual recognition and word reading task. *Psychophysiology* **1985**, *22*, 11–23. [CrossRef]
41. Taylor, M.J.; Eals, M. An event-related potential study of development using visual semantic tasks. *J. Psychophysiol.* **1996**, *10*, 125–139.
42. Squires, N.K.; Squires, K.C.; Hillyard, S.A. Two varieties of long-latency positive waves evoked by unpredictable auditory stimuli in man. *Electroencephalogr. Clin. Neurophysiol.* **1975**, *38*, 387–401. [CrossRef]
43. Polich, J.; Criado, J.R. Neuropsychology and neuropharmacology of P3a and P3b. *Int. J. Psychophysiol* **2006**, *60*, 172–185. [CrossRef]
44. Courchesne, E.; Ganz, L.; Norcia, A.M. Event-related brain potentials to human faces in infants. *Child Dev.* **1981**, *52*, 804–811. [CrossRef]
45. Schulman-Galambos, C.; Galambos, R. Cortical responses from adults and infants to complex visual stimuli. *Electroencephalogr. Clin. Neurophysiol.* **1978**, *45*, 425–435. [CrossRef]
46. Stige, S.; Fjell, A.M.; Smith, L.; Lindgren, M.; Walhovd, K.B. The development of visual P3a and P3b. *Dev. Neuropsychol.* **2007**, *32*, 563–584. [CrossRef]
47. Van Dinteren, R.; Arns, M.; Jongsma, M.L.A.; Kessels, R.P.C. P300 development across the lifespan: A systematic review and meta-analysis. *PLoS ONE* **2014**, *9*, e87347. [CrossRef]
48. Riggins, T.; Scott, L.S. P300 development from infancy to adolescence. *Psychophysiology* **2019**, *57*, e13346. [CrossRef]
49. Courchesne, E. *Chronology of Postnatal Human Brain Development: Event-Related Potential, Positron Emission Tomography, Myelinogenesis, and Synaptogenesis Studies in Event-Related Brain Potentials: Basic Issues and Applications*; Rohrbaugh, J.W., Parasuraman, R., Johnson, R., Jr., Eds.; Oxford University Press: Oxford, UK, 1990; pp. 210–241.
50. Wienke, A.S.; Başar-Eroğlu, C.; Schmiedt-Fehr, C.; Mathes, B. Novelty N2-P3a complex and theta oscillations reflect improving neural coordination within frontal brain networks during adolescence. *Front. Behav. Neurosci.* **2018**, *12*, 218. [CrossRef]
51. Mathes, B.; Khalaidovski, K.; Wienke, A.S.; Schmiedt-Fehr, C.; Basar-Eroglu, C. Maturation of the P3 and concurrent oscillatory processes during adolescence. *Clin. Neurophysiol.* **2016**, *127*, 2599–2609. [CrossRef]
52. Alho, K.; Sainio, K.; Sajaniemi, N.; Reinikainen, K.; Naatanen, R. Electrical brain response of human newborns to pitch change of an acoustic stimulus. *Electroencephalogr. Clin. Neurophysiol.* **1990**, *77*, 151–155. [CrossRef]
53. Lang, A.H.; Earola, O.; Korpilahti, P.; Holopainen, I.; Salo, S.; Aaltonen, O. Practical issues in the clinical application of mismatch negativity. *Ear Hear.* **1995**, *16*, 118–130. [CrossRef]
54. Kraus, N.; McGee, T.; Micco, A.; Carrell, T.; Sharma, A.; Nicol, T. Mismatch negativity in school-age children to speech stimuli that are just perceptibly different. *Electroencephalogr. Clin. Neurophysiol.* **1993**, *88*, 123–130. [CrossRef]
55. Cs ´epe, V. On the origin and development of the mismatch negativity. *Ear Hear.* **1995**, *16*, 91–104. [CrossRef]
56. Jansson-Verkasalo, E.; Cheour, M.; Ceponiene, R.; Boyd, S.; Kushnerenko, H.; Tervaniemi, H. Maturation of MMN. In *Abstracts of the First International Workshop on Mismatch Negativity and Its Clinical Applications*; Tervaniemi, M., Escera, C., Eds.; University of Helsinki: Helsinki, Finland, 1998; p. 60.
57. Ridderinkhof, K.R.; Bashore, T.R. Using event-related brain potentials to draw inferences about human information processing. In *Age Differences in Word and Language Processing*; Allen, P.A., Bashore, T.R., Eds.; Elsevier: Amsterdam, The Netherlands, 1995; pp. 294–313.
58. Neisser, U. *Cognitive Psychology*; Appleton-Century-Crofts: New York, NY, USA, 1967.
59. Özdemir, A.K.; Karakaş, S.; Çakmak, E.D.; Tüfekçi, D.İ.; Arıkan, O. Time-frequency component analyzer and its application to brain oscillatory activity. *J. Neurosci. Methods* **2005**, *145*, 107–125. [CrossRef]
60. Berger, H. Uber des elektroenkephalogram. *Arch. Psychiatry Nervenkr.* **1929**, *87*, 527–570. [CrossRef]
61. Adrian, E.D. Olfactory reactions in the brain of the hedgehog. *J. Physiol.* **1942**, *101*, 459–473. [CrossRef]
62. Mountcastle, V.B. Preface. In *Induced Rhythms in the Brain*; Basar, E., Bullock, T.H., Eds.; Birkhauser: Boston, MA, USA, 1992; pp. 217–231.
63. Başar, E. (Ed.) *Dynamics of Sensory and Cognitive Processing by the Brain*; Springer: London, UK, 1988.
64. Başar, E. Brain function and oscillations I. Brain oscillations. In *Principles and Approaches;* Springer: Berlin, Germany, 1998.
65. Başar, E. Brain function and oscillations II. Integrative brain function. In *Neurophysiology and Cognitive Processes;* Springer: Berlin, Germany, 1999.
66. Karakas, S. A review of theta oscillation and its functional correlates. *Int. J. Psychophysiol.* **2020**, *159*, 82–99. [CrossRef]
67. Karakas, S.; Basar, E. Special issue on models and theories of brain function with special emphasis on cognitive processing. *Int. J. Psychophysiol.* **2006**, *60*, 97–101. [CrossRef]
68. Karakas, S.; Basar, E. Preface: Quiet revolutions in neuroscience. *Int. J. Psychophysiol.* **2006**, *60*, 98–100. [CrossRef]
69. Başar, E. The theory of the whole-brain-work. *Int. J. Psychophysiol.* **2006**, *60*, 133–138. [CrossRef]
70. Johnstone, S.J.; Barry, J.R.; Clarke, A.R. Ten years on: A follow-up of review of ERP research inattention deficit hyperactivity disorder. *Clin. Neurophysiol.* **2013**, *124*, 644–657r. [CrossRef]
71. Benninger, C.; Matthis, P.; Scheffner, D. EEG development of healthy boys and girls: Results of a longitudinal study. *Electroencephalogr. Clin. Neurophysiol.* **1984**, *57*, 1–12. [CrossRef]
72. Clarke, A.; Barry, R.; McCarthy, R.; Selikowitz, M.; Brown, C.R. EEG evidence for a new conceptualization of attention deficit hyperactivity disorder. *Clin. Neurophysiol.* **2002**, *113*, 1036–1044. [CrossRef]

73. Matthis, P.; Scheffner, D.; Benninger, C.; Lipinski, C.; Stolzis, L. Changes in the background activity of the electroencephalogram according to age. *Electroenceph. Clin. Neurophysiol.* **1980**, *49*, 626–635. [CrossRef]
74. Matousek, M.; Petersen, I. Frequency analysis of the EEG in normal children and normal adolescents. In *Automation of Clinical Electroencephalography*; Kellaway, P., Petersen, I., Eds.; Raven: New York, NY, USA, 1973; pp. 75–102.
75. Clarke, A.; Barry, R.; McCarthy, R.; Selikowitz, M. EEG-defined subtypes of children with attention-deficit/hyperactivity disorder. *Clin. Neurophysiol.* **2001**, *112*, 2098–2105. [CrossRef]
76. Klimesch, W.; Schimke, H.; Schwaiger, J. Episodic and semantic memory: An analysis in the EEG theta and alpha band. *Electroencephalogr. Clin. Neurophysiol.* **1994**, *91*, 428–441. [CrossRef]
77. Klimesch, W.; Schack, B.; Schabus, M.; Doppelmayr, M.; Gruber, W.; Sauseng, P. Phase-locked alpha and theta oscillations generate the P1–N1 complex and are related to memory performance. *Brain Res.* **2004**, *19*, 302–316. [CrossRef]
78. Karakas, S.; Bekçi, B.; Çakmak, E.D.; Erzengin, O.U.; Aydın, H.A. Information processing in sleep based on the event-related activity of the brain. *Sleep Biol. Rhythms.* **2007**, *5*, 28–39. [CrossRef]
79. Rechtschaffen, A.; Kales, A. *Manual of Standardised Terminology Techniques and Scoring System for Sleep Stages of Human Subjects*; National Institutes of Health Publication No.204; Government Printing Office: Washington, DC, USA, 1968; pp. 1–12.
80. American Psychiatric Association. *Diagnostic and Statistical Manual of Mental Disorders: DSM-V*, 5th ed.; American Psychiatric Publishing: Washington, DC, USA, 2013.
81. Kinsbourne, M. The control of attention by interaction between the hemispheres. In *Attention and Performance*; Kornblum, S., Ed.; Academic Press: Cambridge, MA, USA, 1973; Volume 4, pp. 239–256.
82. Shaw, P.; Kabani, N.J.; Lerch, J.P.; Eckstrand, K.; Lenroot, R.; Gogtay, N.; Greenstein, D.; Clasen, L.; Evans, A.; Rapoport, J.L.; et al. Neurodevelopmental trajectories of the human cerebral cortex. *J. Neurosci.* **2008**, *28*, 3586–3594. [CrossRef] [PubMed]
83. Shaw, P.; Malek, M.; Watson, B.; Sharp, W.; Evans, A.; Greenstein, D. Development of cortical surface area and gyrification in attention-deficit/hyperactivity disorder. *Biol. Psychiatry* **2012**, *72*, 191–197. [CrossRef]
84. Hervey, A.S.; Epstein, J.N.; Curry, J.F. Neuropsychology of adults with attention deficit/hyperactivity disorder: A meta-analytic review. *Neuropsychology* **2004**, *18*, 485–503. [CrossRef] [PubMed]
85. Schoechlin, C.; Engel, R.R. Neuropsychological performance in adult attention deficit hyperactivity disorder: Meta-analysis of empirical data. *Arch. Clin. Neuropsychol.* **2005**, *20*, 727–744. [CrossRef]
86. Klinkerfuss, G.H.; Lange, P.H.; Weinberg, W.A.; O'Leary, J.L. Electroencephalogram abnormalities of children with hyperkinetic behavior. *Neurology* **1965**, *15*, 883–891. [CrossRef] [PubMed]
87. Bresnahan, S.M.; Barry, R.J. Specificity of quantitative EEG analysis in adults with attention deficit hyperactivity disorder. *Psychiatry Res.* **2002**, *112*, 133–144. [CrossRef] [PubMed]
88. Poil, S.S.; Bollmann, S.; Ghisleni, C.; O'Gorman, R.L.; Klaver, P.; Ball, J.; Eich-Höchli, D.; Brandeis, D.; Michels, L. Age-dependent electroencephalographic changes in attention-deficit/hyperactivity disorder (ADHD). *Clin. Neurophysiol.* **2014**, *125*, 1626–1638. [CrossRef] [PubMed]
89. Hobbs, M.J.; Clarke, A.R.; Barry, R.J.; McCarthy, R.; Selikowitz, M. EEG abnormalities in adolescent females with AD/HD. *Clin. Neurophysiol.* **2007**, *118*, 363–371. [CrossRef] [PubMed]
90. Giertuga, K.; Zakrzewska, M.Z.; Bielecki, M.; Racicka-Pawlukiewicz, E.; Kossuth, M.; Cybulska-Klosowicz, A. Age-related changes in resting-state EEG activity in attention-deficit/hyperactivity disorder: A cross-sectional study. *Front. Hum. Neurosci.* **2017**, *11*, 285. [CrossRef]
91. Chabot, R.J.; Serfontein, G. Quantitative electroencephalographic profiles of children with attention deficit disorder. *Biol. Psychiatry.* **1996**, *40*, 951–963. [CrossRef]
92. Arns, M.; Conners, C.K.; Kraemer, H.C. A decade of EEG theta/beta ratio research in ADHD: A meta-analysis. *J. Atten. Disord.* **2013**, *17*, 374–383. [CrossRef]
93. Hermens, D.F.; Cooper, N.J.; Kohn, M.; Clarke, S.; Gordon, E.; Williams, L.M. Predicting stimulant medication response in ADHD: Evidence from an integrated profile of neuropsychological, psychophysiological and clinical factors. *J. Integr. Neurosci.* **2005**, *4*, 107–121. [CrossRef] [PubMed]
94. Grayson, D.S.; Ray, S.; Carpenter, S.; Iyer, S.; Dias, T.G.; Stevens, C.; Nigg, J.T.; Fair, D.A. Structural and functional rich club organization of the brain in children and adults. *PLoS ONE* **2014**, *9*, e88297. [CrossRef]
95. Posner, J.; Park, C.; Wang, Z. Connecting the dots: A review of resting connectivity MRI studies in attention deficit hyperactivity disorder. *Neuropsychol. Rev.* **2014**, *24*, 3–15. [CrossRef]
96. Buckner, R.L.; Andrews-Hanna, J.R.; Schacter, D.L. The brain's default network: Anatomy, function, and relevance to disease. *Ann. N. Y. Acad. Sci.* **2008**, *1124*, 1–38. [CrossRef]
97. Cao, M.; Shu, N.; Cao, Q.; Wang, Y.; He, Y. Imaging functional and structural brain connectomics in attention-deficit/hyperactivity disorder. *Mol. Neurobiol.* **2014**, *50*, 1111–1112. [CrossRef]
98. Snyder, S.; Hall, J.R. A meta-analysis of quantitative EEG power associated with attention-deficit hyperactivity disorder. *J. Clin. Neurophysiol.* **2006**, *23*, 441–456. [CrossRef]
99. Clarke, A.R.; Barry, R.J.; Dupuy, F.E.; Heckel, L.D.; McCarthy, R.; Selikowitz, M.; Johnstone, S.J. Behavioral difference between EEG-defined subgroups of children with attention-deficit hyperactivity disorder. *Clin. Neurophysiol.* **2011**, *122*, 1333–1341. [CrossRef]

100. Faraone, S.V.; Doyle, A.E.; Mick, E.; Biederman, J. Meta-analysis of the association between the 7-repeat allele of the dopamine D(4) receptor gene and attention deficit hyperactivity disorder. *Am. J. Psychiatry* **2001**, *158*, 1052–1057. [CrossRef]
101. Karakas, S.; Basar-Eroglu, C.; Ozesmi, Ç.; Kafadar, H.; Erzengin, O.U. Gamma response of the brain: A multifunctional oscillation that represents a bottom-up with top-down processing. *Int. J. Psychophysiol.* **2001**, *39*, 137–150. [CrossRef] [PubMed]
102. Karakas, S.; Basar, E. Oscillatory responses of the brain and their cognitive correlates. In *Encyclopedia of Neuroscience*, 3rd ed.; Adelman, G., Smith, B.H., Eds.; Elsevier: San Diego, CA, USA, 2004.
103. Karakaş, S.; Arıkan, O.; Çakmak, E.D.; Bekçi, B.; Doğutepe, E.; Tüfekçi, İ. Early gamma response of sleep is sensory/perceptual in origin. *Int. J. Psychophysiol.* **2006**, *62*, 152–167. [CrossRef] [PubMed]
104. Karakaş, S.; Tüfekçi, İ.; Bekçi, B.; Çakmak, E.D.; Doğutepe, E.; Erzengin, Ö.U.; Özkan, A.; Arıkan, O. Early time-locked gamma response and gender specificity. *Int. J. Psychophysiol.* **2006**, *60*, 225–239. [CrossRef]
105. Karakaş, S.; Bekçi, B.; Erzengin, Ö.U. Early gamma response in human neuroelectric activity is correlated with neuropsychological test scores. *Neurosci. Lett.* **2003**, *340*, 37–40. [CrossRef] [PubMed]
106. Debener, S.; Herrmann, C.S.; Kranczioch, C.; Gembris, D.; Engel, A.K. Topdown attentional processing enhances auditory evoked gamma band activity. *Neuroreport* **2003**, *14*, 683–686. [CrossRef] [PubMed]
107. Landau, A.N.; Esterman, M.; Robertson, L.C.; Bentin, S.; Prinzmetal, W. Different effects of voluntary and involuntary attention on EEG activity in the gamma band. *J. Neurosci.* **2007**, *27*, 11986–11990. [CrossRef]
108. Tiitinen, H.T.; Sinkkonen, J.; Reinikainen, K.; Alho, K.; Lavikainen, J.; Naatanen, R. Selective attention enhances the auditory 40-Hz transient response in humans. *Nature* **1993**, *364*, 59–60. [CrossRef] [PubMed]
109. Gruber, W.R.; Klimesch, W.; Sauseng, P.; Doppelmayr, M. Alpha phase synchronization predicts P1 and N1 latency and amplitude size. *Cereb. Cortex* **2005**, *15*, 371–377. [CrossRef]
110. Hermann, C.S.; Lenz, D.; Junge, S.; Busch, N.A.; Maess, B. Memory-matches evoke human gamma-responses. *BMC Neurosci.* **2004**, *5*, 13. [CrossRef]
111. Hermann, C.S.; Munk, M.H.J.; Engel, A.K. Cognitive functions of gamma-band activity: Memory match and utilization. *Trends Cogn. Sci.* **2004**, *8*, 347–355. [CrossRef]
112. Hermann, C.S.; Mecklinger, A. Gamma activity in human EEG is related to highspeed memory comparisons during object selective attention. *Vis. Cogn.* **2001**, *8*, 593–608. [CrossRef]
113. Sauseng, P.; Klimesch, W.; Gruber, W.R.; Birbaumer, N. Cross-frequency phase synchronization: A brain mechanism of memory matching and attention. *NeuroImage* **2008**, *40*, 308–317. [CrossRef] [PubMed]
114. Yordanova, J.; Kolev, V. Developmental changes in the event-related EEG theta response and P300. *Electroencephalogr. Clin. Neurophysiol.* **1997**, *104*, 418–430. [CrossRef] [PubMed]
115. Yordanova, J.; Kolev, V. Developmental changes in the theta response system: A single sweep analysis. *J. Psychophysiol.* **1998**, *12*, 113–126.
116. Yordanova, J.; Banaschewski, T.; Kolev, V.; Woerner, W.; Rothenberger, A. Abnormal early stages of task stimulus processing in children with attention-deficit hyperactivity disorder—Evidence from event-related gamma oscillations. *Clin. Neurophysiol.* **2001**, *112*, 1096–1108. [CrossRef] [PubMed]
117. Schadow, J.; Lenz, D.; Dettler, N.; Fründ, I.; Herrmann, C.S. Early gamma band responses reflect anticipatory top-down modulation in the auditory cortex. *NeuroImage* **2009**, *47*, 651–658. [CrossRef] [PubMed]
118. Chabot, R.J.; di Michele, F.; Prichep, L. The role of quantitative electroencephalography in child and adolescent psychiatric disorders. *Child Adolesc. Psychiatr. Clin. N. Am.* **2005**, *14*, 21–53. [CrossRef]
119. Martino, A.D.; Ghaffari, M.; Curchack, J.; Reiss, P.; Hyde, C.; Vannucci, M. Decomposing intra-subject variability in children with attention-deficit/hyperactivity disorder. *Biol. Psychiatry* **2008**, *64*, 607–614. [CrossRef] [PubMed]
120. Weissman, D.H.; Roberts, K.C.; Visscher, K.M.; Woldorff, M.G. The neural bases of momentary lapses in attention. *Nat. Neurosci.* **2006**, *9*, 971–978. [CrossRef]
121. Lenz, D.; Krauel, K.; Schadow, J.; Baving, L.; Duzel, E.; Herrmann, C.S. Enhanced gamma-band activity in ADHD patients lacks correlation with memory performance found in healthy children. *Brain Res.* **2008**, *1235*, 117–132. [CrossRef]
122. Lenz, D.; Karuel, K.; Flechtner, H.H.; Schadow, J.; Hinrichs, H.; Hermann, C.S. Altered evoked gamma-band responses reveal impaired early visual processing in ADHD children. *Neuropsychologia* **2010**, *48*, 1985–1993. [CrossRef]
123. Satterfield, J.H.; Schell, A.M.; Nicholas, T. Preferential neural processing of attended stimuli in attention-deficit hyperactivity disorder and normal boys. *Psychophysiology* **1994**, *31*, 1–10. [CrossRef]
124. Van Mourik, R.; Oosterlaan, J.; Heslenfeld, D.J.; Konig, C.E.; Sergeant, J.A. When distraction is not distracting: A behavioral and ERP study on distraction in ADHD. *Clin. Neurophysiol.* **2007**, *118*, 1855–1865. [CrossRef] [PubMed]
125. Frank, Y.; Seiden, J.; Napolitano, B. Visual event-related potentials and reaction time in normal adults, normal children, and children with attention deficit hyperactivity disorder: Differences in short-term memory processing. *Int. J. Neurosci.* **1996**, *88*, 109–124. [CrossRef] [PubMed]
126. Jonkman, L.; Kenemans, J.; Kemner, C.; Verbaten, M.; van Engeland, H. Dipole source localization of event-related brain activity indicative of an early visual selective attention deficit in ADHD children. *Clin. Neurophysiol.* **2004**, *115*, 1537–1549. [CrossRef] [PubMed]
127. Johnstone, S.J.; Barry, R.J. Auditory event-related potentials to a two-tone discrimination paradigm in attention deficit hyperactivity disorder. *Psychiatry Res.* **1996**, *64*, 179–192. [CrossRef] [PubMed]

128. Senderecka, M.; Grabowska, A.; Gerc, K.; Szewczyk, J.; Chmylak, R. Event-related potentials in children with attention deficit hyperactivity disorder: An investigation using an auditory oddball task. *Int. J. Psychophysiol.* **2012**, *85*, 106–115. [CrossRef] [PubMed]
129. Jung, R.; Kornmüller, A.E. Eine methodik der ableitung lokalisierter potential schwankungenaus subcorticalen Hirngebicten. *Arch Psychiatr. Nervenkr.* **1938**, *109*, 1–30. [CrossRef]
130. Grastyan, E.; Karmos, G.; Vereczkey, L.; Kellenyi, L. The hippocampal electrical correlates of the homeostatic regulation of motivation. *Electroencephalogr. Clin. Neurophysiol.* **1966**, *21*, 34–53. [CrossRef]
131. Grastyan, E.; Lissak, K.; Madarasz, I.; Donhoffer, H. Hippocampal electrical activity during the development of conditioned reflexes. *Electroencephalogr. Clin. Neurophysiol.* **1959**, *11*, 409–430. [CrossRef]
132. Aleksanov, S.N.; Vainstein, I.I.; Preobrashenskaya, L.A. Relationship between electrical potentials of the hippocampus, amygdala, and neocortex during instrumental conditioned reflexes. *Neurosci. Behav. Physiol.* **1959**, *16*, 199–207. [CrossRef]
133. Karmos, G.; Grastyan, E. Influence of hippocampal lesions on simple and delayed conditional reflexes. *Acta Physiol. Acad. Sci. Hung.* **1962**, *21*, 215–224. [PubMed]
134. Karmos, G.; Grastyan, E.; Losonczy, H.; Vereczkey, L.; Grosz, J. The possible role of the hippocampus in the organization of the orientation reaction. *Acta Physiol. Acad. Sci. Hung.* **1965**, *26*, 131–141. [PubMed]
135. Buzsaki, G.; Haubenreiser, J.; Grastyan, E.; Czopf, J.; Kellenyi, L. Hippocampal slow wave activity during appetitive and aversive conditioning in the cat. *Electroencephalogr. Clin. Neurophysiol.* **1981**, *51*, 276–290. [CrossRef] [PubMed]
136. O'Keefe, J. Hippocampal neurophysiology in the behaving animal. In *The Hippocampus Book*; Andersen, P., Morris, R., Amaral, D., Bliss, T., O'Keefe, J., Eds.; Oxford University Press: New York, NY, USA, 2007; pp. 474–548.
137. Miller, R. *Cortico-Hippocampal Interplay and the Representation of Contexts in the Brain*; Springer: Berlin/Heidelberg, Germany, 1991.
138. Miller, R. *Discovery and General Behavioural Correlates of the Hippocampal Theta RhyCortico-Hippocampal Interplay and the Representation of Contexts in the Brain*; Springer: Berlin/Heidelberg, Germany; New York, NY, USA, 1991; pp. 60–86.
139. Başar, E.; Basar-Eroglu, C.; Karakas, S.; Schürmann, M. Gamma, alpha, delta and theta oscillations govern cognitive processes. *Int. J. Psychophysiol.* **2001**, *39*, 241–248. [CrossRef] [PubMed]
140. Basar-Eroglu, C.; Basar, E.; Demiralp, T.; Schürmann, M. P300-response: Possible psychophysiological correlates in delta and theta frequency channels: A review. *Int. J. Psychophysiol.* **1992**, *13*, 161–179. [CrossRef]
141. Klimesch, W. EEG alpha, and theta oscillations reflect cognitive and memory performance: A review and analysis. *Brain Res. Rev.* **1999**, *29*, 169–195. [CrossRef]
142. Klimesch, W.; Freunberger, R.; Sauseng, P.; Gruber, W. A short review of slow phase synchronization and memory: Evidence for control processes in different memory systems? *Brain Res.* **2008**, *1235*, 31–44. [CrossRef] [PubMed]
143. Barry, R.; Clark, A.; McCarthy, M.; Selikowitz, M. EEG coherence in attention-deficit/hyperactivity disorder: A comparative study of two DSM-IV types. *Clin. Neurophysiol.* **2002**, *113*, 579–585. [CrossRef]
144. Demiralp, T.; Bayraktaroğlu, Z.; Lenz, D.; Junge, S.; Busch, N.A. Gamma amplitudes are coupled to EEG during the visual-to-theta phase in human perception. *Int. J. Physiol.* **2007**, *64*, 24–30. [CrossRef]
145. Basar, E.; Hari, R.; Lopes Da Silva, F.H.; Schürmann, M. (Eds.) Brain alpha activity—New aspects and functional correlates. *Int. J. Psychophysiol.* **1997**, *26*, 1–482.
146. Basar, E.; Schürmann, M.; Basar-Eroglu, C.; Karakas, S. Alpha oscillations in brain functioning: An integrative theory. *Int. J. Psychophysiol.* **1997**, *26*, 5–29. [CrossRef]
147. Karakas, S.; Erzengin, O.U.; Basar, E. A new strategy involving multiple cognitive paradigms demonstrates that ERP components are determined by the superposition of oscillatory responses. *Clin. Neurophysiol.* **2000**, *111*, 1719–1732. [CrossRef]
148. Schack, B.; Klimesch, W.; Sauseng, P. Phase synchronization between theta and upper alpha oscillations in a working memory task. *Int. J. Psychophysiol.* **2005**, *57*, 105–114. [CrossRef]

brain sciences

MDPI

Article

The Development of Global-Level Categorization: Frequency Tagging EEG Responses

Stefanie Peykarjou [1,2], Stefanie Hoehl [3] and Sabina Pauen [1,*]

1 Department of Psychology, Heidelberg University, 69117 Heidelberg, Germany; stefanie.peykarjou@psychologie.uni-heidelberg.de
2 Department of Developmental and Educational Psychology, Charlotte Fresenius University Wiesbaden, 65185 Wiesbaden, Germany
3 Department of Developmental and Educational Psychology, University of Vienna, 1010 Vienna, Austria; stefanie.hoehl@univie.ac.at
* Correspondence: sabina.pauen@psychologie.uni-heidelberg.de

Abstract: Adults and infants form abstract categories of visual objects, but little is known about the development of global categorization. This study aims to characterize the development of very fast global categorization (living and non-living objects) and to determine whether and how low-level stimulus characteristics contribute to this response. Frequency tagging was used to characterize the development of global-level categorization in $N = 69$ infants (4, 7, 11 months), $N = 22$ children (5–6 years old), and $N = 20$ young adults. Images were presented in an oddball paradigm, with a category change at every fifth position (AAAABAAAABA...). Strong and significant high-level categorization was observed in all age groups, with reduced responses for phase-scrambled control sequences ($R^2 = 0.34$–0.73). No differences between the categorization of living and non-living targets were observed. These data demonstrate high-level visual categorization as living and non-living from four months to adulthood, providing converging evidence that humans are highly sensitive to broad categorical information from infancy onward.

Keywords: categorization; global; EEG; frequency tagging; Fast Periodic Visual Stimulation (FPVS)

1. Introduction

Imagine being a preverbal infant who encounters an object that they have never seen before. It seems most important to decide whether this is a living thing or not, in order to predict whether it will be able to move and act of its own accord. Identifying living entities as such provides the potential evolutionary advantage of treating these more carefully than inanimate objects, independent of the specific type of living or non-living object one encounters. Indeed, under strained circumstances, we can determine whether an image shows a living or non-living object more quickly than identifying which type of object it is [1–3], and infants ontogenetically distinguish broad categories prior to more fine-grained ones [4,5].

Traditionally, the categorization hierarchy is assumed to include a basic level, differentiating everyday object classes such as cats, dogs, cars, and motorcycles at a superordinate level in which these are summarized (e.g., mammals, vehicles), and a subordinate level differentiating category membership in more detail (e.g., Maine Coon or Siamese, Labrador or Beagle). Developmental research has expanded this hierarchy by including an even broader level spanning the living/non-living distinction, the so-called global level [6]. Evidence for such global-level categories comes from fMRI studies in adults [7,8] and behavioral studies in children and infants [8–11]. During infancy, categorization responses develop from broad and inclusive categories to more circumscribed categories, the so-called global-to-basic-level shift [4,11]. Converging evidence for this shift has been found across a variety of paradigms, with slight differences in the developmental timing. While most studies employed gender-balanced designs, no study tested for gender effects on categorization.

Mapping the development of global-level categorization has been aggravated by stark methodological differences between the age groups under study. Research with adults has employed (ultra-)rapid visual categorization, with response times and event related-potentials (ERPs) as dependent variables, and fMRI for evaluating category-specific neural activation patterns [1,3,7]. In contrast, research on the early development of categorization has mainly relied on behavioral paradigms such as matching-to-sample and object sorting tasks in children [10] or habituation–dishabituation measures and sequential touching in infants [9,11,12]. More recently, ERPs have been employed to evaluate infant categorization, but the living/non-living distinction was not directly tested, and paradigms as well as dependent variables differed from prior work with adults [13–15]. It is likely due to these methodological challenges that prior studies have never mapped developmental trajectories over broader periods (e.g., from infants to children, from children to adults).

Recent developments employing frequency tagging in combination with recording EEG provide an opportunity for investigating the development of categorization, as similar methods can be employed across the life-span [16,17]. In this paradigm, changes between categories are embedded into a fixed stimulation sequence and tagged at a specific frequency, defined a priori. EEG responses at this and related frequencies can be employed as an index of categorization in a similar fashion in all age groups. With this approach, it has been demonstrated that infants categorize animal and furniture items from four months onward, but characteristics of adult categorization emerge around eleven months and become more pronounced in preschool children [16]. Adults showed strong categorization responses in occipital-posterior and frontal areas, which were similar for animal and furniture targets. Five-to-six-year-old children showed strong responses in the same areas, and partly enhanced categorization of animal compared to furniture targets. Infants only showed categorization responses in the occipital-posterior region, and an advantage for animal targets emerged by eleven months of age. Phase-scrambled control sequences were employed to evaluate the contribution of low-level visual cues to categorization that are known to be associated with category membership [18]. The categorization responses for adults, children, and eleven-month-olds were much reduced for the phase-scrambled control sequences, indicating categorization based on high-level cues. While there was no statistically significant difference between the responses to the original and phase-scrambled images at four and seven months, significant categorization responses were only observed for original images at these ages.

Thus, high-level visual cues drove categorization responses from eleven months onward, but small yet significant categorization was also evident for stimuli containing only low-level visual cues. Low-level cues refer to local measurements such as orientation, spatial frequency, luminance, and contrast [19], while high-level visual cues refer to the organization of stimulus parts into a global "Gestalt", the configuration of an entity (cf. APA Dictionary of Psychology, n.d.). Several lines of work have demonstrated that among low-level cues, curvilinear features contained in images of living objects are sufficient to elicit global categorization [20,21], but neural differentiation is maintained when controlling for shape differences associated with category membership [7,22]. Existing studies exploring the role of low-level visual cues have recruited gender-balanced samples but have not yet taken participant gender into account in their statistical analyses.

The current study aims to enhance our understanding of high-level categorization, testing the global-level contrast in infants, children, and adults using EEG frequency tagging. By employing a similar approach across age groups and comparing the categorization of original and phase-scrambled control images, these data provide evidence on the trajectory of high- and low-level categorization. To allow for detecting potential differences compared to a more circumscribed categorical contrast, this study closely followed recently published work on animal–furniture categorization [16]. The youngest age group included was four months, the earliest age at which the categorization of objects as living and non-living has been confirmed across studies [8,23], despite some evidence for earlier categorization [5]. Similar to [16], seven- and eleven-month-olds were included to cover potential developmental changes throughout the first year of life. The general paradigm and all hypotheses are based on this prior work. Strong categorization in the frontal and posterior regions was expected for adults and children, with an advantage for categorizing living targets only in children. In infants, categorization was expected only in the posterior region. Based on the evidence for a global-to-basic-level shift [4,11], categorization as living/non-living was hypothesized to emerge even earlier than categorization as animal/furniture item. It was expected to be stronger for the original than phase-scrambled control sequences from at least seven months of age. In contrast, gender was not expected to have a significant impact on categorization performance at any age because the living/non-living distinction is of high relevance irrespective of sex/gender. Moreover, no previous study reported any relevant gender-related differences regarding categorization responses.

Overall, the hypotheses could be confirmed, with strong categorization responses independent of low-level confounds in adults, children, and even infants starting at four months of age.

2. Experiment 1

2.1. Materials and Methods

2.1.1. Participants

Twenty adult subjects were tested (13 females, mean age = 21 years, SD = 2 years) after giving written consent to participate. They received course credit for participation. Five additional participants were tested but excluded due to a technical error (missing triggers at the sequence start). Data were collected in Heidelberg, a mid-sized German university town.

2.1.2. Stimuli/Presentation

Overall, the hypotheses could be confirmed, with strong categorization responses independent of low-level confounds in adults, children, and even infants starting at four months of age. The presentation was similar to recent studies employing frequency tagging [16,24], and full details can be found in Supplementary File S1. Stimulation sequences were displayed at 6 Hz using sinusoidal contrast modulation. At every fifth position, corresponding to 1.2 Hz, the stimulus category changed (6 Hz/5; i.e., AAAABAAAABAAAAB…). EEG amplitude at 1.2 Hz and harmonics (i.e., 2F/5 = 2.4 Hz, 3F/5 = 3.6 Hz…) was used as an index of categorization [17], whereas amplitude at 6 Hz and harmonics was employed to evaluate general attention to the visual display. The schematic stimulation course is illustrated in Figure 1.

Four conditions were presented: living deviant (non-living standard), non-living deviant (living standard), and the corresponding phase-scrambled versions. Per condition, four 60-s sequences were presented, with the order of sequences randomized across participants. Participants were seated at a looking distance of approx. 60 cm from the computer screen and were instructed to watch the presentation passively, but attentively.

2.1.3. EEG Recordings and Analyses

EEG was measured using a BrainProducts actiCap (Gilching, Germany) with 32 active Ag-AgCl electrodes arranged according to the 10-10-system and a right mastoid reference. Sampling rate was set at 250 Hz and the signal was amplified via a BrainAmp amplifier.

2.1.4. EEG Preprocessing and Frequency Domain Analyses

EEG was preprocessed using Letswave (https://www.letswave.org/, accessed on 1 January 2018) and MATLAB 2012b (The Mathworks, Natick, MA, USA) and followed the procedure described in several recent studies [16,24]. A description and flowchart of all steps can be found in Supplementary File S2.

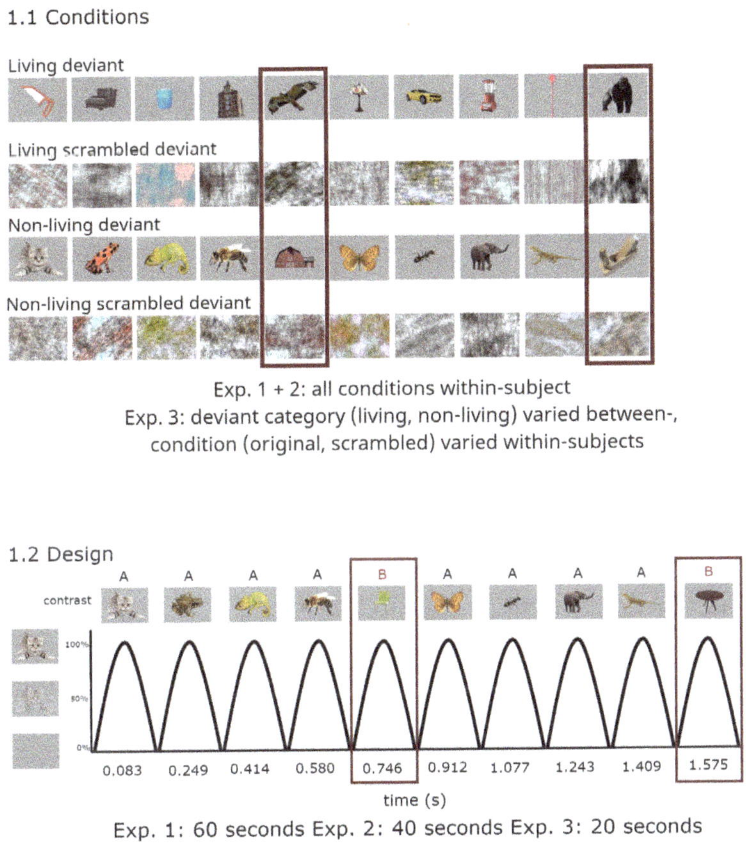

Figure 1. Schematic illustration of the experimental paradigm and conditions. (**1.1**) Four conditions were tested within-subjects in adults and children (Experiments 1 + 2). In infants (Experiment 3), the condition (original, scrambled images) was varied within-, and the deviant category (living, non-living) between-subjects. (**1.2**) Images are presented by sinusoidal contrast modulation at a rate of 6 cycles per second = 6 Hz (1 cycle ≈ 170 ms). Category changes were introduced at fixed intervals of every fifth image (6/5 Hz = 1.2 Hz). Sequence duration was set at 60 s in adults, 40 s in children, and 20 s in infants to accommodate differences in attention span.

To measure the strength of activity, baseline corrected amplitudes (bca) were computed by subtracting the average amplitude of the 24 surrounding bins (12 on each side excluding the immediately adjacent bins, and two extreme bins) from every frequency bin [24]. Bca was employed for comparing the conditions statistically. Z-scores were calculated as the difference between the amplitude at the frequency of interest and the mean amplitude of 24 surrounding bins divided by the standard deviation of those bins [25] and served to identify significant responses. Threshold of significance was placed at Z-score 1.64 ($p < 0.05$, one-tailed). Signal-to-noise ratios (SNRs) were computed by dividing the signal by the amplitude at the neighboring frequency bins and employed to visualize the response patterns.

Electrode clusters were based on previous work [16,22] and visually verified using heatmaps for individual harmonics. For the categorization response, the posterior-occipital cluster consisted of 11 electrodes (P3, P4, Pz, P7, P8, PO9, PO10, O1, O2, Oz, Iz), and the frontal cluster of 10 electrodes (F3, F4, Fz, C3, C4, Cz, FC1, FC2, FC5, FC6). For the base response, which was primarily analyzed as a control of visual attention directed to the different conditions, analyses were limited to occipital electrodes (O1, O2, Oz). Comparisons between conditions were performed using summed baseline-corrected amplitudes (bca) across consecutively significant harmonics (see Supplementary File S1 for details). Responses were averaged across electrodes per cluster for analysis. Group analyses were calculated by averaging the individual amplitude spectra, then computing bca, SNR, and Z-scores on the resulting grand-averaged spectrum.

2.1.5. Statistical Analyses

Conditions were compared using the baseline corrected amplitudes in a JZS Bayes factor repeated measurement analysis of variance (rmANOVA) with default prior scales [26–28]. The Bayes factor rmANOVA provides a more conservative test than the standard rmANOVA and estimates probability for models based on the null and alternative hypotheses, thereby providing scales for interpreting the strength of evidence for H0 and H1. Preliminary analyses did not favor an effect of gender, as all $BFs_{10} < 2.91$, so gender was not considered in the main analyses. The within-subjects ANOVAs were composed as condition (2: original, scrambled) * deviant category (2: animal, furniture). Effect sizes were computed as the increase in R^2 when adding the factor to the null model. Analyses including electrode as a factor can be found in Supplementary File S3, and fully confirm the main analysis.

2.2. Results

2.2.1. Categorization Responses

In the posterior-occipital and the frontal cluster, strong and highly significant categorization responses were obtained for living and non-living deviants (see Figure 2 and Table 1), consistent with rapid categorization in visual and anterior networks. Small but significant categorization responses were also obtained for the phase-scrambled control sequences. This finding reflects neural sensitivity to associations between low-level features with category membership. Consistent with the assumption that Gestalt information and prior knowledge facilitate categorization, responses in the scrambled control conditions were severely reduced compared to the original image sequences.

Table 1. Baseline corrected amplitude (bca) means and standard deviations (SD), Z-score, and signal-to-noise ratio (SNR) ranges for categorization and base rate responses in Experiment 1 (adults). Responses were averaged within electrode clusters. Bca values represent grand-averages across participants and channels, and Z-score and SNR ranges represent individual averages across channels. Z-scores: percentage in parenthesis indicates the portion of participants with a significant response ($Z > 1.64$).

Response	Condition	Deviant Category	Bca Mean	Bca SD	Z-Score Range	SNR Range
Posterior categorization response (harmonics 1–11)	Original	Animal	0.38	0.12	3.05–16.79 (100%)	1.30–2.34
		Furniture	0.41	0.14	3.17–17.59 (100%)	1.25–2.23
	Phase-scrambled	Animal	0.12	0.08	−0.91–9.51 (70%)	0.94–1.48
		Furniture	0.15	0.08	1.14–8.76 (85%)	1.11–1.68
Anterior categorization response (harmonics 1–8)	Original	Animal	0.23	0.11	3.13–16.32 (100%)	1.11–2.06
		Furniture	0.24	0.13	2.97–17.67 (100%)	1.05–2.10
	Phase-scrambled	Animal	0.07	0.07	−0.28–9.46 (70%)	0.87–1.38
		Furniture	0.10	0.08	0.82–8.61 (85%)	0.97–1.60
Base response (harmonics 1–4)	Original	Animal	1.21	0.43	17.42–51.00 (100%)	3.55–15.92
		Furniture	1.26	0.53	19.24–48.20 (100%)	3.78–17.81
	Phase-scrambled	Animal	1.42	0.78	11.91–50.25 (100%)	3.12–19.01
		Furniture	1.37	0.81	12.96–47.78 (100%)	4.13–18.82

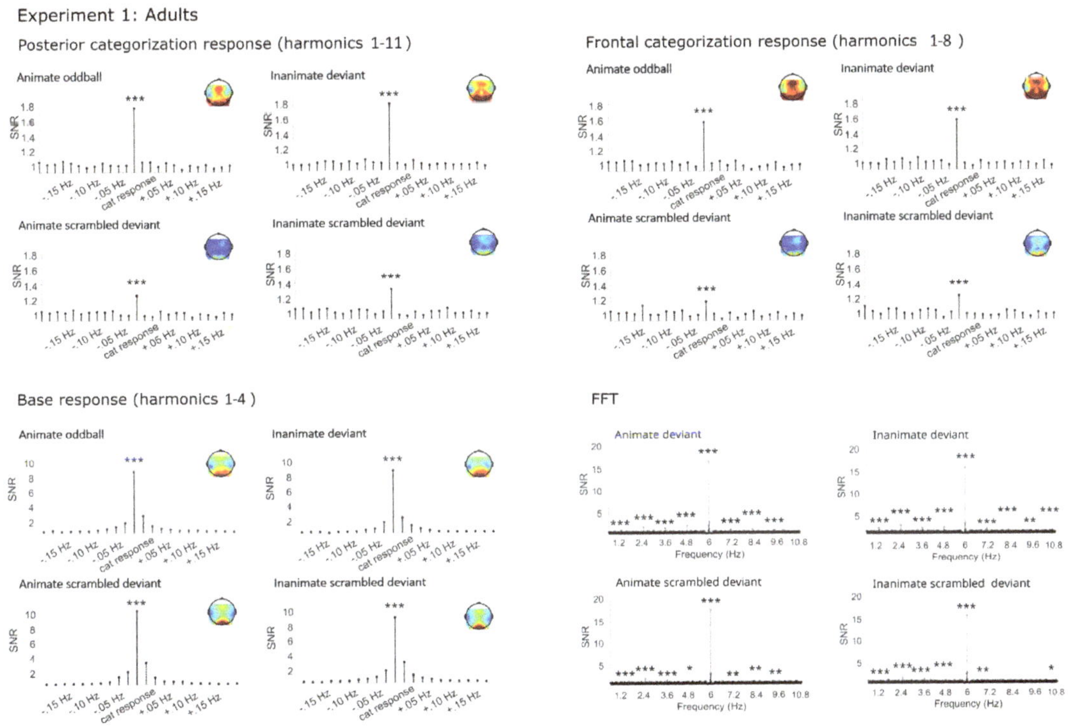

Figure 2. Electroencephalographic responses in Experiment 1 (adults). Signal-to-noise ratios (SNRs) of summed responses for the categorization response at the posterior-occipital and frontal leads and of the base response at the occipital channels. Fast Fourier transformation (FFT) responses at the posterior-occipital channels. Data have been averaged across electrodes per cluster and grand-averaged across participants for display. * = $p < 0.05$, ** = $p < 0.01$, *** = $p < 0.001$.

Categorization Responses in Posterior-Occipital Cluster (Harmonics 1–11)

Significant categorization responses were obtained when averaging across channels in all four conditions (all Zs > 5). A main effect of condition provided extreme evidence for a stronger categorization response in the original than phase-scrambled images, $BF_{10} = 6.00 \times 10^{15}$, $R^2 = 0.64$ (original $M = 0.42$ µV, $SD = 0.16$; scrambled $M = 0.12$ µV, $SD = 0.08$). There was anecdotal evidence against a main effect of category, $BF_{10} = 0.43$, and against an interaction of category and condition, $BF_{10} = 0.43$.

Categorization Response in Anterior Cluster (Harmonics 1–8)

Significant categorization responses were obtained when averaging across channels in all four conditions (all Zs > 4). There was extreme evidence for stronger responses in the original rather than the phase-scrambled conditions, $BF_{10} = 6.61e^9$, $R^2 = 0.71$ (original $M = 0.24$, $SD = 0.09$; scrambled $M = 0.08$, $SD = 0.05$). There were no other effects, with evidence speaking against differences between animal and furniture deviants, $BF_{10} = 0.26$.

2.2.2. Base Frequency (Harmonics 1–4)

Large responses at the base stimulation frequency and its harmonics were observed in all conditions over the medial occipital cortex, with a maximum at O1, O2, Oz. Visual inspection revealed that the amplitude was higher in the phase-scrambled than original conditions. This was verified by the Bayesian ANOVA, $BF_{10} = 18{,}686.70$, $R^2 = 0.46$.

2.3. Summary

Together, these data are consistent with the view that adults quickly and expertly categorize stimuli based on animacy, and that this ability is partly based on low-level cues. To investigate the development of this high-level perceptual categorization, children aged five to six years were tested using the same procedure in Experiment 2.

3. Experiment 2

3.1. Materials and Methods

The materials, procedure, and statistical analysis were similar to Experiment 1. Only deviations will be detailed in the following paragraphs.

3.1.1. Participants

The final sample consisted of N = 22 five-to-six-year-old children (12 females, M age = 75.2 months, SD = 5 months, age range 64–82 months). All children were born full-term (>37 weeks of gestation) and did not report any neurological or visual problems. An additional N = 18 children were tested but not included in the final analyses due to a lack of cooperativeness (N = 2), bad data (too many artifacts, leading to the exclusion of trials, N = 14), or technical errors (N = 2). Relatively high dropout-rates were due to employing four conditions within-subjects, requiring participants to remain seated and attentive for a relatively long period of time (approx. 30 min, see Stimuli/Presentation). Moreover, some participants felt uneasy with the procedure. In accordance with the terms provided by the local ethics committee of Heidelberg University that approved the general procedure, verbal consent was obtained from the children, and written informed consent from their caretakers.

3.1.2. Stimulation

Sequence length was set to 40 s to accommodate the children's need for breaks. Four sequences per condition (16 in total) were presented. During capping, EEG preparation, and after every second FPVS sequence, participants were engaged in a game where they could win stickers to keep them motivated and attentive.

3.1.3. EEG Analysis

On average, the participants watched 14.55 sequences per condition (SD = 1.37), and a mean of 13.23 (SD = 2.09) was kept for analysis after preprocessing (interpolation and sequence exclusion). There were no differences in the number of trials between the four conditions, all $BFs_{10} < 0.62$. Baseline corrections were performed using bins 2–8 (excluding two extreme bins) rather than bins 2–12 as in Experiment 1 to keep the frequency range comparable. Statistical analyses including the electrode as a factor can be found in Supplementary File S3 and largely confirmed the analyses on the averaged electrodes per cluster.

Preliminary analyses spoke against gender effects, all $BFs_{10} < 0.59$, so gender was not considered in the main analysis.

3.2. Results

3.2.1. Categorization Responses

Similar to Experiment 1, strong and highly significant categorization responses were obtained for living and non-living deviants in both the posterior and the frontal cluster (see Figure 3 and Table 2). Again, small significant but strongly reduced categorization responses were obtained in the phase-scrambled control sequences.

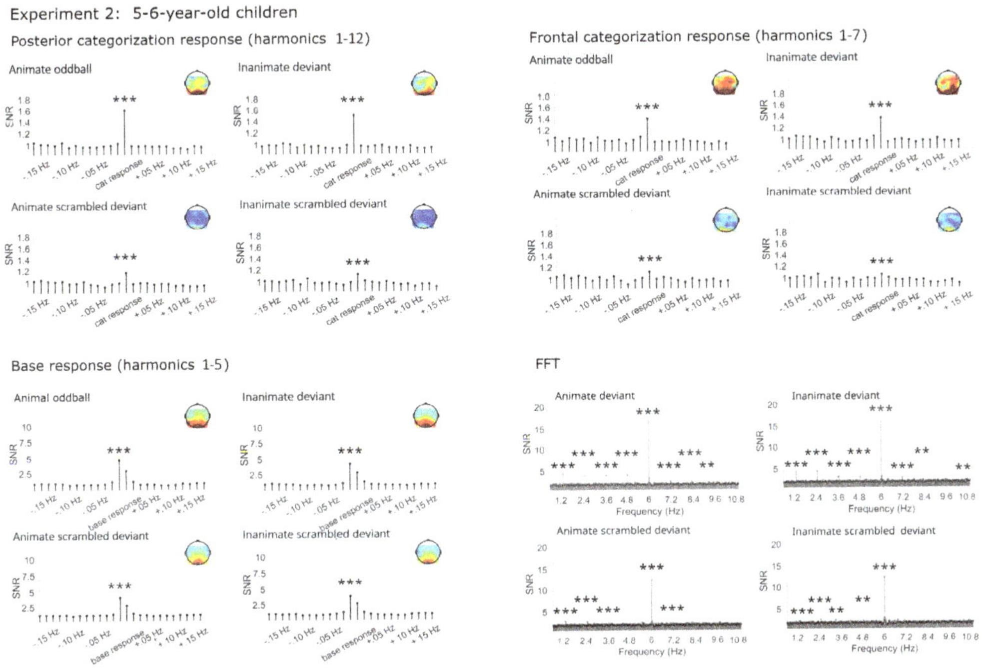

Figure 3. Electroencephalographic responses in Experiment 2 (5–6-year-olds). Signal-to-noise ratios (SNRs) of summed responses for the categorization response at the posterior-occipital and frontal leads and of the base response at the occipital channels. Fast Fourier transformation (FFT) responses at the posterior-occipital channels. Data have been averaged across electrodes per cluster and grand-averaged across participants for display. *** = $p < 0.001$, ** = $p < 0.01$.

Table 2. Baseline corrected amplitude (bca) means and standard deviations (SD), Z-score, and signal-to-noise ratio (SNR) ranges for categorization and base rate responses in Experiment 2 (5–6-year-old children). Responses were averaged within electrode clusters. Bca values represent grand-averages across participants and channels, and Z-score and SNR ranges represent individual averages across channels. Z-scores: percentage in parenthesis indicates the portion of participants with a significant response (Z > 1.64).

Response	Condition	Deviant Category	Bca Mean	Bca SD	Z-Score Range	SNR Range
Posterior categorization response (harmonics 1–12)	Original	Animal	1.63	0.70	2.04–20.52 (100%)	1.19–2.48
		Furniture	1.53	0.63	2.99–14.16 (100%)	1.23–2.34
	Phase-scrambled	Animal	0.47	0.44	−3.28–6.58 (59%)	0.86–1.48
		Furniture	0.36	0.44	−2.97–5.47 (55%)	0.78–1.39
Anterior categorization response (harmonics 1–7)	Original	Animal	0.66	0.51	−0.74–18.94 (81%)	0.95–2.26
		Furniture	0.69	0.52	−1.25–7.28 (77%)	0.86–2.00
	Phase-scrambled	Animal	0.24	0.36	−1.51–11.04 (41%)	0.86–1.58
		Furniture	0.17	0.44	−3.01–6.95 (32%)	0.80–1.59
Base response (harmonics 1–5)	Original	Animal	1.97	0.94	1.65–35.05 (100%)	1.26–7.31
		Furniture	2.12	1.15	2.25–59.05 (100%)	1.58–9.02
	Phase-scrambled	Animal	1.87	1.11	2.96–38.30 (100%)	1.64–7.52
		Furniture	1.85	0.96	1.56–40.62 (95%)	1.24–6.85

Categorization Responses in Posterior-Occipital Cluster (Harmonics 1–12)

Significant categorization responses were obtained when averaging across channels in all four conditions (all Zs > 8). A main effect of condition provided extreme evidence for a stronger categorization response in the original than phase-scrambled images, $BF_{10} = 3.75 \times 10^{14}$, $R^2 = 0.73$ (original $M = 1.58$ µV, $SD = 0.58$; scrambled $M = 0.41$ µV, $SD = 0.27$). There was anecdotal evidence against a main effect of category, or an interaction of category and condition, $BFs_{10} < 0.31$.

Categorization Response in Anterior Cluster (Harmonics 1–7)

Significant categorization responses were obtained when averaging across channels in all four conditions (all Zs > 3). Consistent with the visual inspection, there was moderate evidence for stronger responses in the original than phase-scrambled conditions, $BF_{10} = 9.21$, $R^2 = 0.38$ (original $M = 0.70$, $SD = 0.52$; scrambled $M = 0.24$, $SD = 0.48$). There were no other effects, with the evidence speaking against differences between animal and furniture deviants, $BF_{10} = 0.54$.

3.2.2. Base Frequency (Harmonics 1–5)

In all conditions, large responses at the base stimulation frequency and its harmonics were observed over the medial occipital cortex. Visually, there were no differences between conditions, which was verified by the Bayesian ANOVA, all $BFs_{10} < 0.34$.

3.3. *Summary*

Preschool children showed strong high-level categorization of living and non-living items, similar to the responses of adults recorded in Experiment 1. Next, infants across the first year of life were tested with the same approach to track the development of high-level categorization on the global level.

4. Experiment 3

4.1. *Materials and Methods*

Paradigm and analysis were similar to Experiment 1. Only deviations will be detailed.

4.1.1. Participants

The final sample consisted of $N = 24$ four-month-old (7 females, M age = 4 months, 11 days, SD = 10 days), $N = 24$ seven-month-old (11 females, M age = 7 months, 14 days, SD = 9 days), and $N = 21$ eleven-month-old infants (10 females, M age = 11 months, 17 days, SD = 9 days). All children were born full-term (>37 weeks of gestation) and did not report any neurological or visual problems. An additional $N = 34$ children were tested but not included in the final analyses due to bad data (too many artifacts, leading to the exclusion of trials, $N = 19$), fussiness ($N = 11$), or technical errors ($N = 4$). Written informed consent was obtained from the infants' caretakers.

4.1.2. Stimulation

Sequence length was set to 20 s as customary in infant FPVS studies [15,29], and a mixed design was employed with the condition (original/scrambled) varied within-subjects, and the deviant category (living/non-living) varied between-subjects. A maximum of 16 sequences was presented.

4.1.3. EEG analysis

On average, participants watched 6.43 sequences ($SD = 1.69$) per condition, and a mean of 4.07 ($SD = 2.06$) was kept for analysis after preprocessing (interpolation and sequence exclusion). There was an effect of age group on the number of presented sequences ($BF_{10} = 3.30$), reflecting a higher number of sequences presented at 7 months of age (4 months $M = 5.79$, $SD = 1.83$, 7 months $M = 7.14$, $SD = 1.00$, 11 months $M = 6.21$, $SD = 1.86$). However, no difference in sequences retained after preprocessing was observed, $BF_{10} = 0.58$.

No effects of condition or category were observed for the numbers of presented or retained sequences, all $BFs_{10} < 0.51$. Sequences with more than three bad channels were discarded. Baseline corrections were performed using bins 2–6 (excluding no extreme bins) rather than bins 2–12 as in Experiment 1 to keep the frequency range comparable (±0.2 Hz in Experiments 1 + 2, ±0.25 Hz in Experiment 3) while avoiding narrowing down the range of bins too much. Analyses averaged across electrodes per cluster are reported. Analyses including electrode as a factor can be found in Supplementary File S3. Preliminary analyses provided evidence against gender effects, all $BFs_{10} < 0.83$, so gender was not included in the main statistical analysis.

4.2. Results

4.2.1. Categorization Responses

Similar to Experiment 1, mainly significant categorization responses were obtained for living and non-living deviants in both the posterior and frontal cluster (see Figures 4–6 and Table 3). Here, small significant but strongly reduced categorization responses were also obtained in the phase-scrambled control sequences.

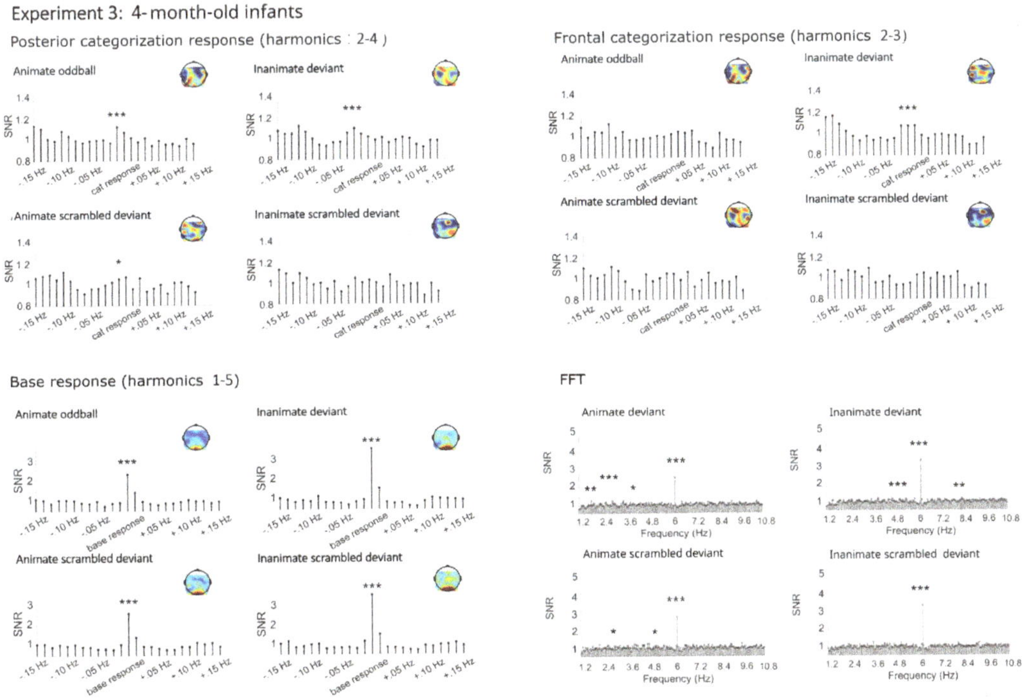

Figure 4. Electroencephalographic responses in Experiment 3 (4-month-old infants). Signal-to- noise ratios (SNRs) of summed responses for the categorization response at the posterior-occipital and frontal leads and of the base response at the occipital channels. Fast Fourier transformation (FFT) responses at the posterior-occipital channels. Data have been averaged across electrodes per cluster and grand-averaged across participants for display. *** = $p < 0.001$, ** = $p < 0.01$, * = $p < 0.05$.

Table 3. Baseline corrected amplitude (bca) means and standard deviations (SD), Z-score and signal-to-noise ratio (SNR) ranges for categorization and base rate responses in Experiment 3 (4-, 7- and 11-month-old infants). Responses were averaged within electrode clusters. Bca values represent grand-averages across participants and channels, and Z-score and SNR ranges represent individual averages across channels. Z-scores: percentage in parenthesis indicates the portion of participants with a significant response (Z > 1.64).

Response	Condition	Deviant Category	Age Group	N	Bca Mean	Bca SD	Z-Score Range	SNR Range
Posterior-occipital categorization response (harmonics 2–4)	Original	Animal	4	11	0.36	0.69	−1.99–5.54 (18%)	0.80–1.41
			7	9	0.37	0.46	−0.48–3.73 (33%)	0.95–1.45
			11	12	0.52	0.43	−0.74–12.54 (67%)	0.92–1.71
		Furniture	4	13	0.09	0.54	−1.33–2.45 (31%)	0.87–1.33
			7	15	0.30	0.34	−1.28–3.70 (40%)	0.87–1.44
			11	9	0.31	0.58	−1.53–3.95 (33%)	0.86–1.71
	Phase-scrambled	Animal	4	11	0.15	0.49	−0.93–2.94 (27%)	0.90–1.30
			7	9	−0.05	0.46	−3.58–1.78 (11%)	0.73–1.18
			11	12	0.08	0.37	−0.91–2.15 (8%)	0.81–1.28
		Furniture	4	13	0.00	0.43	−1.90–1.56 (0%)	0.76–1.20
			7	15	0.08	0.44	−2.01–3.59 (20%)	0.75–1.41
			11	9	0.31	0.46	−0.71–5.03 (33%)	0.93–1.60
Frontal categorization response (harmonics 2–3)	Original	Animal	4	11	−0.02	0.35	−1.38–4.42 (36%)	0.80–1.63
			7	9	0.16	0.27	−1.41–3.55 (11%)	0.77–1.28
			11	12	0.15	0.33	−1.52–6.63 (25%)	0.64–1.73
		Furniture	4	13	0.10	0.40	−2.47–9.91 (31%)	0.73–1.60
			7	15	0.35	0.50	−1.62–7.24 (33%)	0.93–1.71
			11	9	0.21	0.22	−0.01–3.25 (33%)	1.00–1.60
	Phase-scrambled	Animal	4	11	0.08	0.28	−1.31–2.82 (18%)	0.87–1.39
			7	9	0.01	0.20	−1.50–1.10 (0%)	0.81–1.19
			11	12	0.03	0.32	−2.25–3.43 (8%)	0.72–1.32
		Furniture	4	13	−0.10	0.28	−2.80–0.92 (0%)	0.69–1.18
			7	15	0.07	0.28	−1.45–2.78 (13%)	0.80–1.45
			11	9	0.12	0.37	−1.19–3.93 (11%)	1.05–3.47
Base response (harmonics 1–5)	Original	Animal	4	11	1.56	0.86	3.42–13.21 (100%)	1.62–3.54
			7	9	2.05	1.21	−0.62–49.39 (91%)	0.95–4.83
			11	12	2.91	1.74	3.33–35.61 (100%)	1.80–7.30
		Furniture	4	13	2.54	1.18	4.35–23.25 (100%)	1.72–7.47
			7	15	2.14	1.03	3.08–25.25 (100%)	1.55–6.18
			11	9	2.90	1.09	3.83–30.24 (100%)	1.88–8.09
	Phase-scrambled	Animal	4	11	1.90	0.93	1.25–18.02 (91%)	1.22–4.95
			7	9	1.67	1.53	−0.16–25.16 (67%)	0.93–5.98
			11	12	1.38	1.23	−0.03–19.14 (75%)	0.96–3.95
		Furniture	4	13	3.08	1.52	2.04–25.36 (100%)	0.95–4.83
			7	15	2.02	1.04	1.81–27.03 (100%)	1.49–6.80
			11	9	2.77	2.64	1.20–17.98 (89%)	1.33–6.62

Categorization Responses in Posterior-Occipital Cluster (Harmonics 2–4)

Significant categorization responses were obtained when averaging across channels in all conditions except for the animate scrambled deviant (Zs > 2.48; animate scrambled deviant Z = 1.05). A main effect of condition provided moderate evidence for a stronger categorization response in the original rather than the phase-scrambled images, $BF_{10} = 9.15$, $R^2 = 0.27$ (original $M = 0.32$ μV, $SD = 0.51$; scrambled $M = 0.09$ μV, $SD = 0.44$). There was anecdotal evidence against a main effect of age, category, or any interactions, $BFs_{10} < 0.40$.

Categorization Response in Anterior Cluster (Harmonics 2–3)

Significant categorization responses were obtained when averaging across channels only in the original conditions (Zs > 2.20). The Bayesian ANOVA did not provide strong evidence for stronger categorization responses in sequences with original images, $BFs_{10} < 1.31$ (original $M = 0.17$, $SD = 0.40$; scrambled $M = 0.03$, $SD = 0.29$). In particular, there was no evidence for an effect of condition, $BF_{10} = 1.40$, and evidence against an effect of age group, $BF_{10} = 0.19$.

4.2.2. Base Frequency (Harmonics 1–5)

In all conditions, large responses at the base stimulation frequency and its harmonics were observed over the medial occipital cortex (Zs > 27). Visually, there were no differences between conditions. The Bayesian ANOVA also did not provide evidence for any effects, all $BFs_{10} < 1.15$.

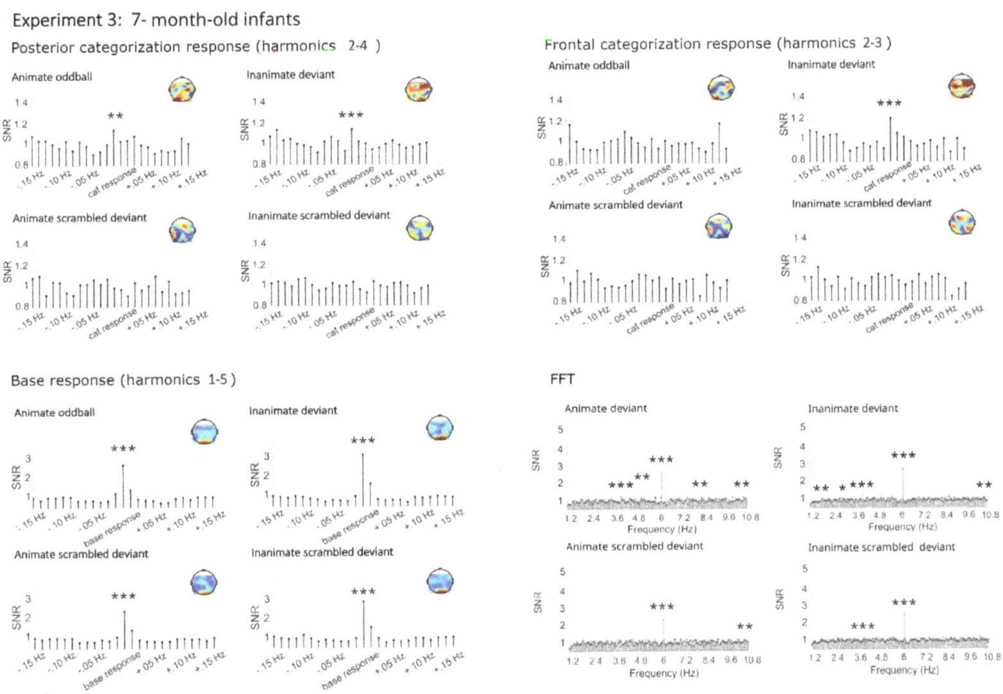

Figure 5. Electroencephalographic responses in Experiment 3 (7-month-old infants). Signal-to-noise ratios (SNRs) of summed responses for the categorization response at the posterior-occipital and frontal leads and of base response at the occipital channels. Fast Fourier transformation (FFT) responses at the posterior-occipital channels. Data have been averaged across electrodes per cluster and grand-averaged across participants for display. *** = p <0.001, ** = $p < 0.01$, * = $p < 0.05$.

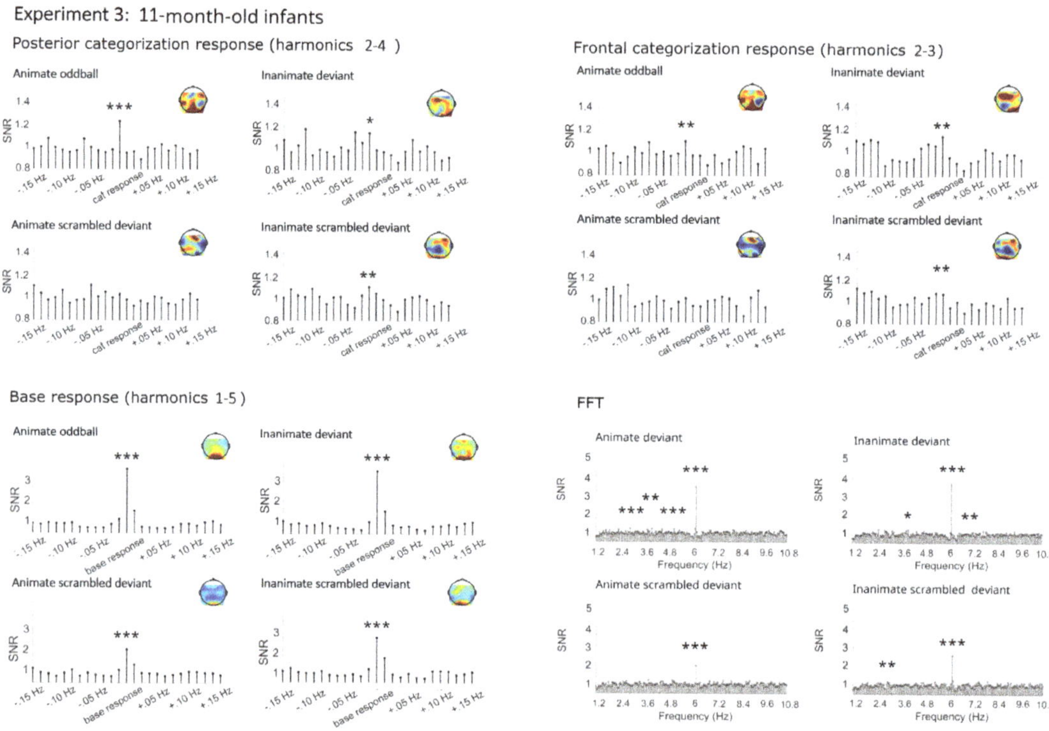

Figure 6. Electroencephalographic responses in Experiment 3 (11-month-old infants). Signal-to-noise ratios (SNRs) of summed responses for the categorization response at the posterior-occipital and frontal leads and of the base response at the occipital channels. Fast Fourier transformation (FFT) responses at the posterior-occipital channels. Data have been averaged across electrodes per cluster and grand-averaged across participants for display. *** = p <0.001, ** = p < 0.01, * = p < 0.05.

4.3. Summary

Four-, seven- and eleven-month-old infants sorted visual stimuli into living and non-living categories. These findings demonstrate sophisticated visual categorization according to animacy from four months onward, with no apparent development throughout the first year of life.

4.4. Control Analyses

As the length of sequences needed to be cut short for developmental populations, control analyses employing only the first 20 s of stimulation were run for Experiments 1 and 2. These control analyses confirm the main results and can be found in Supplementary File S4. Moreover, due to the mixed design, fewer participants were tested per cell than in Experiments 1 and 2. Therefore, another control analysis on a random subset of $N = 10$ from Experiments 1 and 2 can be found in Supplementary S5, which yielded results highly similar to the main analysis reported here.

5. General Discussion

The current study investigated the categorization of broad living and non-living visual items in adults, children, and infants in a frequency tagging EEG paradigm. Strong responses and similar categorization patterns were observed across age groups. In all age groups, categorization responses emerged at the posterior-occipital and frontal leads, and the response strength was independent of the target category (i.e., comparison of sequences with living or non-living items presented at the target positions). Moreover, significant but distinctly reduced categorization responses were evident in the control sequences with images containing only low-level visual information, pointing to the role of high-level Gestalt information for driving categorization from the earliest age studied here (four months).

Adults and preschool children in this study (Experiments 1, 2) showed similar response patterns compared to a recent study testing animal/furniture categorization [16] including similar areas of activation and severely reduced responses for phase-scrambled sequences. Categorization responses were observed in a posterior and frontal cluster, which suggests that both perceptual and higher-order processes may contribute to categorization. While low-level image characteristics were sufficient to elicit categorization in the phase-scrambled conditions, categorization was markedly enhanced for configurational images containing high-level Gestalt information. This is likely due to the differential power-spectra of living and non-living (man-made) objects [18] that are preserved by phase-scrambling. Together, these studies suggest that little development in broad categorization abilities takes place after children reach school-age. Only the influence of the target category varied between studies: in the precursor study, children showed enhanced responses for animal targets, whereas no difference between the target categories was detected in the present case. We can only speculate about this difference between the two tasks. The living and non-living categories were more diverse than animals and furniture items, and were constructed of six sub-categories each. It may be that the enhanced diversity of the non-living category increased the children's attention to these items, thus canceling out potential advantages for detecting animals in the visual environment. Future work is needed to test under which circumstances children show an animacy bias.

Crucially, Experiment 3 investigated categorization across the first year of life in three age groups (4, 7, and 11 months). Based on previous work indicative of a global-to-basic-level shift in categorization early in life [4,11], it was expected that the infants would be able to categorize images based on high-level visual cues on such a broad level from an early age. Indeed, high-level categorization was evident in all three infant age groups, and the evidence spoke against developmental changes. Moreover, significant categorization responses were detected in both the posterior-occipital and a frontal cluster, indicating that both visual and higher-level cognitive processes contribute to the living/non-living categorization from four months onward. This is in contrast to the categorically more constrained animal–furniture contrast, where stronger responses for the original compared to phase-scrambled images were not present prior to eleven months of age, and no frontal activation was detected [16]. No differences in categorization strength for living and non-living targets were observed, similar to the older participants (Experiments 1 and 2). In a previous study, an advantage for animals compared to furniture targets was observed at eleven months, but not at younger ages [16]. Global categorization from four months on is compatible with data from looking-time studies [8,23]. Early broad categorization has also been demonstrated in ERP studies [16,30], whereas the signatures of event-related oscillations have so far only been investigated for more circumscribed categories such as houses or faces [19]. Together, these findings strengthen the assumption of a shift in visual categorization from broader to more fine-grained levels in the first year of life, and demonstrate that categorization is based on high-level visual cues from four months onward.

Despite these markers of high-level categorization, the infants' performance was limited compared to the older participants. While statistical comparisons across age groups were not possible due to differences in the stimulation duration (60 s in adults, 40 s in children, 20 s in infants), qualitative comparisons across experiments indicated less mature categorization in the infants. Effect sizes of high-level categorization were descriptively smaller in the infant participants, and the percentage of infant subjects showing significant categorization responses was more restricted. Moreover, categorization responses were confined to fewer harmonics (2–3 compared to 6–11 in older participants), and no significant response was detected in the first harmonic (1.2 Hz), probably due to a high amount of noise in the low frequency spectrum typical for infant EEG. In contrast, responses to the base stimulation occurred across similar harmonics in all age groups, speaking against general differences in brain responses to frequency tagging associated with age.

To disentangle cognitive from methodological effects, control analyses with sequence durations and participant numbers matched to infant data were performed on the data from the adults and children (Supplementary Files S4 and S5). These demonstrate that differences in the range and number of harmonics remained, even when restricting the available data of the adult and child participants to match the infant data, and significant responses were present in the first harmonic anyway. The major difference compared to the main analysis was that comparatively fewer children showed significant categorization responses when only 20 s of stimulation was analyzed. Based on these control analyses, it seems that differences in elicited categorization harmonics reflect the true differences between age groups, whereas the presence of significant responses in individual averages may be influenced by sequence duration. Of course, it remains a challenge to determine whether other, non-cognitive differences between age groups (such as the presence of movement artefacts) contribute to the distribution of categorization across harmonics.

Employing EEG responses as the dependent variable also allowed us to identify two scalp regions, the occipito-temporal and frontal area, where categorization responses emerged. While source analysis would be necessary to identify the brain regions where these responses were generated, it seems likely that in addition to activations in the visual cortices, a more frontally located region is active. A cortical source analysis of categorization responses would therefore be a promising pathway to further our understanding of developmental trajectories in categorization abilities. An interesting area for future research is whether the frontal activation observed here is related to an activation of cognitive concepts, enabling subjects to form predictions about the prospective behavior of visual objects. Based on findings from behavioral paradigms, such predictions can be expected from at least seven months of age, but earlier ages have not been tested previously [31–33]. The present study employed highly similar methods across ages to evaluate categorization patterns across development, a strength compared to behavioral work that is not suited for investigating subjects across the life-span. Moreover, the role of low-level visual cues to categorization was investigated, with evidence showing that low-level cues suffice, but do not fully explain categorization responses. In all ages tested, high-level visual cues including Gestalt information boosted categorization responses. However, the current study was limited by differences in the stimulation duration and analysis of harmonics between age groups (but see the control analyses in Supplementary Files S4 and S5). Moreover, only two conditions were tested within-subject in the infant participants, precluding statistical analyses across experiments, and unfortunately it is not possible to share original data due to limitations of informed consent. Finally, the dropout rates in developmental populations were relatively high, particularly in preschool children due to limits of attentional span and compliance.

6. Conclusions

Together, the present study demonstrates a high-level categorization of visual objects into broad classes of living and non-living items in infants, children, and adults. While EEG markers associated with categorization underwent qualitative change from infancy to childhood, the categorization patterns did not change with age, and high-level categorization was present from four months until adulthood. Future work is needed to connect categorization performance to cognitive concepts, and investigate how categorization enables humans across their life-span to make predictions about complex properties and the future behavior of living and non-living objects. It seems likely that proficient early identification of living entities enables humans from infancy to form broad predictions about the future behavior of objects and facilitates successful interactions with living and non-living objects.

Contributions

1. Neural correlates revealed the categorization of living and non-living visual objects in infants, children, and adults;
2. At all ages, categorization is based on high-level visual cues;
3. Categorization emerges in the posterior-occipital and frontal areas.

Supplementary Materials: The following supporting information can be downloaded at: https://www.mdpi.com/article/10.3390/brainsci14060541/s1, S1: Methods; S2: Processing steps; S3: Analyses including electrode factor; S4: Analyses for 20-s segments; S5: Analyses with a subset of 10 participants.

Author Contributions: Conceptualization, S.P. (Stefanie Peykarjou), S.H., and S.P. (Sabina Pauen); Formal analysis, S.P. (Stefanie Peykarjou); Investigation, S.P. (Stefanie Peykarjou); Data curation, S.P. (Stefanie Peykarjou); Writing—original draft preparation, S.P. (Stefanie Peykarjou); Writing—review and editing, S.H. and S.P. (Sabina Pauen); Visualization, S.P. (Stefanie Peykarjou); Supervision, S.H. and S.P. (Sabina Pauen). All authors have read and agreed to the published version of the manuscript.

Funding: This work was partly supported by a scholarship awarded to S.P. (Stefanie Peykarjou) by the Fonds de la Recherche Scientifique.

Institutional Review Board Statement: The study was conducted in accordance with the Declaration of Helsinki, and approved by the Ethics Committee of Heidelberg University, Fakultät für Verhaltens- und Empirische Kulturwissenschaften, Heidelberg University, AZ Pey 2020 1/3 (16 October 2020) and AZ Pey 2022 2/1 (7 December 2022).

Informed Consent Statement: In accordance with the terms provided by the local ethics committee, verbal consent was obtained from the children, and written informed consent from the adults and caretakers of the children/infants.

Data Availability Statement: The data necessary to reproduce the analyses presented here are not publicly accessible as the participants' informed consent did not include public data sharing, but are available from the first author upon reasonable request.

Acknowledgments: We are grateful to all participants and to our student assistants who helped during the data collection. We would like to thank Bruno Rossion for inspiring our work and providing highly valuable feedback.

Conflicts of Interest: The authors declare no conflicts of interest.

References

1. Thorpe, S.; Fize, D.; Marlot, C. Speed of Processing in the Human Visual System. *Nature* **1996**, *381*, 520–522. [CrossRef] [PubMed]
2. Mack, M.L.; Palmeri, T.J. The Dynamics of Categorization: Unraveling Rapid Categorization. *J. Exp. Psychol. Gen.* **2015**, *144*, 551. [CrossRef] [PubMed]
3. Poncet, M.; Fabre-Thorpe, M. Stimulus Duration and Diversity Do Not Reverse the Advantage for Superordinate-level Representations: The Animal Is Seen before the Bird. *Eur. J. Neurosci.* **2014**, *39*, 1508–1516. [CrossRef] [PubMed]
4. Pauen, S. The Global-to-Basic Level Shift in Infants' Categorical Thinking: First Evidence from a Longitudinal Study. *Int. J. Behav. Dev.* **2002**, *26*, 492–499. [CrossRef]
5. Quinn, P.C.; Johnson, M.H.; Mareschal, D.; Rakison, D.H.; Younger, B.A. Understanding Early Categorization: One Process or Two? *Infancy* **2000**, *1*, 111–122. [CrossRef] [PubMed]
6. Mandler, J.M.; Bauer, P.J.; McDonough, L. Separating the Sheep from the Goats: Differentiating Global Categories. *Cognit. Psychol.* **1991**, *23*, 263–298. [CrossRef]
7. Proklova, D.; Kaiser, D.; Peelen, M.V. Disentangling Representations of Object Shape and Object Category in Human Visual Cortex: The Animate–Inanimate Distinction. *J. Cogn. Neurosci.* **2016**, *28*, 680–692. [CrossRef] [PubMed]
8. Spriet, C.; Abassi, E.; Hochmann, J.-R.; Papeo, L. Visual Object Categorization in Infancy. *Proc. Natl. Acad. Sci. USA* **2022**, *119*, e2105866119. [CrossRef] [PubMed]
9. Rostad, K.; Yott, J.; Poulin-Dubois, D. Development of Categorization in Infancy: Advancing Forward to the Animate/Inanimate Level. *Infant Behav. Dev.* **2012**, *35*, 584–595. [CrossRef]
10. Wright, K.; Poulin-Dubois, D.; Kelley, E. The Animate–Inanimate Distinction in Preschool Children. *Br. J. Dev. Psychol.* **2015**, *33*, 73–91. [CrossRef]
11. Younger, B.A.; Fearing, D.D. A Global-to-basic Trend in Early Categorization: Evidence from a Dual-category Habituation Task. *Infancy* **2000**, *1*, 47–58. [CrossRef] [PubMed]

12. Poulin-Dubois, D.; Pauen, S. The development of object categories: What, when, and how? In *Handbook of Categorization in Cognitive Science*, 2nd ed.; Cohen, H., Lefebvre, C., Eds.; Elsevier Academic Press: San Diego, CA, USA, 2017; pp. 653–671.
13. Elsner, B.; Jeschonek, S.; Pauen, S. Event-Related Potentials for 7-Month-Olds' Processing of Animals and Furniture Items. *Dev. Cogn. Neurosci.* **2013**, *3*, 53–60. [CrossRef] [PubMed]
14. Grossmann, T.; Gliga, T.; Johnson, M.H.; Mareschal, D. The Neural Basis of Perceptual Category Learning in Human Infants. *J. Cogn. Neurosci.* **2009**, *21*, 2276–2286. [CrossRef] [PubMed]
15. Peykarjou, S.; Wissner, J.; Pauen, S. Categorical Erp Repetition Effects for Human and Furniture Items in 7-month-old Infants. *Infant Child Dev.* **2017**, *26*, e2016. [CrossRef]
16. Peykarjou, S.; Hoehl, S.; Pauen, S. The Development of Visual Categorization Based on High-level Cues. *Child Dev.* **2024**, *95*, e122–e138. [CrossRef] [PubMed]
17. Rossion, B. Understanding Individual Face Discrimination by Means of Fast Periodic Visual Stimulation. *Exp. Brain Res.* **2014**, *232*, 1599–1621. [CrossRef] [PubMed]
18. Torralba, A.; Oliva, A. Statistics of Natural Image Categories. *Netw. Comput. Neural Syst.* **2003**, *14*, 391. [CrossRef]
19. Xie, S.; Hoehl, S.; Moeskops, M.; Kayhan, E.; Kliesch, C.; Turtleton, B.; Köster, M.; Cichy, R.M. Visual Category Representations in the Infant Brain. *Curr. Biol.* **2022**, *32*, 5422–5432. [CrossRef]
20. Yetter, M.; Robert, S.; Mammarella, G.; Richmond, B.; Eldridge, M.A.; Ungerleider, L.G.; Yue, X. Curvilinear Features Are Important for Animate/Inanimate Categorization in Macaques. *J. Vis.* **2021**, *21*, 3. [CrossRef] [PubMed]
21. Zachariou, V.; Del Giacco, A.C.; Ungerleider, L.G.; Yue, X. Bottom-up Processing of Curvilinear Visual Features Is Sufficient for Animate/Inanimate Object Categorization. *J. Vis.* **2018**, *18*, 3. [CrossRef]
22. Stothart, G.; Quadflieg, S.; Milton, A. A Fast and Implicit Measure of Semantic Categorisation Using Steady State Visual Evoked Potentials. *Neuropsychologia* **2017**, *102*, 11–18. [CrossRef] [PubMed]
23. Behl-Chadha, G. Basic-Level and Superordinate-like Categorical Rep resentations in Early Infancy. *Cognition* **1996**, *60*, 105–141. [CrossRef]
24. Dzhelyova, M.; Rossion, B. The Effect of Parametric Stimulus Size Variation on Individual Face Discrimination Indexed by Fast Periodic Visual Stimulation. *BMC Neurosci.* **2014**, *15*, 87. [CrossRef] [PubMed]
25. Liu-Shuang, J.; Norcia, A.M.; Rossion, B. An Objective Index of Individual Face Discrimination in the Right Occipito-Temporal Cortex by Means of Fast Periodic Oddball Stimulation. *Neuropsychologia* **2014**, *52*, 57–72. [CrossRef]
26. Morey, R.D.; Rouder, J.N. *BayesFactor*, version 0.9.12-2; Comprehensive R Archive Network: Vienna, Austria, 2015.
27. Rouder, J.N.; Morey, R.D.; Speckman, P.L.; Province, J.M. Default Bayes Factors for ANOVA Designs. *J. Math. Psychol.* **2012**, *56*, 356–374. [CrossRef]
28. JASP Team JASP 2024. Available online: https://jasp-stats.org/wp-content/uploads/2024/03/Statistical-Analysis-in-JASP-2024.pdf (accessed on 17 May 2024).
29. Bertels, J.; Bourguignon, M.; De Heering, A.; Chetail, F.; De Tiège, X.; Cleeremans, A.; Destrebecqz, A. Snakes Elicit Specific Neural Responses in the Human Infant Brain. *Sci. Rep.* **2020**, *10*, 7443. [CrossRef]
30. Jeschonek, S.; Marinovic, V.; Hoehl, S.; Elsner, B.; Pauen, S. Do Animals and Furniture Items Elicit Different Brain Responses in Human Infants? *Brain Dev.* **2010**, *32*, 863–871. [CrossRef]
31. Träuble, B.; Pauen, S.; Poulin-Dubois, D. Speed and Direction Changes Induce the Perception of Animacy in 7-Month-Old Infants. *Front. Psychol.* **2014**, *5*, 109716. [CrossRef]
32. Pauen, S.; Träuble, B. How 7-Month-Olds Interpret Ambiguous Motion Events: Category-Based Reasoning in Infancy. *Cognit. Psychol.* **2009**, *59*, 275–295. [CrossRef]
33. Poulin-Dubois, D.; Crivello, C.; Wright, K. Biological Motion Primes the Animate/Inanimate Distinction in Infancy. *PLoS ONE* **2015**, *10*, e0116910. [CrossRef]

brain sciences

MDPI

Article

Neurophysiological Oscillatory Mechanisms Underlying the Effect of Mirror Visual Feedback-Induced Illusion of Hand Movements on Nociception and Cortical Activation

Marco Rizzo [1], Laura Petrini [1], Claudio Del Percio [2], Lars Arendt-Nielsen [1,3] and Claudio Babiloni [2,4,*]

[1] Center for Neuroplasticity and Pain (CNAP), SMI®, Department of Health Science and Technology, Aalborg University, 9220 Aalborg, Denmark; marco.rizzo@mssm.edu (M.R.); lap@hst.aau.dk (L.P.); lan@hst.aau.dk (L.A.-N.)
[2] Department of Physiology and Pharmacology "V. Erspamer", Sapienza University of Rome, 00185 Rome, Italy; claudio.delpercio@uniroma1.it
[3] Department of Medical Gastroenterology, Mech-Sense, Aalborg University Hospital, 9220 Aalborg, Denmark
[4] Hospital San Raffaele Cassino, 03043 Cassino, Italy
* Correspondence: claudio.babiloni@uniroma1.it

Abstract: Mirror Visual Feedback (MVF)-induced illusion of hand movements produces beneficial effects in patients with chronic pain. However, neurophysiological mechanisms underlying these effects are poorly known. In this preliminary study, we test the novel hypothesis that such an MVF-induced movement illusion may exert its effects by changing the activity in midline cortical areas associated with pain processing. Electrical stimuli with individually fixed intensity were applied to the left hand of healthy adults to produce painful and non-painful sensations during unilateral right-hand movements with such an MVF illusion and right and bilateral hand movements without MVF. During these events, electroencephalographic (EEG) activity was recorded from 64 scalp electrodes. Event-related desynchronization (ERD) of EEG alpha rhythms (8–12 Hz) indexed the neurophysiological oscillatory mechanisms inducing cortical activation. Compared to the painful sensations, the non-painful sensations were specifically characterized by (1) lower alpha ERD estimated in the cortical midline, angular gyrus, and lateral parietal regions during the experimental condition with MVF and (2) higher alpha ERD estimated in the lateral prefrontal and parietal regions during the control conditions without MVF. These preliminary results suggest that the MVF-induced movement illusion may affect nociception and neurophysiological oscillatory mechanisms, reducing the activation in cortical limbic and default mode regions.

Keywords: high-density electroencephalography (HD-EEG); alpha event-related de/synchronization (ERD/ERS); source analysis; Mirror Visual Feedback (MVF); sensory–motor interaction; pain

1. Introduction

The phenomenon whereby the mirrored movement of one limb is perceived as the simultaneous movement of the opposite limb is known as Mirror Visual Feedback (MVF) [1]. The MVF technique entails placing a mirror perpendicular to the observer's body midline, creating the illusion of viewing the true limb [2,3]. Although the evidence supporting its effectiveness is still inconclusive [4], several investigations in cohorts of patients with chronic pain or motor deficits demonstrated neuroplasticity processes associated with MVF sensory–motor experience [5,6]. Therefore, MVF-based neurorehabilitation strategies were developed to treat chronic pain conditions such as phantom limb pain (PLP) syndrome [3,7,8], post-stroke hemiparesis [9–11], and complex regional pain syndrome [12,13]. These strategies were also successful in facilitating upper-extremities motor recovery [14,15].

Neuroimaging studies in healthy individuals used functional magnetic resonance imaging (fMRI) [16–18], near-infrared spectrometry [19,20], and transcranial magnetic stimulation [21–23] to investigate the brain neural basis of the effects of the MVF procedure. Results showed an MVF-related increase in cortical excitability (i.e., larger motor-evoked potentials) or activation in premotor, primary motor (M1), and primary somatosensory (S1) cortical areas ipsilateral to the moving hand. Other fMRI and TMS studies extended the focus of the investigation to the whole brain, showing the involvement of the visual and parietal posterior cortical areas—strongly connected with the central sensory–motor regions—as responsible for visuomotor transformation processes occurring during MVF-induced illusory experience [10,18,21,23–26]. In this regard, an fMRI study investigated the effects of MVF training in low-limb amputees. After 12 sessions of MVF therapy, patients showed a significant reduction of PLP and increased activity in the bilateral orbitofrontal cortex in response to phantom ankle "imaginary" movement [27]. Interestingly, the phantom ankle movements alone induced activation in the premotor and parietal associative cortical areas rather than the primary sensory–motor cortex [27].

The above neuroimaging results had the advantage of a high spatial resolution but were limited in describing the temporal evolution of the MVF-related cortical activation. For this purpose, electroencephalography (EEG) can probe the cortical activity related to motor and sensory events with less spatial resolution but enhanced temporal resolution (milliseconds) [28,29]. Previous EEG investigations measured lateralized related potentials (as an index of M1 activity) during unilateral hand movement tasks performed with MVF illusion. Those potentials showed clear activation of the M1 contralateral to the hand reflected in the mirror and perceived as moving (indeed, immobile) [30,31].

Other EEG studies considered the modulation of the dominant EEG rhythms at 8–12 Hz (i.e., alpha rhythms) recorded in Rolandic central and posterior scalp regions in the resting-state and psychophysically relaxed conditions as a reflection of cortical inhibition of somatomotor and visual–visuospatial cortical regions, respectively [32,33]. These alpha rhythms reduce in amplitude (the so-called event-related alpha desynchronization, alpha ERD) as a sign of cortical activation during the preparation and execution of voluntary movements [32–35]. Furthermore, central alpha ERD was associated with nociception and sensorimotor interactions (gating) between voluntary hand movements and painful stimuli [35,36].

Previous studies showed consistent alpha ERD (i.e., cortical activation) over frontal and parietal areas during MVF-induced movement illusion [37–39]. Abnormal activity in these areas was also linked to dysesthesia (i.e., abnormal limb perceptions associated with pathological pain like PLP) and emotions due to MVF-induced sensorimotor incongruence [40–43]. Other EEG investigations in healthy participants reported stronger alpha and beta ERD in the bilateral frontal premotor and anterior–posterior cingulate areas when they experienced dysesthesia [42,44] and discomfort sensations [45] induced by sensorimotor incongruence. Notably, dysesthesia is an interesting experimental model as it is characterized by a discomfort like that reported in people with pathological chronic pain [43,46]. Therefore, based on the mentioned investigations in healthy participants, it can be hypothesized that there is a possible neuromodulatory effect of MVF on neurophysiological oscillatory mechanisms generating EEG alpha rhythms in frontal premotor and anterior–posterior cingulate areas.

Keeping in mind the above findings and considerations, we used high-resolution EEG and estimation of alpha ERD cortical sources to shed light on the effects of MVF-induced movement illusion on nociception and frontal premotor and midline cortical areas, including anterior–posterior cingulate and other limbic areas commonly associated with pain processing. These cortical regions may underpin human pain experience related to the affective sense of self, body representation, and interoceptive signals [47–51], dimensions relevant to MVF therapy. Along this line, we tested the novel hypothesis that in healthy adults, the MVF-related illusion of hand movements may affect the nociception and the activation in the premotor and midline cortical areas, as revealed by the alpha ERD. To test this hypothesis, electrical stimuli were applied to the left hand to produce painful or non-painful sensations during unilateral right-hand movements with MVF. As control conditions, the same electrical stimuli were applied during unilateral right-hand and bilateral hand movements without MVF. The alpha ERD was estimated within mathematical sources modeled in the cortical midline and control regions.

2. Materials and Methods

2.1. Subjects

Thirteen healthy adult male volunteers were recruited for this study (mean age = 26.2, SD = ±4.7). The main exclusion criteria included the presence of chronic pain, neurological diseases, current medical treatment, and participation in other studies involving pain stimulation in the past four weeks. The Edinburgh Handedness Inventory [52] confirmed that all the subjects were right-handed. Remarkably, ten subjects participated in our previous MVF study [39]. Experiments were conducted at the Aalborg University (DK). The Scientific Ethical Committee of Region Nordjylland approved the study (N-20190008) and all the participants signed the informed consent form according to the Declaration of Helsinki.

2.2. Experimental Procedure

All participants were seated on a chair with their arms lying symmetrically ahead on a table. Auditory cues of 70 dB, 1000 Hz, and 50 ms of duration [53,54] were used to trigger right index finger movements. These specific auditory parameters were selected to produce an acute short sound, creating a peak in the EEG signal that could be identified and subsequently filtered out. Additionally, previous evidence has demonstrated that this type of passive stimulus does not affect the activity of cortical areas adjacent to the auditory cortex, such as the motor cortex, inferior parietal lobule, and anterior and posterior cingulate cortices [55]. All the subjects underwent three different conditions (80 trials each) in a randomized order. Each condition lasted approx. 13 min, with 10 min breaks between the conditions (the flowchart of the experimental session is represented in Figure 1). The whole session—including EEG net preparation, training, pain threshold measurement, and experiment—lasted approx. 2.5 h. In the experimental condition (Unilateral with Mirror, or UM+), the mirror was placed on the desk, perpendicular to the subject's midsagittal plane, with the reflecting face on the right side (Figure 2). Subjects were trained to move their right index finger in response to the auditory cue while watching the image of the reflected moving hand in the mirror to give the illusion of the simultaneous left index finger movement. The position of the left hand behind the mirror corresponded to the image of the left hand reflected in the mirror. This condition was meant to produce sensory–motor interaction between electrical stimuli applied on the left hand and illusory movements of the (mirrored) left hand. In the control conditions, the mirror was removed from the experimental setting and the left hand was directly visible to the participants. Subjects performed the same unilateral right index finger movements in one control condition (Unilateral without Mirror, UM–) in order to see the effects of electrical stimulation on the left hand (i.e., somatosensory cortex on the right hemisphere) without motor interaction. In the other control condition (Bilateral without Mirror, BM–), subjects simultaneously performed movements of both index fingers. This control condition was meant to investigate the effects of the interaction between electrical stimuli and actual movement of the left hand (i.e., sensory–motor cortex of the right hemisphere). Participants were trained to perform a double extension of their index finger with a slow release toward the table (approx. 1 s). All the participants received brief training to perform the movement correctly and keep their left hand as still as possible during the unilateral conditions. In each condition, an electrical stimulus was delivered on the tip of the left index finger 100 ms after the auditory cue to induce cortical sensory–motor interaction (Figure 3). The interval between the stimuli was fixed at 10 s, a sufficient period to reset the desynchronization of the alpha rhythms [35]. A fixed interval was used to allow predictability of the upcoming auditory and electrical stimuli, optimal for studying the anticipatory alpha ERD/ERS responses.

However, subjects were not informed of this fixed interval to prevent them from using counting strategies.

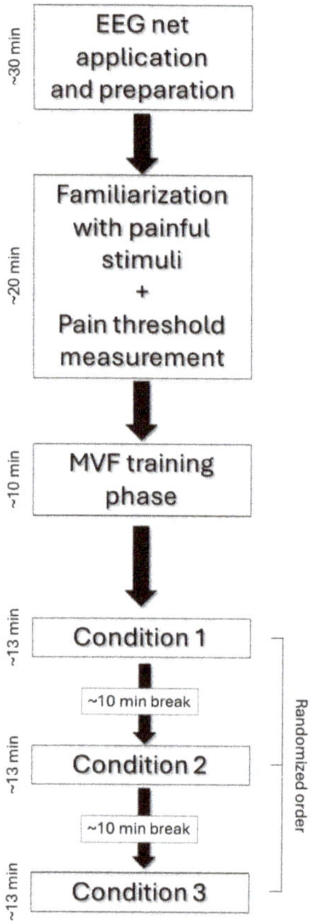

Figure 1. Experimental session flowchart.

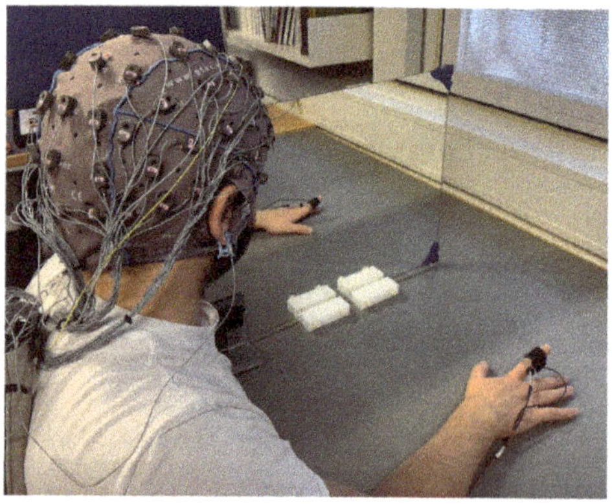

Figure 2. *Cont.*

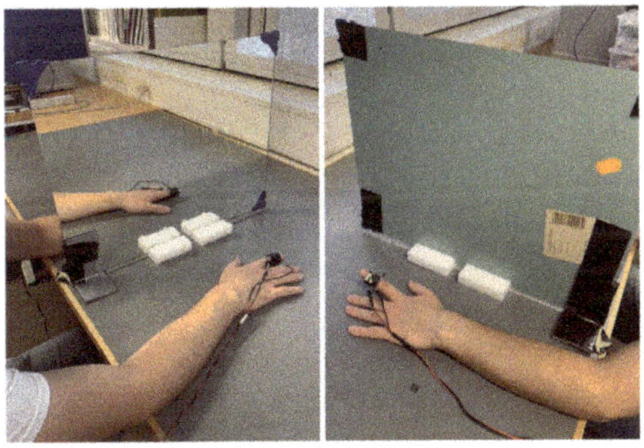

Figure 2. Experimental setup during the Mirror Visual Feedback (MVF) procedure (Unilateral with Mirror or UM+ condition). The mirror is placed in the subject's midsagittal plane to give the illusion of ownership of the left hand. An electrode delivering the stimulations was placed on the index finger of the left hand (bottom right panel). A sham electrode was placed on the right index finger to strengthen the illusory feeling through congruent visual feedback. Movements consisted of double extension (1 s approx.) with a slow release toward the table.

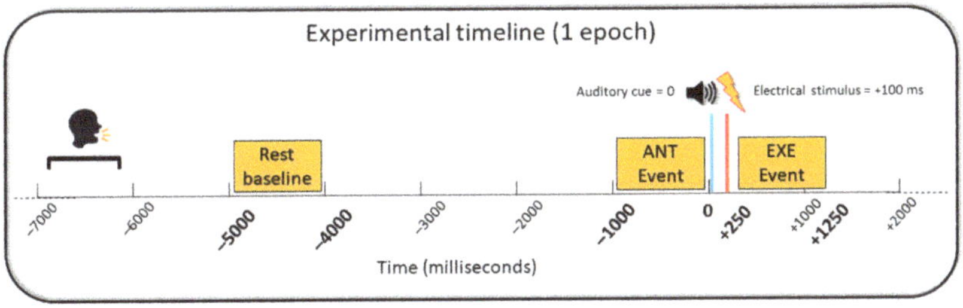

Figure 3. The above figure illustrates the sequence of one trial of the experimental paradigm. In the experiment, a fixed 10 s interstimulus interval was applied. During the offline analysis, 9 s epochs were extracted (from 7 s before the auditory cue to 2 s after the cue). In the above figure, the time is expressed in milliseconds (ms). The 0 (zero) time corresponds to the auditory cue triggering the right index finger movement (blue line). In each trial, electrical stimuli were delivered on the left index finger after 100 ms from the auditory cue (red line). In between the trials, subjects were asked to vocally rate the sensation related to the electrical stimulation on a scale from 0 to 10 (painful > 5; non-painful < 5). The anticipation (ANT) event period was defined as the time interval of 1000 ms before the 0, whereas the execution (EXE) period was defined as the time interval between 250 and 1250 ms after the 0 (yellow boxes). The alpha event-related de/synchronization (ERD/ERS) was calculated for the ANT and EXE periods in relation to a baseline period represented by the yellow box between 5000 and 4000 ms before the 0 (yellow box).

2.3. Electrical Stimulations and Pain Threshold Detection

Electrical stimulations (5 ms) were provided by a voltage-controlled current stimulator (NoxiSTIM; JNI Biomedical; Aalborg, Denmark). Stimuli were delivered on the tip of the left finger using an electrode with a gold pin cathode and a ring anode (Figure 2). An inactive electrode (sham) was placed on the right index finger to achieve congruent Mirror Visual Feedback (Figure 2). A multichannel data acquisition module (NI-6221, National Instruments Corp., Austin, TX, USA) was used to synchronize the electrical stimuli to the auditory cues.

Subjects were asked to evaluate the intensity of electrical stimuli using a Numerical Scale Rating (NRS) from 0 to 10, where 0 represents the absence of any sensory perception and 10 is the "worst imaginable pain". Before the pain threshold measurement, 15 stimuli at random intensities were delivered in a familiarization phase. The method of limits was used to determine the pain threshold for each subject. Starting from 0.5 milliamperes (mA), the current intensity was progressively increased until a rating of 5 out of 10, namely, the minimum current intensity producing a painful sensation [56]. With this procedure, we aimed to reach a fixed intensity that was perceived half the time as painful (NRS > 5) or non-painful (NRS < 5). The same fixed individual pain threshold was used during the three conditions (mean mA = 5.5; SD = ± 3.1). In between the trials (approx. 3–4 s after the cue), subjects were asked to provide a numerical value (decimals included) of the average pain intensity of the last 5 trials. Subjects vocally rated the stimuli intensity, and each value was noted by the experimenter on an Excel spreadsheet. The speaking artefacts were removed in the offline analysis. Since each condition consisted of 80 trials, 16 values per condition were considered in the analysis. Due to the fluctuant subjective perception of fixed stimuli, painful (NRS > 5) and non-painful (NRS < 5) responses were obtained for each condition. If one condition did not show fluctuation in the perceived pain intensity, the single dataset was removed from the analysis. Blocks with fluctuation were defined as blocks presenting a ratio of at least 30%/70% of painful/non-painful trials or vice versa. Moreover, to avoid any sort of habituation or sensitization phenomena, we excluded those blocks that did not show a random distribution between painful and non-painful trials (see RUNS test in Control Analysis and Figure S2 in Supplementary Materials). The final samples considered in the statistical analysis were UM+ = 8 subjects, UM− = 13 subjects, and BM− = 12 subjects. Of interest, the painful epochs (NRS > 5) were 36.5% in the UM− condition, 58.5% in the BM− condition, and 53.9% in the UM+ condition.

2.4. EEG Recording and Preprocessing

Scalp EEG activity was recorded using an active 64-channel system (g.HIamp amplifier, g.tec medical engineering GmbH, Graz, Austria) following the 10–10 international system. Ground and reference electrodes were positioned on the forehead and ear lobes, respectively. Electrode impedance was maintained under 5 kΩ, and data were sampled at 1200 Hz. Eye movements and blinks were detected using Fp1 and Fp2 electrodes and subsequently removed during the offline analysis. Data preprocessing was conducted using the EEGLAB 2021.1 [57] freeware toolbox in MatLab R2018b (MathWorks, Inc., Natick, MA, USA). Following visual inspection, data were filtered with a zero-phase basic FIR filter (passband edges: 0.3–40 Hz, cutoff frequencies: −6 dB, filter order: 846, transition band: 1 Hz), resampled to 256 Hz, and re-referenced to the average of all channels to mitigate reference-related effects from online recording [58]. Data were then segmented into 80 epochs of 9 s each. Artefact removal involved identifying and eliminating artefacts such as blinks, muscle movements, and 50 Hz noise using the Infomax independent component analysis (ICA) algorithm [59]. Approximately 7.9% of artifact-laden epochs and 7.3% of artefact-related components were excluded from the EEG datasets. Subsequently, global EEG spectra were computed using FFT-based methods to determine the individual alpha frequency peak (IAFp). IAFp was defined as the frequency exhibiting the highest peak within the alpha band (8–12 Hz) [60]. EEG data were filtered at the individual alpha frequency band range from IAF–2 Hz to IAF + 2 Hz (mean IAFp = 10.1; SD = ± 1.0) [60,61].

2.5. Cortical Sources Estimation

The freeware "exact Low-Resolution Brain Electromagnetic Tomography" (eLORETA) [62] was utilized for cortical sources estimation using preprocessed EEG data as input. The eLORETA freeware is a functional imaging technique that belongs to a group of standardized linear inverse solution procedures, modelling the 3D distributions of EEG sources within a head volume conductor model, including the scalp, skull, and brain [63,64]. eLORETA addresses the EEG inverse problem by estimating the current density values at the voxel level, offering a brain source space limited to the cortical grey matter, consisting of 6239 voxels (5 mm resolution, with each voxel containing an equivalent current dipole), based on the Montreal Neurological Institute template (MNI152). For each voxel, eLORETA provides information about the MNI coordinates, the lobe, and the Brodmann area (BA).

The input for the source estimation is a spectral power density computed by 62 scalp electrodes, as eLORETA source analysis is a free-reference method [65]. Due to the low spatial resolution characteristic of the EEG techniques, adjacent Brodmann areas were clustered. The clusters included BAs 1–2–3 together with BA 4 (sensory–motor area), BA 24 together with BA 32 (anterior cingulate cortex), BA 44 with BA 45 (inferior frontal gyrus), and BA 9 with BA 46 (superior frontal gyrus or dorsolateral prefrontal cortex).

2.6. Alpha ERD/ERS Calculation

Standard quantification of the alpha ERD/ERS was conducted for both the anticipation and execution stages of the movements [32,36,66–68]. The alpha ERD/ERS was calculated using the formula: ERD/ERS% = $(E - R)/R \times 100$, where E represents the power density at the "Event" period (1 s) and R represents the power density at the "Rest" periods (1 s). The "Rest" period was defined as the interval from 5 s to 4 s before the auditory cues that triggered the movement. The anticipation "Event" was defined as the period of 1 s before the auditory cues. Finally, the execution "Event" was defined as the 1 s period from 250 ms to 1250 ms after the auditory cues, with the alpha ERD peak within this interval considered in the statistical analysis. The first 250 ms post-cue were excluded to remove auditory-evoked potentials (N1-P2 complex) [69] and pain-evoked potentials (N2-P2 complex) [35,70,71]. Negative percentage values represented the alpha ERD, indicating cortical activity, while positive values represented the alpha ERS, indicating cortical inhibition [32,66]. For scalp analysis, electrodes were clustered for each hemisphere at the frontal (F3 and FC3; F4 and FC4) and centro–parietal (C3, CP3, and P3; C4, CP4, and P4) levels. The eLORETA solutions were utilized as input to compute the cortical voxel-level ERD/ERS in the alpha frequency band.

2.7. Statistical Analysis for the ERD/ERS Scalp Distribution

Linear mixed models for repeated measures were used to assess whether the subjective perception of the stimulus intensity depended on the MVF. The fixed factors were Condition (UM+, UM–, and BM–) and Stimulus (pain and no-pain), whereas the random factor was the number of subjects in each group. The models were fitted separately in the left and right hemispheres for the frontal (F3-FC3 and F4-FC4) and centro–parietal (C3-CP3-P3 and C4-CP4-P4) clusters of electrodes, as well as for the anticipation and execution phases of the event. The linear mixed model statistical analysis was carried out using IBM SPSS Statistics 27 (IBM, New York, NY, USA).

2.8. Statistical Analysis for the Alpha ERD/ERS Cortical Sources

To investigate differences in cortical activation between stimuli perceived as painful and non-painful, statistical nonparametric mapping voxel-based wise randomization tests (5000 permutations) were employed for the alpha ERD/ERS eLORETA solutions for each condition (UM+, UM–, and BM–) and stage of the event (Anticipation and Execution). The results derived from this statistical approach are equivalent to those obtained using a statistical parametric method with multiple comparison corrections [72]. This nonparametric permutation approach is appropriate for analyzing low sample sizes with low degrees of freedom [72,73]. The results are shown as voxel-based T statistics maps with corrected $p < 0.05$ (randomization tests correction) [72]. To reduce the type II error (false negatives), an uncorrected significance threshold of $p < 0.001$ was employed. Only cortical areas comprising at least 5 voxels were included in the analysis. Statistical analyses were conducted using the eLORETA statistical toolbox [62]. The significance of the results was statistically confirmed by the effect size (r) for the two-sided Wilcoxon signed rank test, a nonparametric test used to compare paired groups [74,75]. In particular, Wilcoxon's r measured the effect size of the difference between Pain and No pain in the ERS/ERD% corresponding to each's most significant voxel (Table 1). The effect size was calculated for each condition (UM–, BM–, and UM+) and movement stage (ANT and EXE). Wilcoxon's r values range from −1 to 1 and are interpreted as $|0|$-$|0.1|$ = no effect or very small effect,

|0.1|-|0.3| = small effect, |0.3|-|0.5| = medium effect, and |0.5|-|1| = large effect [76]. Wilcoxon's r values are reported in the extreme right column in Table 1. The effect size was calculated using the *effsize* package for RStudio (v. 4.3.1).

Table 1. Voxel-based Student's *t*-test analysis for the eLORETA sources of the alpha ERD at the Brodmann areas (BAs) level. ** Significance threshold of $p < 0.05$ corrected for multiple comparisons; * significance threshold of $p < 0.001$ uncorrected; n.s. not significant. The cluster size denotes the number of voxels that meet the significance threshold for multiple comparisons (uncorrected threshold in brackets). For the most statistically significant voxels, its Montreal Neurological Institute (MNI) coordinates, T value, and effect size (Wilcoxon's r) are reported. Effect size values range from −1 to 1 and were interpreted as |0|-|0.1| = no effect or very small effect, |0.1|-|0.3| = small effect, |0.3|-|0.5| = medium effect, and |0.5|-|1| = large effect. For a better comprehension of the results, it must be noted that the alpha ERD represents a negative percentage indicating decreased power density during the event compared to the baseline. Therefore, when comparing painful versus non-painful epochs, negative T values indicate higher alpha ERD during painful epochs, while positive T values indicate higher alpha ERD during non-painful epochs. In the two control conditions (UM− and BM−), both the anticipation and execution phases of the event are characterized by stronger alpha ERD in the non-painful than painful blocks (as reflected by positive T values) in the sensory–motor and parietal associative areas. In the experimental condition (UM+), the anticipation and execution phases of the event are characterized by stronger alpha ERD in the painful rather than non-painful blocks (negative T values) in those cortical midline regions of the limbic system, default mode network, and attentional network.

			ANTICIPATION						
Conditions Comparisons	BAs	Cluster Size	Region	Hemisphere	MNI Coordinates x	y	z	T Value	Wilcoxon's r
UM− (Pain vs. No pain)	32	5 (9)	Anterior cingulate	L	−5	20	40	4.36 **	−0.87
	40	3 (8)	Inferior parietal lobule	L	−65	−25	20	3.95 **	−0.81
BM− (Pain vs. No pain)	1–2–3, 4	13 (36)	Central gyrus	L/R	60	−30	45	4.61 **	−0.84
	40	113	Inferior parietal lobule	L/R	−65	−40	35	4.46 **	−0.89
	44–45	15	Inferior frontal gyrus	L/R	−60	5	15	3.36 *	−0.81
UM+ (Pain vs. No pain)	30	2 (8)	Posterior cingulate	R	5	−55	5	−4.16 **	0.94
	40	1 (5)	Inferior parietal lobule	R	40	−55	60	−4.73 **	0.93
			EXECUTION						
Conditions Comparisons	BAs	Cluster Size	Region	Hemisphere	MNI Coordinates x	y	z	T Value	Wilcoxon's r
UM− (Pain vs. No pain)	40	5	Inferior parietal lobule	L	−55	−25	15	3.34 n.s.	−0.75
BM− (Pain vs. No pain)	9–46	7 (21)	Superior frontal gyrus	R	55	15	30	4.11 **	−0.85
	32	7	Anterior cingulate	R	15	45	−5	3.67 *	−0.81
	44–45	35	Inferior frontal gyrus	R	60	15	15	3.79 *	−0.9
UM+ (Pain vs. No pain)	10	2 (19)	Medial frontal gyrus	R	10	50	15	−4.31 **	0.94
	19	5	Precuneus	R	30	−75	35	−3.87 *	0.84
	24–32	2 (11)	Anterior cingulate	L/R	5	25	15	−4.66 **	0.85
	39	2 (17)	Angular gyrus	R	55	−65	25	−4.75 **	0.93
	40	3 (10)	Inferior parietal lobule	R	45	−55	45	−4.67 **	0.86

2.9. Control Analysis

Event-related potentials (ERPs) in response to the auditory cues and electrical stimuli were extracted from the Cz electrode. Successively, the waveforms resulting from painful and non-painful blocks were averaged across the subjects for each condition and paired samples *t*-tests were used for statistical comparison. The ERPs considered were the N1-P2 (auditory), P2-N2 complex (electrical stimulus), and P3 (attentional). Furthermore, the randomness of the painful and non-painful epochs was tested by performing the

RUNS test for each subject and condition. Finally, NRS values were compared to control differences in the subjective perception of the stimulus intensity for each condition. All the statistical analyses were performed using IBM SPSS Statistics 27 (IBM, New York, NY, USA). Furthermore, time–frequency maps for C3 and C4 electrodes were plotted to provide an estimation of the whole frequency spectrum (0–40 Hz) for each condition and movement stage (Figure S4). The results from the control analyses are reported in detail in Supplementary Materials.

3. Results

3.1. Scalp Topography

Figure 4 shows the scalp distribution of the alpha ERD/ERS for the three conditions (UM–, BM–, and UM+) at the anticipation (1000 ms before the auditory cue) and execution (peak within 1250 ms after the auditory cue) of the event. The cortical activity patterns resulting from the blocks perceived as painful (NRS > 5) and non-painful (NRS < 5) are thus compared. In the control conditions without MVF (UM– and BM–), results indicate a similar centrally distributed alpha ERD in the anticipation phase of the event. When considering the execution phase, the alpha ERD is more widespread in the frontal–central areas in the non-painful than painful blocks in both conditions. In the experimental MVF condition, the maps show a slightly stronger alpha ERD in the painful than non-painful blocks for both the anticipation and execution phases of the movement. Moreover, the execution phase shows a prominent parietal–central distribution in the painful rather than non-painful blocks.

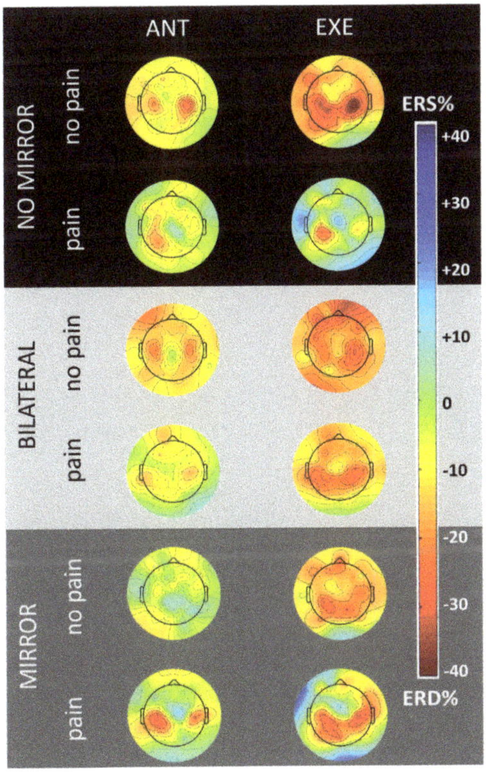

Figure 4. Across subjects' mean 2D maps of the alpha ERD/ERS distribution over the scalp for each condition (Unilateral without Mirror, Bilateral without Mirror, and Unilateral with Mirror), phase of

the event (Anticipation and Execution), and subjective perceived intensity of the stimulus (pain, no-pain). In the maps, the alpha event-related desynchronization (ERD) and synchronization (ERS) are represented with red and blue colors, respectively. For the anticipation phase (ANT), the 1 s interval before the auditory cue is reported as a scalp map. For the execution phase (EXE), the scalp maps are represented by 250 ms time windows after the auditory cue and the interval showing the highest peak of the alpha ERD was reported in the figure. The maps indicate a centrally distributed alpha ERD in the ANT phase and frontocentral alpha ERD in the EXE phase during the two control conditions for both pain and non-pain blocks. In the experimental mirror condition, the maps show a stronger but not statistically significant alpha ERD at the central level in the painful over the non-painful blocks for both stages of the events (ANT and EXE).

However, the linear mixed model analyses performed to assess the interaction between the stimulus intensities and the conditions did not indicate any statistically significant effect for the frontal and central brain regions ($p > 0.05$).

3.2. Control Conditions without MVF

This section of the results considers the differences between painful and non-painful blocks for the control conditions without the MVF illusion (i.e., UM– and BM–). Figures 5 and 6 display the spatial distribution of p values from the parametric statistical maps (Student's t-tests) for the eLORETA solutions of the alpha ERD/ERS, shown in the first and second rows. Table 1 lists the significant T values along with the voxel cluster sizes and their locations.

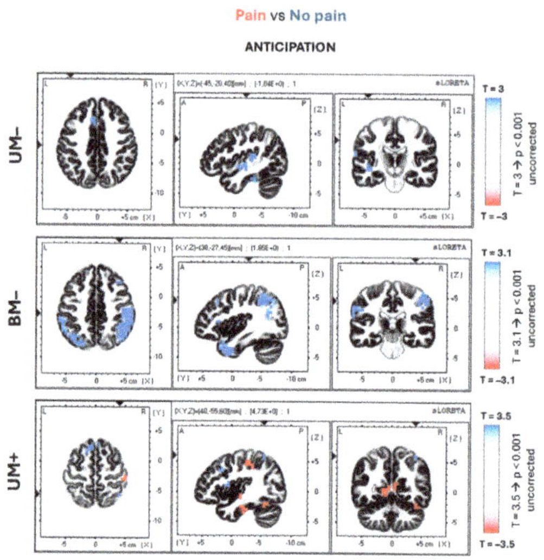

Figure 5. Spatial distribution of the voxel-by-voxel significant p values relative to the Student's t-test for the alpha ERD/ERS eLORETA solutions. The above figure illustrates the comparisons between pain and no-pain blocks for each condition (UM–, BM–, and UM+) at the anticipation phase of the event. The axial, sagittal, and coronal sections are represented. The T values corresponding to the uncorrected significance threshold of $p < 0.001$ are shown on the right side for each condition (the T values change as the sample sizes are different among the groups). In the maps, the red voxels show the areas where the alpha ERD is significantly stronger in the painful than the non-painful blocks (posterior cingulate and inferior parietal lobule). Conversely, the blue voxels show the areas where the alpha ERD is significantly stronger in the non-painful than the painful blocks (sensory–motor and parietal associative cortical areas).

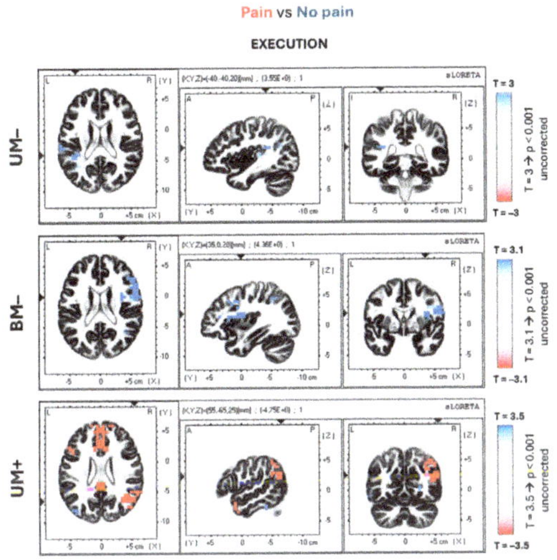

Figure 6. Spatial distribution of the voxel-by-voxel significant p values relative to the Student's t-test for the alpha ERD/ERS eLORETA solutions. The figure illustrates the comparisons between the painful and non-painful blocks for each condition (UM−, BM−, and UM+) at the execution phase of the event. The axial, sagittal, and coronal sections are represented. The T values corresponding to the uncorrected significance threshold of $p < 0.001$ are shown on the right side for each condition (the T values change as the sample sizes are different among the groups). In the maps, the red voxels show the areas where the alpha ERD is significantly stronger in the painful than the non-painful blocks (cortical midline structures of the limbic system, default mode network, and attentional network). Conversely, the blue voxels show the areas where the alpha ERD is significantly stronger in the non-painful than the painful blocks (sensory–motor and parietal associative cortical areas).

For both conditions, the anticipation and execution phases of the event are characterized by a stronger activation (alpha ERD) in the non-painful than painful blocks. In particular, the UM− condition shows this pattern in the left anterior cingulate (BA 32; peak value: T = 4.36; $p < 0.05$ corrected; r = −0.87) and inferior parietal cortex (BA 40; peak value: T = 3.95; $p < 0.05$ corrected; r = −0.81) in the anticipatory phase, and no significant differences in the execution phase. In the BM− conditions, the differences appear in the bilateral sensory–motor (BAs 1–2–3, 4; peak value: 4.61; $p < 0.05$ corrected; r = −0.84), inferior parietal (BA 40; peak value: T = 4.46; $p < 0.05$; r = −0.89) cortical areas, and bilateral inferior frontal gyrus (BAs 44–45; peak value: T = 3.36; $p < 0.001$ uncorrected; r = −0.81), for the anticipation phase (Figure 5). Furthermore, the execution phase of the event (Figure 6) shows those differences in the right superior (BAs 9–46; peak value: T = 4.11; $p < 0.05$ corrected; r = −0.85) and inferior (BAs 44–45; peak value: T = 3.79; $p < 0.001$ uncorrected; r = −0.90) frontal gyri, as well as right anterior cingulate cortex (BA 32; peak value: T = 3.67; $p < 0.001$ uncorrected; r = −0.81).

3.3. Experimental Condition with MVF

The voxel-by-voxel comparison results between painful and non-painful blocks in the experimental MVF (UM+) condition are hereby reported. The spatial distribution of the p values for the Student's t-tests is illustrated in Figures 5 and 6 (bottom rows). The significant T values, as well as the voxel cluster size and localizations, are summarized in Table 1.

Unlike the control conditions, in the UM+ condition, the alpha ERD observed in the painful blocks was stronger than the alpha ERD in the non-painful blocks. During the anticipation phase (Figure 5), the main differences are observed in the right posterior cingulate (BA 30; peak value: T = −4.16; $p < 0.05$ corrected; r = 0.94) and the inferior posterior parietal cortex (BA 40; peak value: T = −4.73; $p < 0.05$ corrected; r = 0.93). During the execution phase (Figure 6), these differences are widespread in the frontal and posterior areas. Results showed significant differences in the right polar frontal (BA 10; peak value: T = −4.31; $p < 0.05$ corrected; r = 0.94) and inferior posterior parietal (BA 40; peal value: T = −4.67; $p < 0.05$ corrected; r = 0.86) cortical areas. The same was true in the right middle temporal gyrus (BA 39; peak value: T = −4.75; $p < 0.05$ corrected; r = 0.93) and precuneus (BA 19; peak value: T = −3.87; $p < 0.001$ uncorrected; r = 0.84), as well as in the bilateral anterior cingulate (BAs 24–32; peak value: T = −4.66; $p < 0.05$ corrected; r = 0.85).

4. Discussion

In the present exploratory study, we hypothesized that the MVF-induced illusion of hand movements may affect the nociception and the activation in those cortical midline areas associated with pain processing. For this purpose, we used an original methodological approach. Electrical stimuli at about the pain threshold were applied to the left hand to produce painful or non-painful sensations during unilateral right-hand movements with the MVF and, as control conditions, during unilateral right and bilateral hand movements without the MVF. The eLORETA freeware estimated the EEG source activity, and the alpha ERD (as a sign of cortical activation) reflected the neurophysiological oscillatory mechanisms underlying the painful and non-painful sensations associated with the MVF-related movement illusion.

The results of a control analysis showed that the vertex N1-P2 and/or N2 peaks of the sensory-evoked potentials exhibited higher amplitude in relation to the painful over the non-painful sensations during both MVF and noMVF conditions (Figure S1 in Supplementary Materials). These results corroborate the reliability of the participants' subjective sensations, in line with previous studies demonstrating that those sensory-evoked potentials can be considered trustworthy neurophysiological signatures of enhanced cortical arousal and pain experience in humans [69–71].

The core results of the present study showed that, compared to the painful sensations, the non-painful sensations during the MVF condition were specifically characterized by lower alpha ERD (cortical activation) estimated in the right cortical midline regions (i.e., medial prefrontal, anterior and posterior cingulate, parietal cortex, and precuneus), angular gyrus, and inferior parietal lobule contralateral to the electric hand stimulation and ipsilateral to the hand movements. In contrast, the non-painful sensations during the control noMVF conditions were specifically characterized by stronger alpha ERD estimated in the left lateral prefrontal and inferior parietal lobule regions ipsilateral to the electric hand stimulation and contralateral to the hand movements. Large effect sizes (i.e., r > 0.8) confirmed the significance of these findings. Although the results of this study are preliminary and need to be interpreted with caution, a functional interpretation is provided in the following section.

4.1. Effects of MVF-Induced Movement Illusion on Cortical Activity and Pain

During the MVF condition, the non-painful over painful sensations were specifically related to lower alpha ERD in the right cortical midline regions of the limbic and default mode network, in the right angular gyrus of the default mode network, and in the right inferior parietal lobule of the ventral attention network. In contrast, the non-painful sensations in the noMVF conditions were characterized by higher alpha ERD in the left lateral prefrontal and inferior parietal lobule of the ventral attention networks.

The current findings emphasize the role of alpha frequency neurophysiological oscillatory mechanisms in modulating the deactivation within cortical regions pertinent to human nociception. These results complement previous neuroimaging research in humans, which highlighted: (1) the involvement of medial prefrontal and anterior–posterior cingulate cortices in processing the affective aspect of pain [77,78]; (2) the activation of cortical midline regions in relation to the unpleasant sensations elicited by painful stimuli [79–81];

and (3) the activation of a posterior cortical subnetwork including the inferior parietal lobe, posterior cingulate cortex, and precuneus, associated with affective and cognitive components of the pain experience, such as pain anticipation and self-awareness [82]. Additionally, the present findings align with previous evidence indicating that sensorimotor incongruence in healthy individuals can lead to alterations in limb perceptions and emotional responses (e.g., dysesthesia) [40,42,45]. These studies have linked dysesthesia to abnormal EEG alpha and beta rhythms observed in parietal, midline premotor, and cingulate cortical areas [44,45]. Notably, such discomfort is similar to that reported in people with pathological chronic pain [43,46].

Although the present findings are preliminary and must be interpreted with caution, they offer—along with the previous neuroimaging results—intriguing insights into the neurophysiological mechanism underlying the experience of pain during MVF illusion. In the current MVF condition, the decreased alpha ERD observed in the right cortical midline limbic regions may be linked to non-painful sensations that diminish the affective sense of self and internal body perception [42,45,47,50,51]. Furthermore, the reduced alpha ERD in the right default mode network regions could be associated with weakened internal representations related to bodily sensations (interoception) [83–85]. Conversely, in the noMVF conditions, the heightened alpha ERD in the left lateral prefrontal and inferior parietal lobule may be associated with non-painful sensations related to significant attentional allocation and cognitive states towards the actual moving hand(s) for "body gnosis", affecting neural signals induced by electrical hand stimulation [86,87]. Unfortunately, EEG techniques do not have the fine spatial resolution necessary to disentangle the different subcomponents of a given midline cortical region belonging to either the limbic or default mode network.

During the present MVF condition, the participants may experience the self-oriented attentional and cognitive states associated with the illusory sense of ownership ("the illusory moving hand belongs to my body") and agency ("I am moving the hand reflected in the mirror") for the moving hand reflected in the mirror. According to previous alpha ERD evidence from our group [39,88], these states may be related to the activation of lateral prefrontal, premotor, and parietal cortical regions in the hemisphere ipsilateral to the true unilateral hand movements and contralateral to the moving hand reflected in the mirror. Notably, in the present experiments, the left inferior parietal lobule was the only region of that cortical network showing an effect related to pain. Specifically, the non-painful (rather than painful) sensations during the MVF condition were associated with lower alpha ERD in the right inferior parietal lobule, as opposed to the higher alpha ERD in the left inferior parietal lobule during the noMVF conditions. It can be speculated that during the MVF condition, the non-painful sensations were also related to a low activation in the right lateral parietal area integrating the visual–somato(nociceptive)–motor information regarding the electrically stimulated left hand.

4.2. Clinical Implications and Future Perspectives

The 64-channel EEG system used in the present study provided a precise delineation of the temporal and spatial evolution of the alpha oscillatory mechanisms during movement preparation and execution, as well as sensory stimuli expectancy and conscious experience during the MVF procedure. Specifically, the present findings suggest that the EEG alpha rhythms may be insightful markers of the neurophysiological underpinning of nociception and sensorimotor interactions during MVF-related rehabilitative processes, such as motor relearning [89–91] and enhanced transcallosal communication [92,93]. If future studies confirm the present results in patients with chronic pain undergoing MVF therapy, even at the individual level, the assessment of the alpha ERD may serve as a neurophysiological biomarker for restored sensory–motor information processing or tracking patients' motor recovery trajectory. An additional aspect of the present findings—potentially relevant in the clinical context—is the strong relationship between nociceptive experience and attentional and internal body representation during MVF experience. The EEG alpha rhythms may be used in future studies as a biomarker indicating the best MVF settings to induce the modulation of EEG alpha source activity in the limbic system. Along this line, the MVF settings using virtual reality technologies may be especially promising [94,95]. Future neurorehabilitation methods may leverage MVF-based virtual reality technologies to provide greater illusory immersion and enhance the range of movement, including asymmetric movements that are otherwise impossible with traditional MVF methods.

Future multimodal EEG–fMRI studies with higher spatial resolution should shed light on the effects of the MVF-related movement illusion on the nociception and the functional connectivity between all cortical and subcortical (e.g., amygdala–hypothalamus, amygdala–septum, hippocampus–mammillary bodies, etc.) components of the limbic system in both healthy participants and patients with chronic pain. A modulation in this connectivity is expected to impact the integration of primary needs, the emotional value of the painful stimuli, and interoceptive signals related to the person's experience in relation to the wellness–illness axis during painful stimulations [48–50]. Furthermore, future studies in both healthy participants and patients with chronic pain may investigate the functional connectivity between the limbic and the default mode networks in connection with the MVF effects on nociception and chronic pain. These studies may suggest new pathophysiological, neurophysiological, and functional neuroanatomical targets and companion biomarkers to integrate MVF therapy and pharmacological analgesic treatments.

4.3. Study Limitations

Due to its exploratory nature, this study has a few significant limitations.

Firstly, this study used a small population of healthy male volunteers to preliminarily test the working hypothesis. As a possible effect of this limited participants' sample, the participants did not clearly exhibit the expected fluctuations in perceived pain intensity. Consequently, the three groups presented different sample sizes, necessitating the use of a more complex mixed-model statistical design. At this early stage of the research, we enrolled only healthy male participants to avoid the potential influence of hormonal fluctuations during the menstrual cycle on pain perception [96]. The present results encourage the planning of future validation studies in a larger sample of participants, including both males and females (taking into account the menstrual cycle in the latter).

Secondly, the phasic electrical stimulations used in this study allowed us to investigate fundamental neurophysiological mechanisms underpinning human nociception and sensorimotor interactions during the MVF experience. The present results encourage the application of the present methodological approach in healthy participants using the model of tonic painful stimulations, which is more relevant in relation to chronic pain and patients with chronic pain conditions typically undergoing MVF-based therapies.

Thirdly, the intrinsic limitations of the EEG technique should be considered in the interpretation of the present results. Although the cortical midline structures present dense connections with the limbic system (e.g., amygdala, thalamus, hippocampus), the involvement of specific subregions of those brain structures could not be probed due to the intrinsically limited spatial resolution of EEG source solutions, even with the present high-resolution techniques (e.g., centimeters).

5. Conclusions

In this exploratory study, we hypothesized that in healthy adults, the MVF-induced movement illusion could alter midline cortical activation, as indicated by the EEG alpha neurophysiological oscillatory mechanisms. In comparison to painful sensations, non-painful sensations were linked to lower alpha ERD in the cortical midline, angular gyrus, and lateral parietal regions during the MVF experimental condition and higher alpha ERD in the lateral prefrontal and parietal regions during the MVF-free control conditions. These preliminary observations suggest that the MVF-induced illusion might influence nociception and neurophysiological mechanisms by reducing activation in cortical limbic and default mode regions, potentially affecting the sense of self, internal body perception, and attention to electrical signals from the stimulated hand. Although preliminary, these findings deserve attention and further verification is essential to develop beneficial approaches in those patients with chronic pain and motor impairments. Future research should use larger samples of healthy participants and patients with chronic pain to achieve generalizable results on both the cortical neurophysiological model of nociception and sensorimotor interactions and MVF neuromodulatory effects on the limbic system in the therapy of patients with chronic pain.

Supplementary Materials: The following supporting information can be downloaded at https://www.mdpi.com/article/10.3390/brainsci14070696/s1: Figure S1: The figure illustrates the mean event-related potentials (ERPs) generated in response to auditory cues and electrical sensory stimuli at the Cz electrode. The differences between painful (red line) and non-painful (black line) blocks are represented for each condition (UM−, BM−, and UM+). Statistical analysis (t-tests) showed significant differences in the N2 peak for the UM- condition as well as in the N1 and P2 peaks for the experimental UM+ condition (* $p < 0.05$).; Figure S2: The graph shows the distribution of the blocks where each subject (Y-axis) perceived the stimuli as painful (NRS > 5, red blocks) or non-painful (NRS < 5, green blocks) in the experimental Unilateral Mirror (UM+) condition. Each block consisted of 5 trials. Therefore, a total of 16 blocks are shown on the X-axis. The RUNS test showed that the red and green blocks are randomly distributed for each subject ($p < 0.05$).; Figure S3: The boxplot displays the extremes, the upper and lower quartiles, the median (line), and the mean (x) of the Numerical Rating Scale (NRS) values provided by the subjects during each condition (a). The ratings were separated into painful (NRS > 5, red) and non-painful (NRS < 5, green). Paired samples t-tests controlled that the painful and non-painful ratings differed significantly in each condition ($p < 0.001$). The table below (b) shows the mean and the standard error (SE) for each condition and perceived intensity.; Figure S4: The figure shows the across-subjects mean event-related spectral power (ERSP) for the pain and no-pain blocks and each condition (*UM−*, *BM−*, and *UM+*) at each frequency (from 0 to 40 Hz). The entire epochs (i.e., from 7 s before the auditory cue to 1 s after the cue) were averaged for each subject and condition. Only the electrodes overlying the left and right sensory-motor cortex (i.e., C3 and C4) were reported. The left panels show the baseline mean power spectrum, whereas the lower panels indicate the low and high mean values (expressed in dB) at each time in the epoch. In the figure, the dark blue colour represents a power reduction in the given frequency, whereas the dark red colour represents a power increase in the frequency.; Table S1: The table shows the mean and standard error values (expressed in mV) for each condition (UM−, BM−, and UM+) for both the painful and non-painful blocks. All the event-related potentials (ERPs: N1, P2, N2, and P3) are reported. The significant differences are shown in Figure S1. Refs. [97–101] are cited in Supplementary Materials.

Author Contributions: Conceptualization, M.R. and C.B.; methodology, M.R. and C.B.; software, M.R. and C.D.P.; formal analysis, M.R. and C.D.P.; investigation, L.P.; resources, L.A.-N.; data curation, M.R.; writing—original draft preparation, M.R.; writing—review and editing, L.P., L.A.-N. and C.B.; supervision, L.P., L.A.-N. and C.B.; project administration, L.A.-N.; funding acquisition, L.A.-N. All authors have read and agreed to the published version of the manuscript.

Funding: This research was funded by Danmarks Grundforskningsfond, Grant/Award Number: DNRD121, and H2020 Marie Skłodowska-Curie Actions, Grant/Award Number: 754465.

Institutional Review Board Statement: The study was conducted in accordance with the Declaration of Helsinki and approved by the Scientific Ethical Committee of Region Nordjylland (protocol code N-20190008 on 11 November 2019).

Informed Consent Statement: Written informed consent was obtained from all subjects involved in this study.

Data Availability Statement: The EEG datasets have been reposited on Figshare at the following site: https://doi.org/10.6084/m9.figshare.24851748.v1 (accessed on 18 December 2023).

Acknowledgments: The authors thank Susanna Lopez for her assistance.

Conflicts of Interest: The authors declare no conflicts of interest.

References

1. Ortiz-Catalan, M.; Guðmundsdóttir, R.A.; Kristoffersen, M.B.; Zepeda-Echavarria, A.; Caine-Winterberger, K.; Kulbacka-Ortiz, K.; Widehammar, C.; Eriksson, K.; Stockselius, A.; Ragnö, C.; et al. Phantom Motor Execution Facilitated by Machine Learning and Augmented Reality as Treatment for Phantom Limb Pain: A Single Group, Clinical Trial in Patients with Chronic Intractable Phantom Limb Pain. *Lancet* **2016**, *388*, 2885–2894. [CrossRef] [PubMed]
2. Ding, L.; Wang, X.; Guo, X.; Chen, S.; Wang, H.; Cui, X.; Rong, J.; Jia, J. Effects of Camera-Based Mirror Visual Feedback Therapy for Patients Who Had a Stroke and the Neural Mechanisms Involved: Protocol of a Multicentre Randomised Control Study. *BMJ Open* **2019**, *9*, e022828. [CrossRef] [PubMed]
3. Ramachandran, V.S.; Rogers-Ramachandran, D. Synaesthesia in Phantom Limbs Induced with Mirrors. *Proc. R. Soc. Lond. B Biol. Sci.* **1997**, *263*, 377–386. [CrossRef]
4. Herrador Colmenero, L.; Perez Marmol, J.M.; Martí-García, C.; Querol Zaldivar, M.D.; Tapia Haro, R.M.; Castro Sánchez, A.M.; Aguilar-Ferrándiz, M.E. Effectiveness of Mirror Therapy, Motor Imagery, and Virtual Feedback on Phantom Limb Pain Following Amputation: A Systematic Review. *Prosthet. Orthot. Int.* **2018**, *42*, 288–298. [CrossRef] [PubMed]
5. Arya, K.N. Underlying Neural Mechanisms of Mirror Therapy: Implications for Motor Rehabilitation in Stroke. *Neurol. India* **2016**, *64*, 38. [CrossRef] [PubMed]
6. Deconinck, F.J.A.; Smorenburg, A.R.P.; Benham, A.; Ledebt, A.; Feltham, M.G.; Savelsbergh, G.J.P. Reflections on Mirror Therapy: A Systematic Review of the Effect of Mirror Visual Feedback on the Brain. *Neurorehabil. Neural Repair* **2015**, *29*, 349–361. [CrossRef] [PubMed]
7. Finn, S.B.; Perry, B.N.; Clasing, J.E.; Walters, L.S.; Jarzombek, S.L.; Curran, S.; Rouhanian, M.; Keszler, M.S.; Hussey-Andersen, L.K.; Weeks, S.R.; et al. A Randomized, Controlled Trial of Mirror Therapy for Upper Extremity Phantom Limb Pain in Male Amputees. *Front. Neurol.* **2017**, *8*, 267. [CrossRef] [PubMed]
8. Thøgersen, M.; Andoh, J.; Milde, C.; Graven-Nielsen, T.; Flor, H.; Petrini, L. Individualized Augmented Reality Training Reduces Phantom Pain and Cortical Reorganization in Amputees: A Proof of Concept Study. *J. Pain* **2020**, *21*, 1257–1269. [CrossRef] [PubMed]
9. Bae, S.H.; Jeong, W.S.; Kim, K.Y. Effects of Mirror Therapy on Subacute Stroke Patients' Brain Waves and Upper Extremity Functions. *J. Phys. Ther. Sci.* **2012**, *24*, 1119–1122. [CrossRef]
10. Bartur, G.; Pratt, H.; Frenkel-Toledo, S.; Soroker, N. Neurophysiological Effects of Mirror Visual Feedback in Stroke Patients with Unilateral Hemispheric Damage. *Brain Res.* **2018**, *1700*, 170–180. [CrossRef]
11. Dohle, C.; Püllen, J.; Nakaten, A.; Küst, J.; Rietz, C.; Karbe, H. Mirror Therapy Promotes Recovery from Severe Hemiparesis: A Randomized Controlled Trial. *Neurorehabil. Neural Repair* **2009**, *23*, 209–217. [CrossRef] [PubMed]
12. McCabe, C.S.; Haigh, R.C.; Ring, E.F.J.; Halligan, P.W.; Wall, P.D.; Blake, D.R. A Controlled Pilot Study of the Utility of Mirror Visual Feedback in the Treatment of Complex Regional Pain Syndrome (Type 1). *Rheumatology* **2003**, *42*, 97–101. [CrossRef] [PubMed]
13. Vladimir Tichelaar, Y.I.G.; Geertzen, J.H.B.; Keizer, D.; Paul van Wilgen, C. Mirror Box Therapy Added to Cognitive Behavioural Therapy in Three Chronic Complex Regional Pain Syndrome Type I Patients: A Pilot Study. *Int. J. Rehabil. Res.* **2007**, *30*, 181. [CrossRef] [PubMed]
14. Bullock, K.; Won, A.S.; Bailenson, J.; Friedman, R. Virtual Reality-Delivered Mirror Visual Feedback and Exposure Therapy for FND: A Midpoint Report of a Randomized Controlled Feasibility Study. *J. Neuropsychiatry Clin. Neurosci.* **2020**, *32*, 90–94. [CrossRef] [PubMed]
15. Zhang, J.J.; Fong, K.N.K. Enhancing Mirror Visual Feedback with Intermittent Theta Burst Stimulation in Healthy Adults. *Restor. Neurol. Neurosci.* **2019**, *37*, 483–495. [CrossRef] [PubMed]
16. Merians, A.S.; Tunik, E.; Fluet, G.G.; Qiu, Q.; Adamovich, S.V. Innovative Approaches to the Rehabilitation of Upper Extremity Hemiparesis Using Virtual Environments. *Eur. J. Phys. Rehabil. Med.* **2009**, *45*, 123–133. [PubMed]
17. Shinoura, N.; Suzuki, Y.; Watanabe, Y.; Yamada, R.; Tabei, Y.; Saito, K.; Yagi, K. Mirror Therapy Activates Outside of Cerebellum and Ipsilateral M1. *NeuroRehabilitation* **2008**, *23*, 245–252. [CrossRef] [PubMed]
18. Numata, K.; Murayama, T.; Takasugi, J.; Monma, M.; Oga, M. Mirror Observation of Finger Action Enhances Activity in Anterior Intraparietal Sulcus: A Functional Magnetic Resonance Imaging Study. *J. Jpn. Phys. Ther. Assoc.* **2013**, *16*, 1–6. [CrossRef] [PubMed]
19. Imai, I.; Takeda, K.; Shiomi, T.; Taniguchi, T.; Kato, H. Sensorimotor Cortex Activation during Mirror Therapy in Healthy Right-Handed Subjects: A Study with Near-Infrared Spectroscopy. *J. Phys. Ther. Sci.* **2008**, *20*, 141–145. [CrossRef]
20. Inagaki, Y.; Seki, K.; Makino, H.; Matsuo, Y.; Miyamoto, T.; Ikoma, K. Exploring Hemodynamic Responses Using Mirror Visual Feedback with Electromyogram-Triggered Stimulation and Functional Near-Infrared Spectroscopy. *Front. Hum. Neurosci.* **2019**, *13*, 60. [CrossRef]
21. Fukumura, K.; Sugawara, K.; Tanabe, S.; Ushiba, J.; Tomita, Y. Influence of Mirror Therapy on Human Motor Cortex. *Int. J. Neurosci.* **2007**, *117*, 1039–1048. [CrossRef] [PubMed]
22. Aziz-Zadeh, L.; Maeda, F.; Zaidel, E.; Mazziotta, J.; Iacoboni, M. Lateralization in Motor Facilitation during Action Observation: A TMS Study. *Exp. Brain Res.* **2002**, *144*, 127–131. [CrossRef] [PubMed]
23. Garry, M.I.; Loftus, A.; Summers, J.J. Mirror, Mirror on the Wall: Viewing a Mirror Reflection of Unilateral Hand Movements Facilitates Ipsilateral M1 Excitability. *Exp. Brain Res.* **2005**, *163*, 118–122. [CrossRef] [PubMed]

24. Fritzsch, C.; Wang, J.; dos Santos, L.F.; Mauritz, K.-H.; Brunetti, M.; Dohle, C. Different Effects of the Mirror Illusion on Motor and Somatosensory Processing. *Restor. Neurol. Neurosci.* **2014**, *32*, 269–280. [CrossRef] [PubMed]
25. Hamzei, F.; Läppchen, C.H.; Glauche, V.; Mader, I.; Rijntjes, M.; Weiller, C. Functional Plasticity Induced by Mirror Training: The Mirror as the Element Connecting Both Hands to One Hemisphere. *Neurorehabil. Neural Repair* **2012**, *26*, 484–496. [CrossRef] [PubMed]
26. Matthys, K.; Smits, M.; Van der Geest, J.N.; Van der Lugt, A.; Seurinck, R.; Stam, H.J.; Selles, R.W. Mirror-Induced Visual Illusion of Hand Movements: A Functional Magnetic Resonance Imaging Study. *Arch. Phys. Med. Rehabil.* **2009**, *90*, 675–681. [CrossRef] [PubMed]
27. Seidel, S.; Kasprian, G.; Furtner, J.; Schöpf, V.; Essmeister, M.; Sycha, T.; Auff, E.; Prayer, D. Mirror Therapy in Lower Limb Amputees—A Look Beyond Primary Motor Cortex Reorganization. *RöFo Fortschritte Auf Dem Geb. Röntgenstrahlen Bildgeb. Verfahr.* **2011**, *183*, 1051–1057. [CrossRef]
28. Kim, J.H.; Lee, B.-H. Mirror Therapy Combined With Biofeedback Functional Electrical Stimulation for Motor Recovery of Upper Extremities after Stroke: A Pilot Randomized Controlled Trial. *Occup. Ther. Int.* **2015**, *22*, 51–60. [CrossRef]
29. Light, G.A.; Williams, L.E.; Minow, F.; Sprock, J.; Rissling, A.; Sharp, R.; Swerdlow, N.R.; Braff, D.L. Electroencephalography (EEG) and Event-Related Potentials (ERPs) with Human Participants. *Curr. Protoc. Neurosci.* **2010**, *52*, 6.25.1–6.25.24. [CrossRef]
30. Debnath, R.; Franz, E.A. Perception of Hand Movement by Mirror Reflection Evokes Brain Activation in the Motor Cortex Contralateral to a Non-Moving Hand. *Cortex J. Devoted Study Nerv. Syst. Behav.* **2016**, *81*, 118–125. [CrossRef]
31. Touzalin-Chretien, P.; Dufour, A. Motor Cortex Activation Induced by a Mirror: Evidence from Lateralized Readiness Potentials. *J. Neurophysiol.* **2008**, *100*, 19–23. [CrossRef]
32. Pfurtscheller, G.; Lopes da Silva, F.H. Event-Related EEG/MEG Synchronization and Desynchronization: Basic Principles. *Clin. Neurophysiol. Off. J. Int. Fed. Clin. Neurophysiol.* **1999**, *110*, 1842–1857. [CrossRef]
33. Babiloni, C.; Del Percio, C.; Arendt-Nielsen, L.; Soricelli, A.; Romani, G.L.; Rossini, P.M.; Capotosto, P. Cortical EEG Alpha Rhythms Reflect Task-Specific Somatosensory and Motor Interactions in Humans. *Clin. Neurophysiol.* **2014**, *125*, 1936–1945. [CrossRef]
34. Neuper, C.; Wörtz, M.; Pfurtscheller, G. ERD/ERS Patterns Reflecting Sensorimotor Activation and Deactivation. *Prog. Brain Res.* **2006**, *159*, 211–222. [CrossRef] [PubMed]
35. Babiloni, C.; Capotosto, P.; Brancucci, A.; Del Percio, C.; Petrini, L.; Buttiglione, M.; Cibelli, G.; Romani, G.L.; Rossini, P.M.; Arendt-Nielsen, L. Cortical Alpha Rhythms Are Related to the Anticipation of Sensorimotor Interaction between Painful Stimuli and Movements: A High-Resolution EEG Study. *J. Pain* **2008**, *9*, 902–911. [CrossRef]
36. Babiloni, C.; Capotosto, P.; Del Percio, C.; Babiloni, F.; Petrini, L.; Buttiglione, M.; Cibelli, G.; Marusiak, J.; Romani, G.L.; Arendt-Nielsen, L.; et al. Sensorimotor Interaction between Somatosensory Painful Stimuli and Motor Sequences Affects Both Anticipatory Alpha Rhythms and Behavior as a Function of the Event Side. *Brain Res. Bull.* **2010**, *81*, 398–405. [CrossRef] [PubMed]
37. Bartur, G.; Pratt, H.; Dickstein, R.; Frenkel-Toledo, S.; Geva, A.; Soroker, N. Electrophysiological Manifestations of Mirror Visual Feedback during Manual Movement. *Brain Res.* **2015**, *1606*, 113–124. [CrossRef] [PubMed]
38. Lee, H.-M.; Li, P.-C.; Fan, S.-C. Delayed Mirror Visual Feedback Presented Using a Novel Mirror Therapy System Enhances Cortical Activation in Healthy Adults. *J. NeuroEngineering Rehabil.* **2015**, *12*, 56. [CrossRef]
39. Rizzo, M.; Petrini, L.; Del Percio, C.; Lopez, S.; Arendt-Nielsen, L.; Babiloni, C. Mirror Visual Feedback during Unilateral Finger Movements Is Related to the Desynchronization of Cortical Electroencephalographic Somatomotor Alpha Rhythms. *Psychophysiology* **2022**, *59*, e14116. [CrossRef]
40. Foell, J.; Bekrater-Bodmann, R.; McCabe, C.S.; Flor, H. Sensorimotor Incongruence and Body Perception: An Experimental Investigation. *Front. Hum. Neurosci.* **2013**, *7*, 310. [CrossRef]
41. McCabe, C.S.; Haigh, R.C.; Halligan, P.W.; Blake, D.R. Simulating Sensory–Motor Incongruence in Healthy Volunteers: Implications for a Cortical Model of Pain. *Rheumatology* **2005**, *44*, 509–516. [CrossRef] [PubMed]
42. Katayama, O.; Osumi, M.; Kodama, T.; Morioka, S. Dysesthesia Symptoms Produced by Sensorimotor Incongruence in Healthy Volunteers: An Electroencephalogram Study. *J. Pain Res.* **2016**, *9*, 1197–1204. [CrossRef]
43. McCabe, C.S.; Cohen, H.; Blake, D.R. Somaesthetic Disturbances in Fibromyalgia Are Exaggerated by Sensory–Motor Conflict: Implications for Chronicity of the Disease? *Rheumatology* **2007**, *46*, 1587–1592. [CrossRef] [PubMed]
44. Katayama, O.; Nishi, Y.; Osumi, M.; Takamura, Y.; Kodama, T.; Morioka, S. Neural Activities behind the Influence of Sensorimotor Incongruence on Dysesthesia and Motor Control. *Neurosci. Lett.* **2019**, *698*, 19–26. [CrossRef] [PubMed]
45. Nishigami, T.; Nakano, H.; Osumi, M.; Tsujishita, M.; Mibu, A.; Ushida, T. Central Neural Mechanisms of Interindividual Difference in Discomfort during Sensorimotor Incongruence in Healthy Volunteers: An Experimental Study. *Rheumatology* **2014**, *53*, 1194–1199. [CrossRef] [PubMed]
46. Daenen, L.; Nijs, J.; Roussel, N.; Wouters, K.; Van Loo, M.; Cras, P. Sensorimotor Incongruence Exacerbates Symptoms in Patients with Chronic Whiplash Associated Disorders: An Experimental Study. *Rheumatology* **2012**, *51*, 1492–1499. [CrossRef] [PubMed]
47. Araujo, H.F.; Kaplan, J.; Damasio, A. Cortical Midline Structures and Autobiographical-Self Processes: An Activation-Likelihood Estimation Meta-Analysis. *Front. Hum. Neurosci.* **2013**, *7*, 548. [CrossRef] [PubMed]
48. Cavanna, A.E.; Trimble, M.R. The Precuneus: A Review of Its Functional Anatomy and Behavioural Correlates. *Brain* **2006**, *129*, 564–583. [CrossRef] [PubMed]

49. Goffaux, P.; Girard-Tremblay, L.; Marchand, S.; Daigle, K.; Whittingstall, K. Individual Differences in Pain Sensitivity Vary as a Function of Precuneus Reactivity. *Brain Topogr.* **2014**, *27*, 366–374. [CrossRef]
50. Northoff, G.; Bermpohl, F. Cortical Midline Structures and the Self. *Trends Cogn. Sci.* **2004**, *8*, 102–107. [CrossRef]
51. Northoff, G.; Heinzel, A.; de Greck, M.; Bermpohl, F.; Dobrowolny, H.; Panksepp, J. Self-Referential Processing in Our Brain—A Meta-Analysis of Imaging Studies on the Self. *NeuroImage* **2006**, *31*, 440–457. [CrossRef] [PubMed]
52. Oldfield, R.C. The Assessment and Analysis of Handedness: The Edinburgh Inventory. *Neuropsychologia* **1971**, *9*, 97–113. [CrossRef] [PubMed]
53. Kida, T.; Wasaka, T.; Inui, K.; Akatsuka, K.; Nakata, H.; Kakigi, R. Centrifugal Regulation of Human Cortical Responses to a Task-Relevant Somatosensory Signal Triggering Voluntary Movement. *NeuroImage* **2006**, *32*, 1355–1364. [CrossRef] [PubMed]
54. Williams, L.M.; Simms, E.; Clark, C.R.; Paul, R.H.; Rowe, D.; Gordon, E. The Test-Retest Reliability of a Standardized Neurocognitive and Neurophysiological Test Battery: "Neuromarker". *Int. J. Neurosci.* **2005**, *115*, 1605–1630. [CrossRef] [PubMed]
55. Justen, C.; Herbert, C. Snap Your Fingers! An ERP/sLORETA Study Investigating Implicit Processing of Self- vs. Other-Related Movement Sounds Using the Passive Oddball Paradigm. *Front. Hum. Neurosci.* **2016**, *10*, 465. [CrossRef] [PubMed]
56. Hu, L.; Cai, M.M.; Xiao, P.; Luo, F.; Iannetti, G.D. Human Brain Responses to Concomitant Stimulation of Aδ and C Nociceptors. *J. Neurosci.* **2014**, *34*, 11439–11451. [CrossRef] [PubMed]
57. Delorme, A.; Makeig, S. EEGLAB: An Open Source Toolbox for Analysis of Single-Trial EEG Dynamics Including Independent Component Analysis. *J. Neurosci. Methods* **2004**, *134*, 9–21. [CrossRef] [PubMed]
58. Zhang, L.; Wang, P.; Zhang, R.; Chen, M.; Shi, L.; Gao, J.; Hu, Y. The Influence of Different EEG References on Scalp EEG Functional Network Analysis During Hand Movement Tasks. *Front. Hum. Neurosci.* **2020**, *14*, 367. [CrossRef] [PubMed]
59. Bell, A.; Sejnowski, T. An Informatio-Maximization Approach to Blind Separation and Blind Deconvolution. *Neural Comput.* **1995**, *7*, 1129–1159. [CrossRef]
60. Klimesch, W. EEG Alpha and Theta Oscillations Reflect Cognitive and Memory Performance: A rKlimesch, W. EEG Alpha and Theta Oscillations Reflect Cognitive and Memory Performance: A Review and Analysis. *Brain Res. Rev.* **1999**, *29*, 169–195. [CrossRef]
61. Corcoran, A.W.; Alday, P.M.; Schlesewsky, M.; Bornkessel-Schlesewsky, I. Toward a Reliable, Automated Method of Individual Alpha Frequency (IAF) Quantification. *Psychophysiology* **2018**, *55*, e13064. [CrossRef] [PubMed]
62. Pascual-Marqui, R.D. Discrete, 3D Distributed Linear Imaging Methods of Electric Neuronal Activity. Part 1: Exact, Zero Error Localization. *arXiv* **2007**, arXiv:0710.3341.
63. Pascual-Marqui, R.D.; Esslen, M.; Kochi, K.; Lehmann, D. Functional Imaging with Low-Resolution Brain Electromagnetic Tomography (LORETA): A Review. *Methods Find. Exp. Clin. Pharmacol.* **2002**, *24*, 91–95. [PubMed]
64. Pascual-Marqui, R.D.; Lehmann, D.; Koenig, T.; Kochi, K.; Merlo, M.C.G.; Hell, D.; Koukkou, M. Low Resolution Brain Electromagnetic Tomography (LORETA) Functional Imaging in Acute, Neuroleptic-Naive, First-Episode, Productive Schizophrenia. *Psychiatry Res. Neuroimaging* **1999**, *90*, 169–179. [CrossRef] [PubMed]
65. Peng, W.; Hu, Y.; Mao, Y.; Babiloni, C. Widespread Cortical α-ERD Accompanying Visual Oddball Target Stimuli Is Frequency but Non-Modality Specific. *Behav. Brain Res.* **2015**, *295*, 71–77. [CrossRef] [PubMed]
66. Pfurtscheller, G.; Neuper, C.; Flotzinger, D.; Pregenzer, M. EEG-Based Discrimination between Imagination of Right and Left Hand Movement. *Electroencephalogr. Clin. Neurophysiol.* **1997**, *103*, 642–651. [CrossRef] [PubMed]
67. Pfurtscheller, G.; Aranibar, A. Evaluation of Event-Related Desynchronization (ERD) Preceding and Following Voluntary Self-Paced Movement. *Electroencephalogr. Clin. Neurophysiol.* **1979**, *46*, 138–146. [CrossRef]
68. Pfurtscheller, G.; Neuper, C. Event-Related Synchronization of Mu Rhythm in the EEG over the Cortical Hand Area in Man. *Neurosci. Lett.* **1994**, *174*, 93–96. [CrossRef] [PubMed]
69. Lightfoot, G. Summary of the N1-P2 Cortical Auditory Evoked Potential to Estimate the Auditory Threshold in Adults. *Semin. Hear.* **2016**, *37*, 1–8. [CrossRef]
70. Mouraux, A.; Iannetti, G.D.; Plaghki, L. Low Intensity Intra-Epidermal Electrical Stimulation Can Activate Aδ-Nociceptors Selectively. *Pain* **2010**, *150*, 199–207. [CrossRef]
71. Fabrizi, L.; Verriotis, M.; Williams, G.; Lee, A.; Meek, J.; Olhede, S.; Fitzgerald, M. Encoding of Mechanical Nociception Differs in the Adult and Infant Brain. *Sci. Rep.* **2016**, *6*, 28642. [CrossRef]
72. Nichols, T.E.; Holmes, A.P. Nonparametric Permutation Tests for Functional Neuroimaging: A Primer with Examples. *Hum. Brain Mapp.* **2001**, *25*, 1–25. [CrossRef]
73. Bian, Y.; Qi, H.; Zhao, L.; Ming, D.; Guo, T.; Fu, X. Dynamic Visual Guidance with Complex Task Improves Intracortical Source Activities during Motor Imagery. *NeuroReport* **2019**, *30*, 645–652. [CrossRef]
74. Wilcoxon, F. Individual Comparisons by Ranking Methods. In *Breakthroughs in Statistics: Methodology and Distribution*; Kotz, S., Johnson, N.L., Eds.; Springer: New York, NY, USA, 1992; pp. 196–202. ISBN 978-1-4612-4380-9.
75. Tomczak, M.; Tomczak, E. The Need to Report Effect Size Estimates Revisited. *An Overview of Some Recommended Measures of Effect Size.* **2014**, *1*, 19–25.
76. Cohen, J. *Statistical Power Analysis for the Behavioral Sciences*, 2nd ed.; Routledge: New York, NY, USA, 1988; ISBN 978-0-203-77158-7.
77. Peng, K.; Steele, S.C.; Becerra, L.; Borsook, D. Brodmann Area 10: Collating, Integrating and High Level Processing of Nociception and Pain. *Prog. Neurobiol.* **2018**, *161*, 1–22. [CrossRef]

78. Tolomeo, S.; Christmas, D.; Jentzsch, I.; Johnston, B.; Sprengelmeyer, R.; Matthews, K.; Douglas Steele, J. A Causal Role for the Anterior Mid-Cingulate Cortex in Negative Affect and Cognitive Control. *Brain* **2016**, *139*, 1844–1854. [CrossRef] [PubMed]
79. De Ridder, D.; Vanneste, S. Burst and Tonic Spinal Cord Stimulation: Different and Common Brain Mechanisms. *Neuromodulation J. Int. Neuromodulation Soc.* **2016**, *19*, 47–59. [CrossRef] [PubMed]
80. Kulkarni, B.; Bentley, D.E.; Elliott, R.; Youell, P.; Watson, A.; Derbyshire, S.W.G.; Frackowiak, R.S.J.; Friston, K.J.; Jones, A.K.P. Attention to Pain Localization and Unpleasantness Discriminates the Functions of the Medial and Lateral Pain Systems. *Eur. J. Neurosci.* **2005**, *21*, 3133–3142. [CrossRef] [PubMed]
81. Price, D.D. Psychological and Neural Mechanisms of the Affective Dimension of Pain. *Science* **2000**, *288*, 1769–1772. [CrossRef]
82. Fauchon, C.; Faillenot, I.; Quesada, C.; Meunier, D.; Chouchou, F.; Garcia-Larrea, L.; Peyron, R. Brain Activity Sustaining the Modulation of Pain by Empathetic Comments. *Sci. Rep.* **2019**, *9*, 8398. [CrossRef]
83. Broyd, S.J.; Demanuele, C.; Debener, S.; Helps, S.K.; James, C.J.; Sonuga-Barke, E.J.S. Default-Mode Brain Dysfunction in Mental Disorders: A Systematic Review. *Neurosci. Biobehav. Rev.* **2009**, *33*, 279–296. [CrossRef] [PubMed]
84. Gusnard, D.A.; Raichle, M.E. Searching for a Baseline: Functional Imaging and the Resting Human Brain. *Nat. Rev. Neurosci.* **2001**, *2*, 685–694. [CrossRef] [PubMed]
85. Raichle, M.E.; MacLeod, A.M.; Snyder, A.Z.; Powers, W.J.; Gusnard, D.A.; Shulman, G.L. A Default Mode of Brain Function. *Proc. Natl. Acad. Sci. USA* **2001**, *98*, 676–682. [CrossRef] [PubMed]
86. Rusconi, E.; Pinel, P.; Dehaene, S.; Kleinschmidt, A. The Enigma of Gerstmann's Syndrome Revisited: A Telling Tale of the Vicissitudes of Neuropsychology. *Brain* **2010**, *133*, 320–332. [CrossRef]
87. Rusconi, E.; Walsh, V.; Butterworth, B. Dexterity with Numbers: rTMS over Left Angular Gyrus Disrupts Finger Gnosis and Number Processing. *Neuropsychologia* **2005**, *43*, 1609–1624. [CrossRef] [PubMed]
88. Rizzo, M.; Del Percio, C.; Petrini, L.; Lopez, S.; Arendt-Nielsen, L.; Babiloni, C. Cortical Sources of Electroencephalographic Alpha Rhythms Related to the Anticipation and Experience of Mirror Visual Feedback-Induced Illusion of Finger Movements. *Psychophysiology* **2023**, *60*, e14281. [CrossRef]
89. Carino-Escobar, R.; Carrillo-Mora, P.; Valdes, R.; Rodríguez Barragán, M.; Hernandez-Arenas, C.; Quinzaños-Fresnedo, J.; Galicia Alvarado, M.; Cantillo-Negrete, J. Longitudinal Analysis of Stroke Patients' Brain Rhythms during an Intervention with a Brain-Computer Interface. *Neural Plast.* **2019**, *2019*, 7084618. [CrossRef]
90. Gandolfi, M.; Formaggio, E.; Geroin, C.; Storti, S.F.; Boscolo Galazzo, I.; Waldner, A.; Manganotti, P.; Smania, N. Electroencephalographic Changes of Brain Oscillatory Activity after Upper Limb Somatic Sensation Training in a Patient with Somatosensory Deficit after Stroke. *Clin. EEG Neurosci.* **2015**, *46*, 347–352. [CrossRef] [PubMed]
91. Jochumsen, M.; Rovsing, C.; Rovsing, H.; Cremoux, S.; Signal, N.; Allen, K.; Taylor, D.; Niazi, I.K. Quantification of Movement-Related EEG Correlates Associated with Motor Training: A Study on Movement-Related Cortical Potentials and Sensorimotor Rhythms. *Front. Hum. Neurosci.* **2017**, *11*, 604. [CrossRef]
92. Carson, R.G. Neural Pathways Mediating Bilateral Interactions between the Upper Limbs. *Brain Res. Brain Res. Rev.* **2005**, *49*, 641–662. [CrossRef]
93. Yavuzer, G.; Selles, R.; Sezer, N.; Sütbeyaz, S.; Bussmann, J.B.; Köseoğlu, F.; Atay, M.B.; Stam, H.J. Mirror Therapy Improves Hand Function in Subacute Stroke: A Randomized Controlled Trial. *Arch. Phys. Med. Rehabil.* **2008**, *89*, 393–398. [CrossRef] [PubMed]
94. Martini, M. Real, Rubber or Virtual: The Vision of "One's Own" Body as a Means for Pain Modulation. A Narrative Review. *Conscious. Cogn.* **2016**, *43*, 143–151. [CrossRef] [PubMed]
95. Matamala-Gomez, M.; Donegan, T.; Bottiroli, S.; Sandrini, G.; Sanchez-Vives, M.V.; Tassorelli, C. Immersive Virtual Reality and Virtual Embodiment for Pain Relief. *Front. Hum. Neurosci.* **2019**, *13*, 279. [CrossRef] [PubMed]
96. Yıldırım, E.; Güntekin, B.; Hanoğlu, L.; Algun, C. EEG Alpha Activity Increased in Response to Transcutaneous Electrical Nervous Stimulation in Young Healthy Subjects but Not in the Healthy Elderly. *PeerJ* **2020**, *8*, e8330. [CrossRef] [PubMed]
97. Heinrichs-Graham, E.; Kurz, M.J.; Gehringer, J.E.; Wilson, T.W. The Functional Role of Post-Movement Beta Oscillations in Motor Termination. *Brain Struct. Funct.* **2017**, *222*, 3075–3086. [CrossRef] [PubMed]
98. Körmendi, J.; Ferentzi, E.; Weiss, B.; Nagy, Z. Topography of Movement-Related Delta and Theta Brain Oscillations. *Brain Topogr.* **2021**, *34*, 608–617. [CrossRef] [PubMed]
99. Pellegrino, G.; Tomasevic, L.; Herz, D.M.; Larsen, K.M.; Siebner, H.R. Theta Activity in the Left Dorsal Premotor Cortex during Action Re-Evaluation and Motor Reprogramming. *Front. Hum. Neurosci.* **2018**, *12*, 364. [CrossRef] [PubMed]
100. Saleh, M.; Reimer, J.; Penn, R.; Ojakangas, C.L.; Hatsopoulos, N.G. Fast and Slow Oscillations in Human Primary Motor Cortex Predict Oncoming Behaviorally Relevant Cues. *Neuron* **2010**, *65*, 461–471. [CrossRef]
101. Schramm, S.; Albers, L.; Ille, S.; Schröder, A.; Meyer, B.; Sollmann, N.; Krieg, S.M. Navigated Transcranial Magnetic Stimulation of the Supplementary Motor Cortex Disrupts Fine Motor Skills in Healthy Adults. *Sci. Rep.* **2019**, *9*, 17744. [CrossRef]

brain sciences

MDPI

Article

Predicted Brain Age in First-Episode Psychosis: Association with Inexpressivity

Dean F. Salisbury [1,*], Brian M. Wulf [1], Dylan Seebold [1], Brian A. Coffman [1], Mark T. Curtis [1,†] and Helmet T. Karim [1,2]

1 Department of Psychiatry, University of Pittsburgh, Pittsburgh, PA 15213, USA
2 Department of Bioengineering, University of Pittsburgh, Pittsburgh, PA 15260, USA
* Correspondence: salisburyd@upmc.edu
† Current address: Department of Psychological and Brain Sciences, Washington University in St. Louis, St. Louis, MO 63130, USA.

Abstract: Accelerated brain aging is a possible mechanism of pathology in schizophrenia. Advances in MRI-based brain development algorithms allow for the calculation of predicted brain age (PBA) for individuals. Here, we assessed PBA in 70 first-episode schizophrenia-spectrum individuals (FESz) and 76 matched healthy neurotypical comparison individuals (HC) to determine if FESz showed advanced aging proximal to psychosis onset and whether PBA was associated with neurocognitive, social functioning, or symptom severity measures. PBA was calculated with BrainAgeR (v2.1) from T1-weighted MR scans. There were no differences in the PBAs between groups. After controlling for actual age, a "younger" PBA was associated with higher vocabulary scores among all individuals, while an "older" PBA was associated with more severe negative symptom "Inexpressivity" component scores among FESz. Female participants in both groups had an elevated PBA relative to male participants. These results suggest that a relatively younger brain age is associated with a better semantic memory performance. There is no evidence for accelerated aging in FESz with a late adolescent/early adult onset. Despite a normative PBA, FESz with a greater residual PBA showed impairments in a cluster of negative symptoms, which may indicate some underlying age-related pathology proximal to psychosis onset. Although a period of accelerated aging cannot be ruled out with disease course, it does not occur at the time of the first episode.

Keywords: brain age; first-episode psychosis; cognitive function; schizophrenia

1. Introduction

Since the time of Kraepelin [1], psychotic disorders have been suggested to involve a progressive degenerative course, as reflected in his new term dementia praecox, or progressive dementia. Later, Bleuler [2] objected to the idea of an invariable dementing course, citing a number of cases that showed full or partial recovery, and renamed the syndrome schizophrenia (split mind). (See Kendler, Tabb, and Wright [3] for a history of the mind versus biology debate in the development of psychiatry as a medical specialty.) For most of the 20th century, there was little post-mortem evidence of neurodegeneration or brain volume reductions in schizophrenia, with the exception of well-documented increases in ventricular size, e.g., [4].

The dominant model became a static perinatal lesion (whether from the genetic misprogramming of cortical differentiation, insults in the third trimester such as maternal infection or famine, and birth complications, etc.), which laid the foundation for abnormalities in brain function during post-pubescent brain maturation, e.g., [5,6]. Still, several lines of evidence have suggested a possible progressive change in young adulthood in the disease. For example, Benes [7] showed that reduced interneurons and increased dopaminergic fibers interacted during late myelination to produce neurodevelopmental abnormalities in the prefrontal cortices. Building on the initial speculation of Feinberg [8], McGlashan and Hoffman [9] suggested that late neurodevelopmental synaptic pruning might play a role in schizophrenia-related pathology. Waddington, Scully, and Youseff [10] suggested that the early developmental origins of a disorder are not antithetical to active progressive disease during adulthood, pointing to the Alzheimer's-like plaques and neurofibrillary tangles that develop at relatively young ages in Down's syndrome, a genetic disorder. They suggested some brain areas were more susceptible to processes during normal aging in schizophrenia that appear as progressive lesions but reflect accelerated aging.

In the 1990s and early 2000s, increasing evidence suggested that there was a demonstrable decrease in gray matter (GM) around the time of the emergence of psychotic symptoms. These findings were made possible by advances in MR imaging, where 1 mm^3 voxels (volume pixels) could now be recorded. Volumetric work in several laboratories demonstrated the presence of subtle progressive GM loss in individuals following the emergence of psychosis. For example, Gur et al. [11] showed progressive frontal cortex loss, Thompson et al. [12] showed progressive loss in the parietal, temporal, and frontal cortices, Lieberman et al. [13] showed progressive ventricle increases, Cahn et al. [14] showed the progressive loss of whole brain GM and increased ventricle volume, Kasai et al. [15,16] showed progressive left superior temporal gyrus loss following first psychotic break, Ho et al. [17] showed increased sulcal cerebral spinal fluid after first break, and Bachmann et al. [18] showed progressive frontal and temporal GM loss. In addition, several studies indicated that, among individuals at clinical high-risk for developing psychosis (CHR), individuals that transitioned to psychosis showed progressive cortical GM loss. For example, Pantelis et al. [19] showed progressive loss in the temporal and frontal cortices, and Job et al. [20] showed the largest progressive loss of the left temporal lobe in genetic high-risk individuals that transitioned to psychosis. All of these studies indicate progressive GM loss, likely greatest in the temporal and frontal cortices, respectively, around the emergence of psychosis, both before and after.

To date, there is no known mechanism for this progression. Several candidates have been proposed, such as a failure to stop adolescent synaptic pruning, e.g., [9], glutamate-induced excitotoxicity and dendritic regression short of apoptosis, e.g., [21], and oxidative stress, e.g., [22]. Referring back to Kraepelin's idea of a progressive dementing disorder, a plethora of papers have posited that schizophrenia (the most severe psychotic disorder) is a consequence of accelerated aging, e.g., [23,24]. However, while some evidence is consistent with accelerated aging, it remains an unproven hypothesis with some conflicting findings (see review [25]).

A promising methodology for shedding light on this hypothesis of accelerated aging is comparing the "brain ages" derived from neurotypical individuals to those of individuals with psychosis. In a recent meta-analysis, Ballester et al. [26] suggested an increased brain age of around 3 years in schizophrenia sampled across illness durations. For example, Shahab et al. [27] reported a brain age gap (BAG) of ~8 years in a group of younger individuals with schizophrenia with about 11 years of illness duration (mean age ~33 years), but no gap in an older group of individuals with schizophrenia with about 40 years of illness duration (mean age ~64). A second sample with about 15 years of illness duration (mean age ~45) showed an approximately 6-year age gap. Likewise, Shnack et al. [28] suggested that individuals with psychosis were approximately 3 years older in terms of BAG than neurotypicals, with accelerated loss in the first few years that levelled off after approximately 5 years of illness. By contrast, a large meta-analysis [29] of 2803 individuals with schizophrenia versus 2598 neurotypical individuals against an independently developed brain age model indicated a ~3.5 years BAG in schizophrenia, although both groups were older than that predicted by the model. This meta-analysis found no effect of illness duration or age at onset (nor symptom severity), suggesting that this increased brain age occurred prior to the emergence of psychosis. On the other hand, Kaufmann et al. [30] found an increased brain age in schizophrenia (n = 1100) and schizophrenia-spectrum illnesses (n = 300), but not in CHR individuals (n = 94). It is of primary importance whether these BAGs are present at first psychosis, within 1 year from the onset of psychosis. As detailed above, significant GM loss occurs after psychosis onset, e.g., [12,15,16]. Hajek et al. [31] showed a greater brain age in first-episode schizophrenia-spectrum individuals, albeit younger (~2.5 years) than in the larger cross-sectional studies described above. Chung et al. [32] examined CHR individuals to test early versus late neurodevelopmental issues in those that transitioned to psychosis. They found that all CHR individuals (236 non-converters and 39 converters to psychosis) had a greater brain age than neurotypical individuals (n = 109). Among the converters, younger individuals had a significantly greater BAG (12–17), but older individuals (18–21) did not, and only the older individuals showed an increased trajectory of loss. These findings suggest both early and late neurodevelopmental deficits that are reflected in age of onset, with an early neurodevelopmental deficit in younger individuals and no accelerated aging in older individuals prior to psychosis onset. Of some importance, however, the meta-analysis of Ballester et al. [26] found a greater BAG with age, but also a greater brain age in first-episode psychosis. Thus, while the data support progressively accelerated aging after first psychosis, it is possible that accelerated aging exists before the emergence of psychosis.

In the current study, we examine the brain ages of 98 individuals within 6 months of their first clinical contact for a suspected psychotic disorder against 76 neurotypical individuals. If a significantly greater BAG was detected at first psychotic episode, then either neurodevelopmental abnormalities or the initiation of accelerated aging were likely to have occurred prior to the emergence of psychosis. By contrast, if a greater BAG was not detected in this large first-episode sample, then the acceleration of aging probably occurs post-psychosis onset. For consistency with the current literature, our first analysis determined brain age and BAG and compared them between groups. However, BAG derived from subtracting actual age from predicted brain age (PBA) has been criticized for being linearly dependent on actual age [33], such that any group differences in BAG may reflect differences in the distribution of age in the groups. Hence, our main comparisons comprised regression analyses on PBA, which are more statistically robust.

2. Materials and Methods

2.1. Participants and Study Design

We recruited participants as part of ongoing first-episode psychosis (FEP) studies in the Salisbury laboratory. The samples comprised 76 psychiatrically well participants as healthy controls (HC) and 91 individuals with FEP. The HCs were recruited through fliers and online advertisements in the local area. The HCs performed Axis I and Axis II screening and reported no first-degree relatives with a psychiatric disorder. The FEP individuals were recruited from the UPMC Western Psychiatric Hospital inpatient and outpatient services. Among the 91 FEP individuals, 70 received baseline diagnoses within the schizophrenia spectrum. This first-episode schizophrenia-spectrum (FESz) group comprised 47 individuals with diagnoses of schizophrenia (paranoid: n = 27; undifferentiated: n = 15; and residual: n = 5), 6 of schizoaffective disorder, 5 of schizophreniform disorder, and 12 of psychotic disorder Not Otherwise Specified. All FESz individuals had less than one year of lifetime antipsychotic medication exposure and less than one year since either their first clinical contact (emergency room visit or hospitalization) or first psychological or pharmacological treatment for psychosis (Table 1). No participant had neurological comorbidities, a history of concussion or head injury with sequelae, or a history of alcohol or drug addiction in the last 5 years. All participants performed a urine drug screen prior to scanning, although a history of alcohol and marijuana use was not a rule out. All participants underwent magnetic resonance imaging (MRI), cognitive testing, and an assessment of their demographic and clinical measures. All procedures were in accordance with the Declaration of Helsinki, and were approved by the University of Pittsburgh Institutional Review Board. All participants provided written informed consent and were paid for their participation. Data were collected under 3 protocols: Tim Trio MRI—PRO14010185, Cortical Cells, Circuits, Connectivity and Cognition in Psychosis, initial approval 3 March 2014;

Prisma MRI—PRO16030020, Attentional Modulation of Sensory Signals in First Episode Psychosis, initial approval 24 June 2016; merged with STUDY19030077, Pathology and Pathophysiology in Early Psychosis, initial approval 28 June 2017.

Table 1. Demographic, neuropsychological functioning, and social functioning measures.

	FESz (N = 70)	HC (N = 76)	Statistic *	*p*-Value	Cohen's d
Age	22.5 (4.5)	23.6 (5.1)	$t_{144} = 1.37$	0.172	0.227
PBA	25.6 (7.0)	26.5 (6.8)	$t_{144} = 0.84$	0.201	0.139
BAG	3.1 (5.3)	3.0 (5.2)	$t_{144} = 0.16$	0.877	0.026
Sex (F/M)	18/52	29/47	$X^2 = 2.59$	0.108	
pSES	41.0 (13.8)	47.7 (11.6)	$t_{144} = 3.19$	**0.002**	**0.532**
WASI vocabulary	51.1 (9.7)	51.6 (6.5)	$t_{144} = 0.34$	0.738	0.055
WASI matrix reasoning	52.3 (8.7)	58.5 (5.7)	$t_{144} = 4.58$	**<0.001**	**0.759**
MATRICS TMT	40.0 (13.0)	48.4 (9.9)	$t_{142} = 4.41$	**<0.001**	**0.735**
MATRICS processing speed	39.2 (14.0)	52.4 (8.4)	$t_{142} = 6.91$	**<0.001**	**1.153**
MATRICS attention	36.4 (13.5)	47.8 (8.9)	$t_{142} = 6.00$	**<0.001**	**1.00**
MATRICS working memory	39.4 (13.5)	48.6 (8.7)	$t_{142} = 4.94$	**<0.001**	**0.825**
MATRICS verbal learning	43.6 (11.2)	52.7 (9.2)	$t_{142} = 5.37$	**<0.001**	**0.897**
MATRICS visual learning	39.7 (12.7)	46.1 (7.7)	$t_{142} = 3.71$	**<0.001**	**0.619**
MATRICS reasoning	44.2 (11.5)	51.2 (7.2)	$t_{142} = 4.47$	**<0.001**	**0.745**
MATRICS social cognition	41.6 (14.1)	54.5 (8.4)	$t_{142} = 6.72$	**<0.001**	**1.125**
MATRICS composite	34.9 (15.6)	50.6 (7.1)	$t_{142} = 7.90$	**<0.001**	**1.323**
SFS Engagement/Withdrawal	102.5 (11.8)	110.7 (10.4)	$t_{142} = 4.42$	**<0.001**	**0.737**
SFS Interpersonal Interaction	118.2 (18.5)	141.8 (8.0)	$t_{142} = 10.03$	**<0.001**	**1.674**
SFS Recreation	110.0 (13.4)	112.7 (9.7)	$t_{142} = 1.38$	0.170	0.231
SFS Employment	115.5 (7.7)	120.9 (4.3)	$t_{142} = 5.21$	**<0.001**	**0.870**
SFS Independence Performance	100.3 (11.5)	106.7 (9.0)	$t_{142} = 3.74$	**<0.001**	**0.624**
SFS Independence Competence	112.3 (11.8)	120.0 (7.1)	$t_{142} = 4.82$	**<0.001**	**0.803**
SFS Prosocial	108.4 (12.6)	112.1 (9.3)	$t_{142} = 2.01$	**0.046**	**0.335**
GF:R current	5.5 (2.2)	9.0 (0.1)	$t_{144} = 13.50$	**<0.001**	**2.237**
GF:R lowest in last year	5.4 (2.2)	8.9 (0.3)	$t_{144} = 14.19$	**<0.001**	**2.350**
GF:R highest in last year	7.4 (1.5)	9.0 (0.2)	$t_{144} = 9.66$	**<0.001**	**1.600**
GF:S current	5.4 (1.7)	9.0 (0.2)	$t_{144} = 18.10$	**<0.001**	**2.998**
GF:S lowest in last year	5.2 (1.8)	8.9 (0.3)	$t_{144} = 18.22$	**<0.001**	**3.017**
GF:S highest in last year	7.3 (1.2)	9.0 (0.1)	$t_{144} = 12.10$	**<0.001**	**2.004**

Note: Significant group differences are **bolded**. * differences in d.f. indicate missing data. For pSES, 2 individuals were in foster/state care, and no value was entered. PBA = Predicted Brain Age. BAG = Brain Age Gap. pSES = parental SocioEconomic Status. TMT = Trail Making Test.

The groups were matched for age, sex distribution, and premorbid IQ, estimated with the vocabulary subtest of the Wechsler Abbreviated Scale of Intelligence (WASI [34]). Diagnoses were based on the Structured Clinical Interview for DSM-IV interview [35] and chart review, and symptoms were rated using the Brief Psychiatric Rating Scale (BPRS [36]), the Scale for the Assessment of Negative Symptoms (SANS, [37]), and the Scale for the Assessment of Positive Symptoms (SAPS, [38]). Symptom clusters based on the SANS and SAPS items were calculated based on the hierarchical cluster analysis by Longenecker et al. [39]. The participants also performed the Matrix Reasoning test of the WASI and underwent MATRICS Consensus Cognitive Battery testing (MCCB [40]) to assess the current effects of their psychosis on their cognition. Social functioning was assessed with the Social Functioning Scale (SFS) and the Global Functioning: Social and Role scales (GF:S and GF:R [41]). All assessments were conducted by an expert diagnostician independent from the imaging laboratories.

2.2. MRI Acquisition

All participants underwent MR scanning at the MR Research Center of the University of Pittsburgh. Axial T1-weighted images were used to calculate PBA. In total, 61 individuals (36 FESz, 25 HC) underwent scanning on a 3T Siemens Tim Trio using a 32-channel phase array head coil. Sagittal T1-weighted anatomical MR images were obtained with a multi-echo 3D MPRAGE sequence (TR/multi-echo TE/TI = 2530/1.74, 3.6, 5.46, 7.32/1260 ms, flip angle = 7°, field of view (FOV) = 220 × 220 mm, 1 mm isotropic voxel size, 176 slices, GRAPPA acceleration factor = 2). A total of 85 individuals (34 FESZ and 51 HC) underwent scanning on a 3T Siemens MAGNETOM Prisma scanner using a 32-channel phase array head coil. Sagittal T1-weighted anatomical MR images were obtained with a 3D MPRAGE sequence [TR/TE/TI = 2400/2.22/1000 ms, flip angle = 7°, field of view (FOV) = 256 × 240 mm, 0.8 mm isotropic voxel size, 208 slices, GRAPPA acceleration factor = 2]. The participant distributions between scanners differed (chi-squared (df = 1) = 5.15, $p = 0.023$). Consequently, the scanner was controlled in subsequent analyses. The participants underwent approximately 1 h of scanning, but only the T1-weighted scans are presented here.

2.3. Data Availability

Data for individuals scanned on the Prisma are available on the US NIMH NDA database. All raw data are available upon request to the first author.

2.4. Brain Age Estimation

Brain age estimation was conducted using BrainAgeR (v2.1), an open-access algorithm for predicting brain age from raw T1-weighted scans (https://github.com/james-cole/brainageR, access dates: 1 May 2022 through 21 June 2022, [42]). Briefly, this algorithm conducted preprocessing and then prediction from a sample of 3377 healthy individuals (18–92 years old)—there was no additional training performed on the data from this study. In the preprocessing steps, structural scans were segmented and normalized to MNI space using SPM12 (https://www.fil.ion.ucl.ac.uk/spm/software/spm12, access dates: 1 May 2022 through 21 June 2022). Gray matter, white matter, and cerebrospinal fluid were first masked using a threshold of 0.3 from a mean image template generated from 200 scans (a subset of the 3377 individuals that included 20 individuals from 10 scanners). In the prediction steps, the algorithm was applied to the new data. This algorithm used principal components analysis to first retain 80% of the variance in brain volumes and estimated the ages using a Gaussian Process Regression. This resulted in 435 principal components that were then applied to our data to predict the ages. On independent data, this model has performed well.

2.5. Statistical Analysis

Brain age was predicted in R (v4.1.3). We first evaluated the performance of the brain age algorithm by evaluating the MAE and ensuring that this was approximately 5 years in the HCs. Other analyses were conducted in SPSS v28. Demographics, neuropsychological performances, and social functioning were compared using *t*-tests or chi-squared tests where appropriate. Correlations between age, PBA, and BAG are reported for each sample. Kolmogorov–Smirnov Z was used to test for similarity of age distributions between the samples.

Next, regression analyses between the PBAs (dependent variable) and the groups were conducted. Age, sex, scanner, and pSES were entered as control variables. Because pSES was not associated with PBA, subsequent regression models were adjusted for age, sex, and scanner. Subsequent regressions were conducted between PBA and the following sets of variables separately: (1) Premorbid IQ (WASI Vocabulary t-score); (2) MATRICS fluid cognitive measures sensitive to psychosis; (3) social functioning measures; and (4) for FESz only, symptom measures (SAPS and SANS component scores). Finally, for FESz, premorbid IQ and any significant symptom associates were included to ensure that symptoms but not

IQ were associated with PBA. We examined Probability–Probability plots for normality and report the Variance Inflation Factor (VIF) for each model tested.

3. Results

3.1. Demographics, Neuropsychological, and Social Functioning

All means (s.d.), tests of significance, and effect sizes (Cohen's d) are presented in Table 1. The groups did not differ in age, sex distribution, or WASI Vocabulary t-scores. FESZ had lower pSES than the HCs. Consistent with the effects of psychosis on cognition, the WASI Matrix Reasoning, MATRICS Composite Scores, and all MATRICS subtests were lower in FESZ. FESZ scored lower on all subscales of the SFS except Recreation. The current GF:S and GF:R, as well as the highest and lowest in the last year on both domains, were lower in FESZ. The clinical measures for FESz are presented in Table 2.

Table 2. Clinical Measures in FESZ participants.

Test	Score
BPRS Total score	47.20 (9.53)
SANS/SAPS-derived components	
Auditory Hallucinations	3.37 (3.21)
Unusual Perceptions	0.93 (1.33)
Delusions	5.95 (3.87)
Thought Disorder	0.23 (0.34)
Inattention	2.17 (2.10)
Inexpressivity	7.43 (5.76)
Apathy/Asociality	7.57 (3.94)
Medicated/Unmedicated	54/16
Medication dosage (CPZ equivalents)	222.40 (122.17)
Mean days from first clinical contact to scan	53.26 (71.2)
Median days from first contact to scan	27
Mean months DUP	23.17 (38.53)
Median months DUP	6.89

3.2. Predicted Brain Age and Brain Age Gap

We found an MAE of 4.2 years in the HCs, indicating a good performance for the model and an MAE of 4.4 years in the whole sample. As mentioned above, age did not differ between the groups. The Kolmogorov–Smirnov test for distributions indicated that both groups were similarly distributed ($Z = 0.974$, $p = 0.299$). PBA and BAG did not differ significantly between the groups (Table 1). However, age correlated with PBA in both the HCs ($r75 = 0.65$, $p < 0.001$) and FESZ ($r69 = 0.65$, $p < 0.001$, Figure 1). In contrast, BAG did not correlate significantly with age in either the HCs ($r75 = -0.131$, $p = 0.261$) or FESZ ($r90 = 0.006$, $p = 0.962$).

3.2.1. Regression Analyses

In our initial regression model, PBA was not associated with group (ß = -0.040, $t = 0.616$, $p = 0.539$, Table 3). For the following results, only significant associations are presented in the text. The full model results are presented in Table 4.

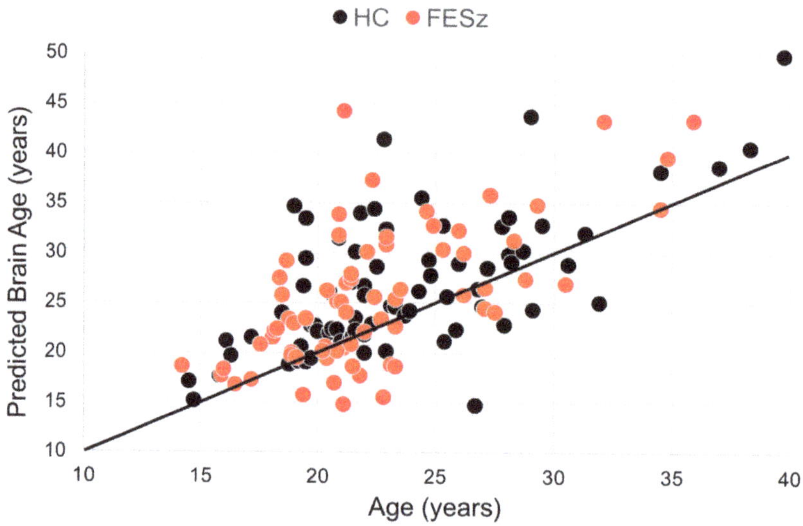

Figure 1. Correlations between predicted brain age and age. The solid line indicates perfect model prediction. Note the tendency for predicted ages to be older in both groups.

Table 3. Predicted brain age (PBA) regression with group.

Variable	B	ß	t-Statistic	*p*-Value	VIF
Age	0.886	0.624	9.703	**<0.001**	1.080
Scanner	1.902	0.136	2.111	**0.037**	1.096
Sex	−2.865	−0.196	3.126	**0.002**	1.031
pSES	−0.016	−0.031	0.478	0.633	1.085
Group (FESz vs. HC)	−0.278	−0.04	0.616	0.539	1.133

Note: The B represents unstandardized coefficients, while ß is a standardized coefficient. Significant effects are bolded. VIF = Variance Inflation Factor. pSES = parental socio-economic status. FESz = first-episode schizophrenia-spectrum individuals. HC = healthy comparison individuals.

Table 4. Full model results of predicted brain age (PBA) associations with neuropsychological, social functioning, and clinical measures.

Variable	B	ß	t-Statistic	*p*-Value	VIF
Premorbid IQ model					
Age	0.887	0.623	10.093	**<0.001**	1.072
Scanner	1.573	0.113	1.833	0.069	1.074
Sex	−2.692	−0.184	3.074	**0.003**	1.004
WASI vocabulary t-score	−0.138	−1.630	−2.725	**0.007**	1.005
MATRICS battery model					
Age	0.969	0.684	10.381	**<0.001**	1.207
Scanner	1.678	0.121	1.911	0.058	1.121
Sex	−2.626	−0.180	−2.891	**0.004**	1.079
Trail Making Test	−0.076	−0.136	−1.231	0.221	3.392
Speed of Processing	0.087	0.167	1.228	0.221	5.144
Attention and Vigilance	−0.036	−0.065	−0.761	0.448	2.011
Working Memory	−0.032	−0.055	−0.564	0.574	2.697
Verbal Learning	0.047	0.077	0.940	0.349	1.850
Visual Learning	−0.058	−0.092	−1.146	0.254	1.810
Reasoning and Problem Solving	−0.108	−0.157	−1.923	0.057	1.859
Social Cognition	0.013	0.025	0.366	0.715	1.282

Table 4. *Cont.*

Variable	B	ß	t-Statistic	*p*-Value	VIF
Social Functioning Model					
Age	0.829	0.584	7.973	**<0.001**	1.379
Scanner	2.180	0.158	2.381	**0.019**	1.137
Sex	−1.972	−0.136	−1.908	0.059	1.296
SFS Social Engagement/Withdrawal	0.056	0.097	1.307	0.193	1.411
SFS Interpersonal Interaction	−0.062	−0.169	−1.871	0.064	2.093
SFS Recreation	−0.047	−0.080	−1.100	0.328	1.403
SFS Employment Occupation	0.054	0.053	0.698	0.486	1.508
SFS Independence Performance	0.068	0.108	1.282	0.202	1.825
SFS Independence Competence	0.007	0.011	0.129	0.898	1.752
SFS Prosocial	0.029	0.048	0.640	524	1.543
GF:R Highest in Last Year	−0.416	−0.080	−0.674	0.502	3.657
GF:S Highest in Last Year	0.448	0.082	0.662	0.509	3.936
Symptom Models (FESz only)					
Age	1.041	0.669	7.582	**<0.001**	1.117
Scanner	0.729	0.053	0.588	0.559	1.155
Sex	−4.763	−0.301	−3.353	**0.001**	1.159
Auditory Hallucinations	0.019	0.009	0.088	0.930	1.378
Unusual Perceptions and Behaviors	−0.773	−0.147	−1.556	0.125	1.285
Delusions	0.262	0.145	1.355	0.181	1.653
Thought Disorder	0.863	0.043	0.438	0.663	1.355
Inattention	0.343	0.103	0.997	0.323	1.545
Inexpressivity	0.312	0.258	2.472	**0.016**	1.568
Apathy and Asociality	−0.150	−0.085	−0.857	0.395	1.406
Age	1.038	0.667	8.006	**<0.001**	1.091
Scanner	0.900	0.065	0.769	0.444	1.091
Sex	−4.035	−0.255	−3.053	**0.003**	1.066
WASI Vocabulary t-score	−0.128	−0.178	−2.073	**0.042**	1.121
Inexpressivity	0.254	0.210	2.532	**0.014**	1.054

Note: The B represents unstandardized coefficients, while ß is a standardized coefficient. Significant effects are in bolded. VIF = Variance Inflation Factor. WASI = Wechsler Abbreviated Scale of Intelligence. MATRICS = Measurement and Treatment Research to Improve Cognition in Schizophrenia (MATRICS) cognitive battery. SFS = Social Functioning Scale. GF:R = Global Functioning: Role. GF:S = Global Functioning: Social. FESz = first-episode schizophrenia-spectrum individuals. Emotional Inexpressivity is derived from hierarchical principal component analysis in Longenecker et al. (2022) [39]. WASI = Wechsler Abbreviated Scale of Intelligence.

3.2.2. Neuropsychological Performance

First, the WASI Vocabulary subtest estimates of premorbid IQ were tested. Among all participants, there was a significant association between premorbid IQ and PBA above and beyond the effects of age (ß = −0.163, t = 2.725, $p = 0.007$, Figure 1). In the model examining the MATRICS subtests (sensitive to the state effects of psychosis), there were no significant associations between PBA and neuropsychological tests. (See Table 4 for full results).

3.2.3. Social Functioning

In the initial model with all social functioning measures, where no associations were detected, the GF:R Current and GF:R Lowest in the last year had VIFs over 30, and the GF:S Current and GF:S Lowest in the last year had VIFs over 16. They were removed from the model. In this second model, there were no significant associations between PBA and social functioning. (See Table 4 for full results).

3.2.4. Symptoms

In FESZ only, the current symptom severity on the SANS/SAPS hierarchical clusters was regressed against PBA. In this model, above and beyond the effects of age, there was a significant association between an older PBA and poorer emotional inexpressivity (ß = 0.258, t = 2.472, p = 0.016, Figure 2). In a follow-up model in FESz, including the WASI Vocabulary subtest estimates of premorbid IQ, both IQ (ß = −0.178, t = 2.073, p = 0.042) and emotional inexpressivity (ß = 0.210, t = 2.532 p = 0.014) remained significantly associated with PBA. (See Table 4 for full results).

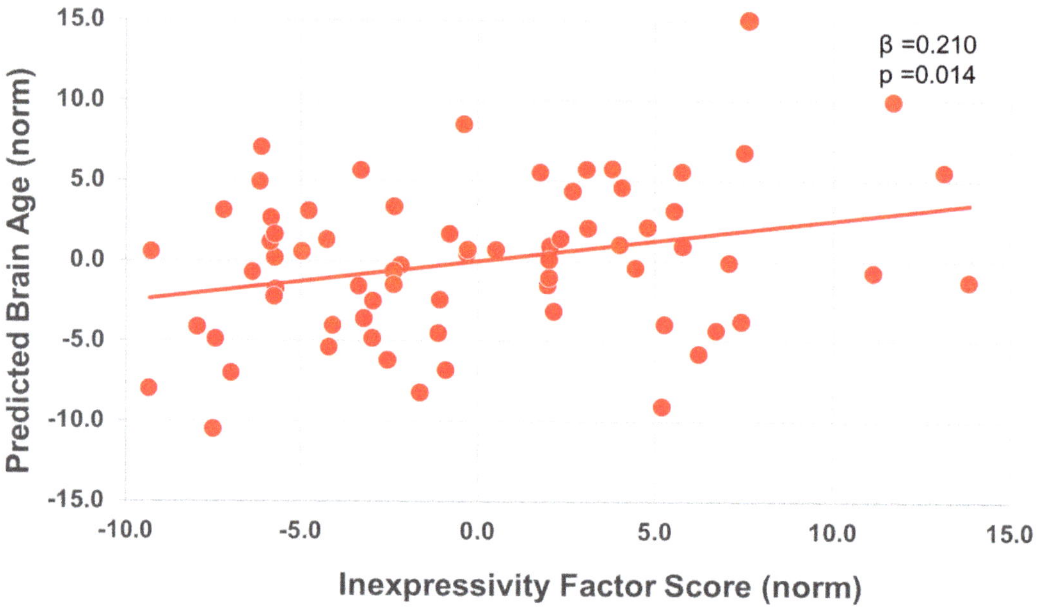

Figure 2. Relationship between predicted brain age and inexpressivity negative symptom component in first-episode schizophrenia-spectrum individuals. Partial regression plot of normalized values.

4. Discussion

Predicted brain age (PBA) and brain age gap (BAG) were examined in first-episode schizophrenia-spectrum individuals (FESz), as well as PBA's associations with neuropsychological, social functioning, and symptom measures. FESz were tested relatively close to their first clinical contact for suspected psychosis, with a median of 27 days until scan. PBA and BAG were not elevated in FESz. Although the groups did not differ in age or age distribution, a more conservative regression model was used to assess for any PBA differences, adjusting for age, sex, and scanner. There was no association between PBA and group. These negative findings were associated with small effect sizes (Table 1), suggesting that the lack of difference was robust and not due to an underpowered study. In terms of accelerated aging prior to the onset of psychosis, these findings suggest little evidence for brain reduction in these individuals prior to the onset of psychosis. This is consistent with the findings of Chung et al. [32], where "older" CHR individuals (18–21) that converted to psychosis did not show an increased BAG, while younger individuals that converted to psychosis did. Chung et al. [32] also suggested an accelerated trajectory of gray matter loss in older individuals. Here, we did not see a correlation between BAG and age among FESz individuals, which would support a progressive course. This could be attributed to the relatively stable scan times among the FESz with respect to onset. However, the duration of untreated psychosis (DUP) did not correlate with PBA or BAG, nor did days from first contact to scan (all r values < 0.082). Thus, we can reasonably conclude that there was little evidence in this sample of 70 FESz compared to a matched group of 76 HCs for accelerated brain aging proximal to first clinical contact for psychosis. However, it is entirely possible that these FESz individuals were entering a post-psychosis period of accelerated aging.

Only longitudinal follow-up testing of the cohort will provide definitive answers as to the individual trajectory of any accelerated brain aging in FESz.

Among all individuals, above and beyond the effects of age, a younger PBA was associated with higher estimates of premorbid IQ (WASI vocabulary performance). Vocabulary performance relies on verbal semantic memory, which is relatively robust against the effects of psychosis, unlike the domains assessed on the MATRICS, which are sensitive to the state effects of psychosis on cognition. The association between PBA and WASI Vocabulary performance was largely similar in FESz and HCs. Although perhaps not remarkable, possessing a younger brain age for one's age has benefits for this cognitive domain among all individuals.

Within FESz, there were generally no associations between PBA and symptom severity above and beyond the effects of age. The exception to this was the negative symptom cluster of emotional inexpressivity. This cluster comprises SANS and SAPS items, including Relationships with Friends/Peers, Ability to Feel Intimacy/Closeness, Persecutory Delusions, Impersistence at Work/School, Recreational Interests and Activities, and Physical Anergia [39]. Although at first glance, these may seem disparate, they tend to vary together among individuals, suggesting sensitivity to pathology in an as of yet unknown common neural substrate. The results here suggest, once again beyond the effects of age itself, that an older PBA is associated with an increased severity of this negative symptom constellation. Importantly, including WASI Vocabulary scores in the final model indicated that the association with negative symptoms persisted and was not due to premorbid IQ.

Caveats include this being a relatively small sample of FESz, but this is <1% of the population once in their lives. Multi-center studies and mega-analyses of harmonized MRI data from individuals within 1 year of clinically significant psychotic symptoms would be beneficial. There are, at present, no longitudinal scans to examine PBA or BAG trajectory among these individuals. These data would be vitally important for determining whether late adolescent/early adulthood-onset FESz show an accelerated trajectory of aging. PBA models may suffer from regression to the mean errors (see [33]). Our groups were very well matched for age, and we performed alternative regressions to ameliorate such confounds. Samples were scanned on two different scanners. Adjusting for the scanner in the models may remove critical variance. We continue to collect participants on the newer scanner, but harmonization may be useful, albeit to the lowest common denominator. Parental socioeconomic status differed between the FESz and HCs. Although the models suggested there was no meaningful association, future samples should be matched to avoid potential confounds. The Variance Inflation Factor (VIF) was unacceptably high for the Current and Lowest in the last year Global Functioning scores. Although no associations were found, this suggests collinearity among the GF:S, GF:R, and SFS measures that may have affected the model sensitivity. However, the analysis of social functioning without these measures did not reveal any associations between social functioning and PBA. We chose to use a publicly available brain age model. It is possible that it was not optimal for our dataset, but we felt it important to use a validated model. Training a model on the HC group and testing group differences would be 'double-dipping', wherein we build a model on the controls and then test if there are differences between these groups on the model built on controls. The BrainAgeR model is a well-established and widely used model with a good performance, thus, we used this model to improve the generalizability of our results. In addition, while the MAE was less than 5 years, it is possible that a tighter-fitting model would be better. However, some previous evidence suggests that this may not be ideal, as this overfits pathological processes and prevents the detection of true clinical effects [43]. Future work developing a validated model with large sample sizes is necessary. Likewise, sex was associated with PBA, either significantly or at the trend level, in all regression models. Although females ($n = 47$, 22.8 ± 4.5 years) and males ($n = 99$, 23.2 ± 5.0 years) did not differ in age ($t_{144} = 0.461$, $p = 0.645$), the PBA was significantly greater in females than males (females: 27.8 ± 7.4 years, males: 25.3 ± 6.5 years, $t_{144} = 2.047$, $p = 0.042$). ANOVA with sex as a random factor and group as a fixed factor found no interaction between sex and group on PBA, so this is unlikely to reflect a differential sex effect in FESz. There is evidence that, even in healthy adults, this is a period of continued neural gray and white matter changes that differs between men and women [44]. The future development of brain age models that better account for sex would be beneficial. Among our HCs and FESz, we did not assess birth traumas, inflammatory markers, or specific genetic risk factors. These may be uncontrolled confounds. Future work with more detailed assessments of these measures is warranted. Finally, medications may affect cognitive performance, for better or for worse. However, we are ethically obligated to provide treatment, and a controlled wash-out period was not feasible.

In summary, PBA was associated in all individuals with a better semantic verbal memory, above and beyond the effects of actual age. There was no evidence for accelerated brain aging in these late adolescent/early adult-onset FESz. However, the negative symptom cluster of reduced emotional expressivity, along with several behavioral domains, was associated in FESz with a greater PBA, even when adjusted for actual age and premorbid IQ. Future work will examine these individuals during the early course of their psychotic illness to detect any increased trajectory of brain aging.

Author Contributions: Conceptualization, H.T.K. and D.F.S.; methodology, H.T.K., B.A.C. and D.F.S.; formal analysis, H.T.K., D.F.S., B.A.C., M.T.C. and B.M.W.; investigation, D.F.S., B.A.C. and M.T.C.; resources, D.F.S. and H.T.K.; data curation, M.T.C. and D.S.; writing—original draft preparation, D.F.S.; writing—review and editing, D.F.S., B.A.C. and H.T.K.; visualization, D.F.S., H.T.K. and B.A.C.; supervision, D.F.S. and H.T.K.; project administration, D.F.S., H.T.K. and D.S.; funding acquisition, D.F.S. and H.T.K. All authors have read and agreed to the published version of the manuscript.

Funding: This research was funded by NIH grant numbers P50 MH103204 and R01 MH108568.

Institutional Review Board Statement: The study was approved by the University of Pittsburgh Institutional Review Board. Data was collected under 3 protocols: Tim Trio MRI—PRO14010185, Cortical Cells, Circuits, Connectivity and Cognition in Psychosis, initial approval 3 March 2014; Prisma MRI—PRO16030020, Attentional Modulation of Sensory Signals in First Episode Psychosis, initial approval 24 June 2016; merged with STUDY19030077, Pathology and Pathophysiology in Early Psychosis, initial approval 28 June 2017.

Informed Consent Statement: Informed consent was obtained from all subjects involved in the study.

Data Availability Statement: Portions of the data are available in the US NIMH NDA database. The original data was obtained elsewhere, are available from the first author upon appropriate request.

Conflicts of Interest: The authors declare no conflicts of interest.

References

1. Kraepelin, E. *Psychiatrie: Ein Lehrbuch fur Studirende und Aerzte*; Verlag Von Johann Ambrosius Barth: Leipzig, Germany, 1899.
2. Bleuler, E. *Dementia Praecox, or the Group of Schizophrenias*; Zinkin, J., Translator; International Universities Press: New York, NY, USA, 1911.
3. Kendler, K.S.; Tabb, K.; Wright, J. The emergence of psychiatry: 1650–1850. *Am. J. Psychiatry* **2022**, *179*, 329–335. [CrossRef]
4. Moore, M.T.; Nathan, D.; Elliott, A.R.; Laubach, C. Encephhalographic studies in schizophrenia (dementia praecox): Report of Sixty Cases. *Am. J. Psychiatry* **1933**, *89*, 801–810. [CrossRef]
5. Weinberger, D.R. Implications of normal brain development for the pathogenesis of schizophrenia. *Arch. Gen. Psychiatry* **1987**, *44*, 660–669. [CrossRef]
6. Lipska, B.K. Using animal models to test a neurodevelopmental hypothesis of schizophrenia. *J. Psychiatry Neurosci.* **2004**, *29*, 282–286.
7. Benes, F.M. Evidence for neurodevelopment disturbances in anterior cingulate cortex of post-mortem schizophrenic brain. *Schizophr. Res.* **1991**, *5*, 187–188. [CrossRef]
8. Feinberg, I. Schizophrenia: Caused by a fault in programmed synaptic elimination during adolescence? *J. Psychiatr. Res.* **1982**, *17*, 319–334. [CrossRef]
9. McGlashan, T.H.; Hoffman, R.E. Schizophrenia as a disorder of developmentally reduced synaptic connectivity. *Arch. Gen. Psychiatry* **2000**, *57*, 637–648. [CrossRef]
10. Waddington, J.L.; Scully, P.J.; Youssef, H.A. Developmental trajectory and disease progression in schizophrenia: The conundrum, and insights from a 12-year prospective study in the Monaghan 101. *Schizophr. Res.* **1997**, *23*, 107–118. [CrossRef]

11. Gur, R.E.; Cowell, P.; Turetsky, B.I.; Gallacher, F.; Cannon, T.; Bilker, W.; Gur, R.C. A follow-up magnetic resonance imaging study of schizophrenia: Relationship of neuroanatomical changes to clinical and neurobehavioral measures. *Arch. Gen. Psychiatry* **1998**, *55*, 145–152. [CrossRef]
12. Thompson, P.M.; Vidal, C.; Giedd, J.N.; Gochman, P.; Blumenthal, J.; Nicolson, R.; Toga, A.W.; Rapoport, J.L. Mapping adolescent brain change reveals dynamic wave of accelerated gray matter loss in very early-onset schizophrenia. *Proc. Natl. Acad. Sci. USA* **2001**, *98*, 11650–11655. [CrossRef]
13. Lieberman, J.; Chakos, M.; Wu, H.; Alvir, J.; Hoffman, E.; Robinson, D.; Bilder, R. Longitudinal study of brain morphology in first episode schizophrenia. *Biol. Psychiatry* **2001**, *49*, 487–499. [CrossRef]
14. Cahn, W.; Pol, H.E.H.; Lems, E.B.; van Haren, N.E.; Schnack, H.G.; van der Linden, J.A.; Schothorst, P.F.; van Engeland, H.; Kahn, R.S. Brain volume changes in first-episode schizophrenia: A 1-year follow-up study. *Arch. Gen. Psychiatry* **2002**, *59*, 1002–1010. [CrossRef]
15. Kasai, K.; Shenton, M.E.; Salisbury, D.F.; Hirayasu, Y.; Lee, C.U.; Ciszewski, A.A.; Yurgelun-Todd, D.; Kikinis, R.; Jolesz, F.A.; McCarley, R.W. Progressive decrease of left superior temporal gyrus gray matter volume in patients with first-episode schizophrenia. *Am. J. Psychiatry* **2003**, *160*, 156–164. [CrossRef]
16. Kasai, K.; Shenton, M.E.; Salisbury, D.F.; Hirayasu, Y.; Onitsuka, T.; Spencer, M.H.; Yurgelun-Todd, D.A.; Kikinis, R.; Jolesz, F.A.; McCarley, R.W. Progressive decrease of left Heschl gyrus and planum temporale gray matter volume in first-episode schizophrenia: A longitudinal magnetic resonance imaging study. *Arch. Gen. Psychiatry* **2003**, *60*, 766–775. [CrossRef]
17. Ho, B.C.; Andreasen, N.C.; Nopoulos, P.; Arndt, S.; Magnotta, V.; Flaum, M. Progressive structural brain abnormalities and their relationship to clinical outcome: A longitudinal magnetic resonance imaging study early in schizophrenia. *Arch. Gen. Psychiatry* **2003**, *60*, 585–594. [CrossRef]
18. Bachmann, S.; Bottmer, C.; Pantel, J.; Schröder, J.; Amann, M.; Schad, M.E.L.R. MRI-morphometric changes in first-episode schizophrenic patients at 14 months follow-up. *Schizophr. Res.* **2004**, *67*, 301–303. [CrossRef]
19. Pantelis, C.; Velakoulis, D.; McGorry, P.D.; Wood, S.J.; Suckling, J.; Phillips, L.J.; Yung, A.R.; Bullmore, E.T.; Brewer, W.; Soulsby, B.; et al. Neuroanatomical abnormalities before and after onset of psychosis: A cross-sectional and longitudinal MRI comparison. *Lancet* **2003**, *361*, 281–288. [CrossRef]
20. Job, D.E.; Whalley, H.C.; Johnstone, E.C.; Lawrie, S.M. Grey matter changes over time in high risk subjects developing schizophrenia. *Neuroimage* **2005**, *25*, 1023–1030. [CrossRef]
21. Olney, J.W. Excitotoxic amino acids and neuropsychiatric disorders. *Annu. Rev. Pharmacol. Toxicol.* **1990**, *30*, 47–71. [CrossRef]
22. Okusaga, O.O. Accelerated aging in schizophrenia patients: The potential role of oxidative stress. *Aging Dis.* **2014**, *5*, 256–262. [CrossRef]
23. Kirkpatrick, B.; Messias, E.; Harvey, P.D.; Fernandez-Egea, E.; Bowie, C.R. Is schizophrenia a syndrome of accelerated aging? *Schizophr. Bull.* **2008**, *34*, 1024–1032. [CrossRef]
24. Harvey, P.D.; Rosenthal, J.B. Cognitive and functional deficits in people with schizophrenia: Evidence for accelerated or exaggerated aging? *Schizophr. Res.* **2018**, *196*, 14–21. [CrossRef]
25. Nguyen, T.T.; Eyler, L.T.; Jeste, D.V. Systemic biomarkers of accelerated aging in schizophrenia: A critical review and future directions. *Schizophr. Bull.* **2018**, *44*, 398–408. [CrossRef] [PubMed]
26. Ballester, P.L.; Romano, M.T.; de Azevedo Cardoso, T.; Hassel, S.; Strother, S.C.; Kennedy, S.H.; Frey, B.N. Brain age in mood and psychotic disorders: A systematic review and meta-analysis. *Acta Psychiatr. Scand.* **2022**, *145*, 42–55. [CrossRef]
27. Shahab, S.; Mulsant, B.H.; Levesque, M.L.; Calarco, N.; Nazeri, A.; Wheeler, A.L.; Foussias, G.; Rajji, T.K.; Voineskos, A.N. Brain structure, cognition, and brain age in schizophrenia, bipolar disorder, and healthy controls. *Neuropsychopharmacology* **2019**, *44*, 898–906. [CrossRef]
28. Schnack, H.G.; Van Haren, N.E.; Nieuwenhuis, M.; Hulshoff Pol, H.E.; Cahn, W.; Kahn, R.S. Accelerated brain aging in schizophrenia: A longitudinal pattern recognition study. *Am. J. Psychiatry* **2016**, *173*, 607–616. [CrossRef] [PubMed]
29. Constantinides, C.; Han, L.K.; Alloza, C.; Antonucci, L.A.; Arango, C.; Ayesa-Arriola, R.; Banaj, N.; Bertolino, A.; Borgwardt, S.; Bruggemann, J.; et al. Brain ageing in schizophrenia: Evidence from 26 international cohorts via the ENIGMA Schizophrenia consortium. *Mol. Psychiatry* **2023**, *28*, 1201–1209. [CrossRef] [PubMed]
30. Kaufmann, T.; van der Meer, D.; Doan, N.T.; Schwarz, E.; Lund, M.J.; Agartz, I.; Alnæs, D.; Barch, D.M.; Baur-Streubel, R.; Bertolino, A.; et al. Common brain disorders are associated with heritable patterns of apparent aging of the brain. *Nat. Neurosci.* **2019**, *22*, 1617–1623. [CrossRef]
31. Hajek, T.; Franke, K.; Kolenic, M.; Capkova, J.; Matejka, M.; Propper, L.; Uher, R.; Stopkova, P.; Novak, T.; Paus, T.; et al. Brain age in early stages of bipolar disorders or schizophrenia. *Schizophr. Bull.* **2019**, *45*, 190–198. [CrossRef]
32. Chung, Y.; Addington, J.; Bearden, C.E.; Cadenhead, K.; Cornblatt, B.; Mathalon, D.H.; McGlashan, T.; Perkins, D.; Seidman, L.J.; Tsuang, M.; et al. Use of machine learning to determine deviance in neuroanatomical maturity associated with future psychosis in youths at clinically high risk. *JAMA Psychiatry* **2018**, *75*, 960–968. [CrossRef]
33. Butler, E.R.; Chen, A.; Ramadan, R.; Le, T.T.; Ruparel, K.; Moore, T.M.; Satterthwaite, T.D.; Zhang, F.; Shou, H.; Gur, R.C.; et al. Pitfalls in brain age analyses. *Hum. Brain Mapp.* **2021**, *42*, 4092–4101. [CrossRef] [PubMed]
34. Wechsler, D. *Wechsler Abbreviated Scale of Intelligence (WASI)*; APA: Washington, DC, USA, 1999.

35. First, M.B.; Gibbon, M. The Structured Clinical Interview for DSM-IV Axis I Disorders (SCID-I) and the Structured Clinical Interview for DSM-IV Axis II Disorders (SCID-II). In *Comprehensive Handbook of Psychological Assessment, Vol. 2. Personality Assessment*; Hilsenroth, M.J., Segal, D.L., Eds.; John Wiley & Sons, Inc.: Hoboken, NJ, USA, 2004.
36. Overall, J.E.; Gorham, D.R. The Brief Psychiatric Rating Scale. *Psychol. Rep.* **1962**, *10*, 799–812. [CrossRef]
37. Andreasen, N.C. *Scale for the Assessment of Negative Symptoms (SANS)*; The University of Iowa: Iowa City, IA, USA, 1981.
38. Andreasen, N.C. *The Scale for the Assessment of Positive Symptoms (SAPS)*; The University of Iowa: Iowa City, IA, USA, 1984.
39. Longenecker, J.M.; Haas, G.L.; Salisbury, D.F. Hierarchical symptom components in early psychosis. *Schizophr. Bull.* **2022**, *48*, 893–901. [CrossRef] [PubMed]
40. Nuechterlein, K.H.; Green, M.F.; Kern, R.S.; Baade, L.E.; Barch, D.M.; Cohen, J.D.; Essock, S.; Fenton, W.S.; Frese, F.J., III; Gold, J.M.; et al. The MATRICS Consensus Cognitive Battery, part 1: Test selection, reliability, and validity. *Am. J. Psychiatry* **2008**, *165*, 203–213. [CrossRef] [PubMed]
41. Cornblatt, B.A.; Auther, A.M.; Niendam, T.; Smith, C.W.; Zinberg, J.; Bearden, C.E.; Cannon, T.D. Preliminary findings for two new measures of social and role functioning in the prodromal phase of schizophrenia. *Schizophr. Bull.* **2007**, *33*, 688–702. [CrossRef] [PubMed]
42. Cole, J.H.; Ritchie, S.J.; Bastin, M.E.; Hernández, V.; Muñoz Maniega, S.; Royle, N.; Corley, J.; Pattie, A.; Harris, S.E.; Zhang, Q.; et al. Brain age predicts mortality. *Mol. Psychiatry* **2018**, *23*, 1385–1392. [CrossRef] [PubMed]
43. Bashyam, V.M.; Erus, G.; Doshi, J.; Habes, M.; Nasrallah, I.M.; Truelove-Hill, M.; Srinivasan, D.; Mamourian, L.; Pomponio, R.; Fan, Y.; et al. MRI signatures of brain age and disease over the lifespan based on a deep brain network and 14 468 individuals worldwide. *Brain* **2020**, *143*, 2312–2324. [CrossRef]
44. Bethlehem, R.A.I.; Seidlitz, J.; White, S.R.; Vogel, J.W.; Anderson, K.M.; Adamson, C.; Adler, S.; Alexopoulos, G.S.; Anagnostou, E.; Areces-Gonzalez, A.; et al. Brain charts for the human lifespan. *Nature* **2022**, *604*, 525–533. [CrossRef]

brain sciences

MDPI

Review

The Necessity of Taking Culture and Context into Account When Studying the Relationship between Socioeconomic Status and Brain Development

Julie M. Schneider [1], Mohammad Hossein Behboudi [2] and Mandy J. Maguire [2,3,*]

1 Department of Communication Sciences and Disorders, Louisiana State University, 72 Hatcher Hall, Field House Drive, Baton Rouge, LA 70803, USA; juschnei@lsu.edu

2 Callier Center for Communication Disorders, The University of Texas at Dallas, 1966 Inwood Road, Dallas, TX 75235, USA; behboudi@utdallas.edu

3 Center for Children and Families, The University of Texas at Dallas, 800 W Campbell Road, Dallas, TX 75080, USA

* Correspondence: mjm053000@utdallas.edu

Abstract: Decades of research has revealed a relationship between childhood socioeconomic status (SES) and brain development at the structural and functional levels. Of particular note is the distinction between income and maternal education, two highly correlated factors which seem to influence brain development through distinct pathways. Specifically, while a families' income-to-needs ratio is linked with physiological stress and household chaos, caregiver education influences the day-to-day language environment a child is exposed to. Variability in either one of these environmental experiences is related to subsequent brain development. While this work has the potential to inform public policies in a way that benefits children, it can also oversimplify complex factors, unjustly blame low-SES parents, and perpetuate a harmful deficit perspective. To counteract these shortcomings, researchers must consider sociodemographic differences in the broader cultural context that underlie SES-based differences in brain development. This review aims to address these issues by (a) identifying how sociodemographic mechanisms associated with SES influence the day-to-day experiences of children, in turn, impacting brain development, while (b) considering the broader cultural contexts that may differentially impact this relationship.

Keywords: socioeconomic status (SES); EEG; brain development; culture

1. Introduction

1.1. Historical Perspectives Linking Socioeconomic Status (SES) and Brain Development

For decades, socioeconomic status (SES) was used as a covariate in studies of brain development in children, but starting around 2009, researchers became more interested in studying SES in and of itself as a predictor, or correlate, of differences in brain development [1]. This change was due to the understanding that SES can represent an individual's access to various resources, which can directly impact brain development [2–8], educational attainment [9], physical well-being [10], mental health [11], and cognitive development [1]. As research has expanded to examine the impact of SES on brain development, researchers have acknowledged that SES is a complex construct, defined by correlated but not identical factors, such as parental educational attainment, occupational status, neighborhood quality, and family income. Recent work has made incredible strides in teasing apart the complexity of these relationships by studying both SES and brain structure and function from multiple perspectives. Through this work, researchers have begun to fill in gaps in our understanding of the complex relationship between an individual's accumulated environmental experiences, their brain development, and how these differences in brain function and structure might manifest behaviorally. Despite these astounding efforts, most research examining the associations between SES and brain development are largely limited to WEIRD (Western, Educated, Industrial, Rich, and Democratic) countries, and often fail to consider the larger cultural context and belief systems that shape child cognitive and brain development. Here, we (1) review what is known about the relationship between SES and brain development based on magnetic resonance imaging (MRI) and resting state electroencephalography (EEG) studies; (2) pinpoint the unique influences of maternal education and income on this trajectory; and (3) review what questions remain to be answered, including taking a more in-depth, culturally responsive approach to understanding how a child's day-to-day life experiences differentially relate to brain development.

1.2. The Importance of Studying the Association between SES and Brain Development

Brain maturation is a prolonged process that continues throughout childhood and adolescence. This extended period allows for a long window of experience-dependent plasticity, where early experiences can significantly shape brain development. During early development, the cortical thickness of the brain, linked to synaptic pruning and myelination, undergoes a rapid decrease before plateauing in early adulthood. Meanwhile, the surface area of the cortex undergoes a substantial expansion through early adolescence followed by a contraction through middle childhood [12–14]. Related to brain volume, white matter volume undergoes a rapid increase throughout adolescence and into young adulthood due to myelination before becoming reduced in older adulthood [15–17]. On the other hand, gray matter volume exhibits a rapid increase during early childhood, reflecting the growth and branching of dendrites and synaptogenesis before peaking in early adolescence, followed by a gradual volume thinning and density reduction due to synaptic pruning [15,16,18]. Studies using structural MRI have identified global differences in these developmental changes which are correlated to SES [1,6,7,19,20]. For example, coming from a lower SES background is often linked to less cortical gray matter, which researchers posit is due to less gray matter formation early in development, or accelerated thinning [5,20–23]. Meanwhile, children from higher SES backgrounds exhibit greater global and local efficiency of the structural connectome, indicative of enhanced neural communication speed and efficiency [24]. Specific areas of the brain, including the prefrontal cortex (PFC) [1,6,7,23,25–27], the left anterior and posterior superior temporal gyrus (STG) [28], and the left inferior frontal gyrus (IFG) [29,30], are sensitive to environmental experiences and support the development of language, attention, executive function, emotion regulation, and memory [5,31]. Thus, these differences in brain development have been the focus of much SES-related research.

In addition to studies examining SES-related differences in brain *structure*, recent research has also sought to clarify SES-related differences in brain *function*. Specifically, SES has been linked to resting state brain function, as measured by both resting-state functional MRI (rsfMRI) and electroencephalography (rsEEG). rsfMRI and rsEEG allow for the investigation of spontaneous signal fluctuations in the brain when an individual is awake but not engaged in a specific cognitive task [32]. These spontaneous signal fluctuations are organized in multiple highly specific functional anatomical networks (resting state networks, RSNs) [33,34] which fluctuate at frequencies between 0.01 and 0.1 Hz, and strongly overlap with sensory–motor, visual, auditory, attention, language, and default networks [32,35]. These low-frequency fluctuations are thought to highly overlap between rsfMRI and rsEEG, as fluctuations in the band-limited power of the local field potential measured using EEG fluctuate at approximately the same frequency as the BOLD signal [36]. Importantly, both measures are thought to measure the brain's "readiness" to predict and integrate new experiences, a critical component of cognitive development, as it allows children to generate and test predictions about their environment [37]. These measures are often used in diagnostic research, as they provide unique insights into brain development, related to changes in functional networks.

Research utilizing rsfMRI has suggested that differences in functional brain development may be sensitive to neighborhood-and family-level-SES [38,39]. In these studies, coming from a lower SES background has been associated with reduced functional connectivity in the default mode network, sensorimotor network [40], and between the hippocampus and amygdala [41]. For example, children and adolescents living in more advantaged neighborhoods demonstrate a stronger positive association between age and functional connectivity compared to youth from more disadvantaged neighborhoods—a pattern which the authors interpret as being related to faster functional brain development among higher SES children [42]. Taken together, structural and functional fMRI studies suggest that coming from a higher SES background is associated with enhanced gray matter formation and functional connectivity, ultimately leading to more efficient cortical networks in adulthood [43].

While much of the research investigating SES-related differences in brain function has been accomplished using rsfMRI, rsEEG has recently gained substantial momentum. One clear advantage of rsEEG is that it is portable, less expensive, and significantly less invasive than MRI, allowing for research in diverse low-income environments. As a result, rsEEG research is more inclusive of international populations and children reared in extreme poverty/adverse conditions than is often observed in MRI research. In rsEEG data, researchers focus on how the overall power or amplitude of specific frequency bands shifts over the course of development. The bands are not universally defined, but are generally categorized as delta (0.5–4 Hz), theta (4–8 Hz), alpha (8–12 Hz), beta (12–30 Hz), and gamma (30–50 or higher Hz) [44]. The typical development trajectory of rsEEG involves an overall decrease in EEG power (average across all frequencies) from late childhood to early adulthood [45,46]. This reduction in EEG power is correlated with a decrease in the volume of gray matter, which occurs during the process of synaptic pruning. During this process, the quantity of synapses that evoke EEG signals diminishes, consequently leading to a decrease in the amplitude of the EEG signal [47–49].

When examining the developmental trajectories of rsEEG within distinct frequency bands, it is observed that this overall power reduction affects the low frequencies more consistently, with the power of absolute delta and theta frequencies decreasing and the power of alpha and beta frequencies slightly increasing [50–55]. The mechanism behind this different effect on the power of low- and high-frequency oscillations during development is relatively understudied [46]. It has been proposed that low frequencies originate from synchronous local neural activity between cortical neurons, and a reduction in synapses involved in this activity could lead to a significant loss of power [48]. In contrast, the beta frequency band, which is associated with asynchronous activity, resulting in lower EEG power, could potentially see an increase in power with the elimination of synapses, if the pruned synapses are activated in conflict with the dominant adjacent synapses. The alpha frequency band is more influenced by structural changes in the thalamus or thalamocortical relays than changes in cortical gray matter. This is supported by other studies, which show that increasing thalamocortical connectivity is associated with increased alpha activity in development [56].

Studies of school-aged children from low-SES backgrounds, defined as coming from very low-income households and/or having an illiterate mother, display a different pattern in EEG power distribution, with higher values of overall Absolute Power (AP) relative to children from higher SES backgrounds. This phenomenon has been interpreted as evidence of a maturational lag in brain development [53,57]. Studies that analyzed frequency-specific power show higher delta and theta AP in children from low-SES households or high-risk children in Mexico [58,59]. In higher frequencies, like alpha and gamma, there is an increase in AP at frontal regions for children from higher SES backgrounds compared to children from lower SES backgrounds. Researchers have also identified consistent SES-related differences in higher frequency bands in studies conducted in the US and Europe [60–62], with children from lower SES homes exhibiting significantly lower frontal gamma power compared to their higher SES peers [61]. In these studies, income status, maternal education, and maternal occupation are all associated with these reductions in gamma power over the frontal channels [61]. This reduced gamma activity is also observed in children with poorer language outcomes and attention [63–65]. Thus, in children from low-SES backgrounds, reduced gamma-band activity over the frontal areas may be an early indicator of a greater subsequent risk for poor language outcomes.

While absolute power is most commonly utilized in rsEEG studies, some argue that relative power (RP—the proportion of frequency-band power compared to total power of all frequency-bands) better addresses issues related to cortical thickness over the course of development [51]. Moreover, although RP findings are closely tied to AP findings, the developmental changes in high-frequency RP are more pronounced than in AP. There are well-established trends in rsEEG RP changes throughout development: there is a reduction in delta and theta RP, and a corresponding increase in alpha and beta RP with advancing age [3,48,58,59,66–69]. Specifically, children in high-risk situations in Mexico (based on maternal education, more than 4 siblings, and sharing a home with families with similar characteristics) demonstrate differences in lower frequency ranges, including higher delta and theta RP in frontal brain regions, and less alpha RP in posterior brain regions compared to children in more economically advantageous environments [58,59]. These rsEEG differences observed between children in high and low-risk situations diminishes with age, although differences in frontal theta and occipital/left temporal alpha bands persist at 6 years of age [58,59].

It is critical to acknowledge that, while these studies demonstrate a link between SES and brain development, it is the larger environmental context associated with SES which shapes brain development. Specifically, factors which are intimately tied to an individual's SES, such as enhanced exposure to stress and reduced environmental enrichment, are likely the mechanisms which drive brain development and subsequent cognitive outcomes [19,70–72]. Across both rsMRI and rsEEG studies, SES is thought to exert its influence on neural development and cognitive outcomes through two pathways which are often correlated.

The first pathway suggests that mothers with lower educational attainment provide less linguistically diverse vocabulary input to their children in the home environment, in turn affecting the development of the language processing system, including the left inferior frontal gyrus, visual word form area, and perisylvian cortex. Specifically, studies have shown that a lack of diverse vocabulary input negatively impacts child vocabulary development through language-specific neural regions, such as the left inferior frontal gyrus [29].

The second pathway is related to lower income, which is thought to result in a more stressful and less structured home environment, due to caregivers' difficulty accessing resources (food, shelter, transportation, etc.). This in turn affects the developing child's cognitive control system, including the hippocampus, prefrontal cortex, and amygdala. For example, when children experience higher physiological stress there is a negative impact on working memory, an aspect of cognitive control, through working memory prefrontal neural activity [73–76].

Despite evidence of distinct pathways, maternal education and income are highly correlated across households: individuals with a college degree make nearly 250% more than those with less than a ninth-grade level of education and 130% more than those with a high school diploma [77]. Despite this correlation, few studies have investigated the differential or overlapping effects of each construct on child cognitive development. In one study investigating the effects of both maternal education and income on child development, Khanam and Nghiem (2016) found that both constructs predicted children's literacy and math scores, however, only income predicted differences in receptive vocabulary and spatial perception (as measured by matrix reasoning) [78]. Other studies have also provided evidence that income and maternal education influence language development [79–81]; however, each of these studies emphasizes the mediating role of other environmental factors, such as parenting style, physical environment, health, and stress. Of note, researchers have found independent, non-overlapping effects of each construct, with income having a stable effect over the course of development on children's later success, while maternal education has more of an influence during infancy [81]. Thus, by concentrating on a single measure of SES, we are oversimplifying existing theoretical models of the relationship between SES and child development as the mechanisms related to SES are largely ignored. In the next sections, we provide more detailed evidence for the overlapping, yet unique influences of maternal education and income on children's environmental experiences, brain development, and cognitive outcomes.

1.3. Pathway 1. Maternal Education, Language Experiences, and Brain Development

A wealth of research has established, that, on average, higher levels of maternal education are related to maternal intellectual quotient (IQ) [82] and the quality of language input in the home [83]. Studies examining maternal IQ and child cognitive outcomes have specified that the home environment predicts child cognitive outcomes, above and beyond maternal IQ, and thus, variability in children's environments related to the quality of language input is a stronger predictor of child outcomes [84,85]. Specifically, higher maternal education is linked to children hearing more diverse words [86], more complex syntax [87–89], and experiencing more conversational turns [29,90–94]; all of which are linked to positive language outcomes. However, the reasons why maternal education influences the quantity and quality of language interactions is less clear [95]. One hypothesis is that higher levels of education results in greater knowledge of child development, including an understanding of the benefits a rich language environment can have on child outcomes (see Figure 1). Supporting this claim, parents' knowledge about child development mediates the relationship between SES and language input [92].

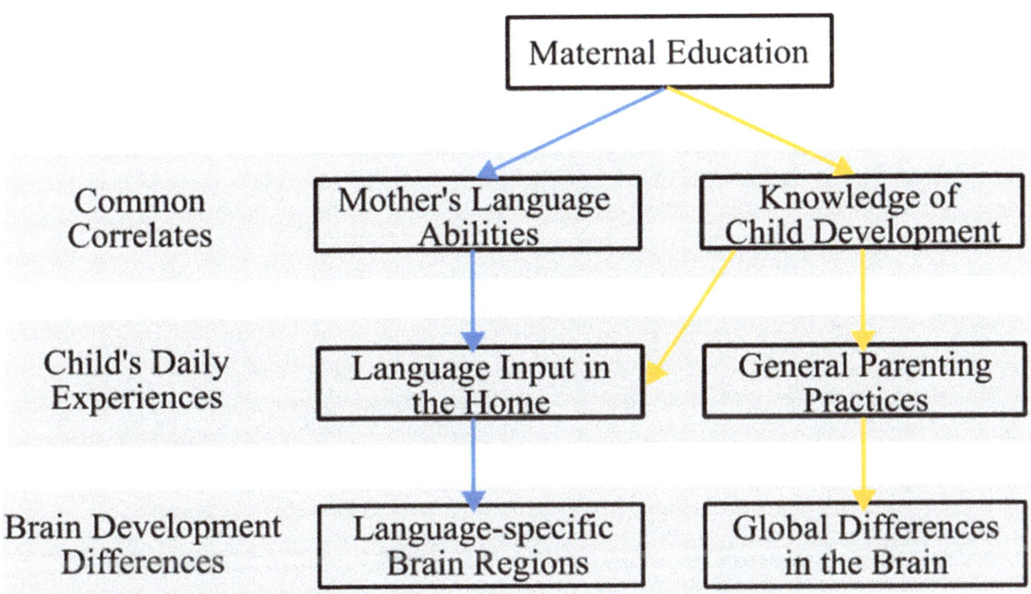

Figure 1. Theoretical model depicting the different pathways by which maternal education influences children's daily experiences and subsequent brain development. Blue and yellow arrows distinguish different pathways by which maternal education relates to brain development.

An alternative hypothesis is that higher levels of maternal education results in an increase in lexical richness and grammatical complexity of mothers' speech (see Figure 1). Supporting this claim, mothers with higher levels of educational attainment talk more, use a richer vocabulary, and produce longer utterances than those with lower educational attainment who had similar beliefs about children's language abilities [96]. Further, for Spanish–English bilingual children, the effects of maternal education are often only observed in the language in which the mother received her highest level of education, indicating the mother's experiences in the education system influenced her language abilities in each language, which then influenced how she uses each language with her child [97]. In both cases, maternal education results in a higher quantity and quality of language used by the parent when speaking to the child. In the first argument, these differences in the language environment are due to a better knowledge of child development in general, likely bleeding into other areas of the child's daily experiences, while in the second, these effects are specific to the language the child hears from the mother. The difference between these arguments is subtle, but may have important differential impacts on brain development.

It appears to be these daily language interactions that lead to differences in the development of language-specific brain regions in children whose mothers have lower levels of educational attainment. Specifically, structural and functional differences in brain development related to SES are due to a deprivation of meaningful language stimulation which causes early proliferation and pruning, resulting in inefficiencies in language outcomes. For example, Noble et al. [5,76] demonstrated that children from lower SES homes have decreased volume in language-specific regions (left inferior frontal gyrus (LIFG) and left superior temporal gyrus (LSTG)), which they attribute to accumulating language deprivation over children's lifetimes; however, children's language environments were not directly measured. Expanding this line of work to the examination of brain stiffness and organization, Schneider, McIlvain, and Johnson [98] reported that changes in neural tissue composition are sensitive to malleable aspects of the environment, specifically language input, whereas tissue organization is more strongly associated with vocabulary outcome. Importantly though, these findings were independent of maternal education, indicating that language input, above and beyond maternal education, relates to the structural organization of children's brains.

Related to brain function, studies have indicated that there are SES-based differences in the recruitment of neural regions that are necessary for language processing. Raizada et al. [99] reports that child language skills mediate the relationship between SES and functional brain activation in the bilateral IFG. More specifically, Romeo et al. [94] reported that it is conversational turns between caregivers and children that contributes to differences in left IFG activation and subsequent vocabulary development. Similar to Schneider, McIlvain, and Johnson [98], these results were independent of maternal education, indicating that differences in children's environments are more likely to predict functional development and language outcomes than SES in and of itself.

Maternal education and language input have further been shown to influence the development and maturation of cortical networks [4,60,62,100,101]. Generally speaking, measures of high-quality language input have been associated with lower relative power in low-frequency bands (theta) and higher relative power in high-frequency bands (beta, gamma) in infant EEG studies [100,101]. These same frequencies are also predictive of children's language outcomes: decreases in lower frequencies (theta) and increases in higher frequencies (alpha, beta, gamma) are positively associated with children's expressive and receptive vocabulary skills. Despite a clear association between rsEEG, maternal education, and language processes, only a few studies have studied the inter-relationship between these variables. Cantiani et al. [60] found that children from high SES homes had more positive central gamma power at 6 months and a longer mean length of utterance at 24 months of age. These studies provide foundational evidence supporting the inter-relationship between maternal education, rsEEG, and language skills. More recently, Schneider et al. (under review) reported unique pathways by which environmental experiences related to SES influence cognitive and neural development in school-aged children. Specifically, changes in the gamma frequency were linked to both maternal education and vocabulary outcome, but not income.

While most research examining the link between SES and brain development has focused primarily on the correlations between each construct, recent work has sought to consider pathways by which SES-related differences in brain structure and function emerge. Several recent studies have revealed that adult–child conversational turns—that is, contingent, responsive verbal interactions—mediate links between SES and brain structure [94,102] and function [29] in language-supporting regions. Caregiver self-reports of cognitive stimulation in the home further mediate the link between SES and cortical thickness [103]. These structural and functional MRI studies suggest that cognitive and linguistic stimulation in the home may be the pathway by which socioeconomic differences in brain development emerge and impact language and cognitive outcomes [104].

1.4. Pathway 2. Income, Stress Related Experiences, and Brain Development

The income-to-needs (ITN) ratio, a commonly reported proxy for income, is determined by dividing the total income of the household by the poverty threshold, which is adjusted according to the size of the family for the specific year the data is collected [105]. Families with a higher ITN ratio generally possess greater economic resources, which has been shown to be important for child brain development [2,104,106] (see Figure 2). Most notably, increasing low-SES families' income via a poverty reduction intervention has been associated with greater rsEEG power in the mid-to high-frequency bands [107].

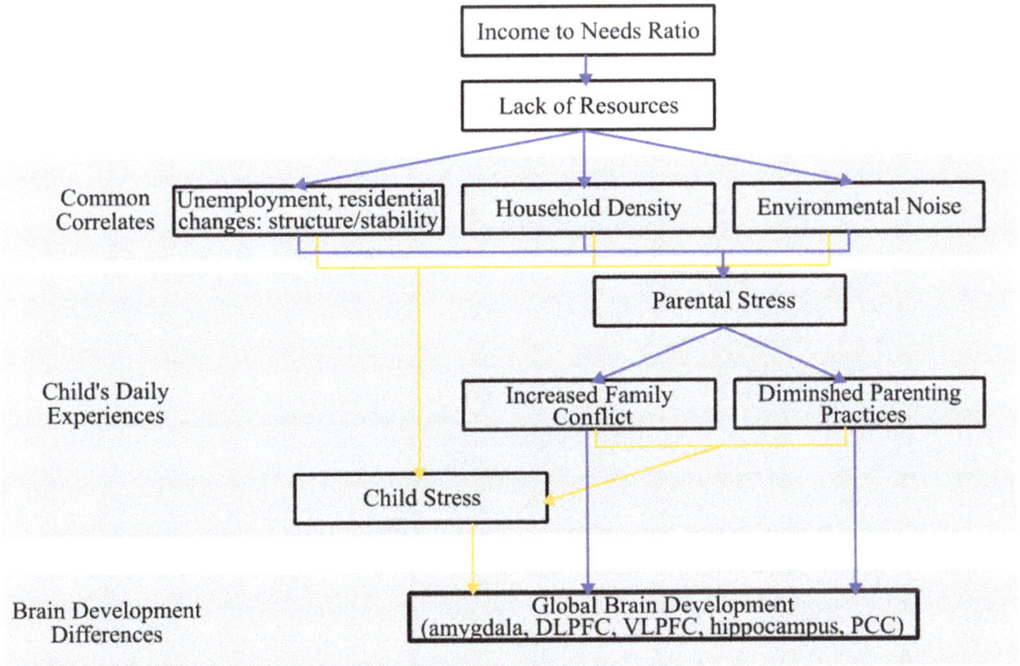

Figure 2. Theoretical model depicting different pathways by which income to needs ratio influences children's daily experiences and subsequent brain development. Blue and yellow arrows distinguish different pathways by which income relates to brain development.

The impacts of income are most pronounced during the preschool and early school years, particularly when low income is a persistent condition, and when the level of poverty is severe [108]. Despite the impact of income on brain development, it is crucial to understand that these adverse neural outcomes are not directly caused by income, but rather mediated by limited access to prenatal care, maternal nutrition, perceived stress, and psychological distress, associated with living in a lower income environment [109–115].

Numerous studies have delved into the intricate relationship between income and stress, highlighting the multifaceted nature of this relationship (see Figure 2). A study by Attar et al. [116] revealed that low-income fourth-grade students in the Chicago metropolitan area experienced a 35% increase in stressful life events and challenges within a year compared to their middle-income peers. This is because residing in an economically disadvantaged neighborhood can serve as a persistent stressor, as these communities face higher rates of unemployment and frequent residential changes. These same neighborhoods are also characterized by limited resources and heightened crime rates which cause increased physiological and psychological stress [108,117–119]. In addition, families with a lower ITN ratio usually live in a more crowded household and noisier environment, which leads to heightened physiological and psychological stress [120–122]. Furthermore, parental emotional distress, often triggered by low income or indebtedness, can result in increased inter-family conflict and the diminished efficacy of parenting practices. Both of these factors have cascading negative effects on children's behavioral, health, and cognitive outcomes [111,123–125].

Evidence derived from empirical research indicates that exposure to stressful events and environments is a significant contributing factor to the observed disparities in brain structure and function among children from lower-income households [106]. Stressors common in low-income families significantly impact the amygdala, dorsolateral, and ventrolateral prefrontal cortex (DLPFC and VLPFC, respectively) [126,127]. These brain regions play a crucial role in regulating emotions, controlling cognition, executing functions, and promoting goal-oriented behaviors [128–130]. According to a report by Kim et al. [128], exposure to chronic stressors during childhood serves as a mediator between low-income conditions in childhood and decreased activity in the DLPFC and VLPFC in adults. This suggests that the stressors associated with low-income environments can have long-term effects on brain function. The hippocampus, central to many cognitive functions and emotional processes, also exhibits developmental differences due to stress from low-income households, which explains variations in long-term memory, learning, control of neuroendocrine functions, and the modulation of emotional behavior [131]. It has been reported that lower chronic stress is associated with higher hippocampal volume, which is associated with higher memory performance [108,131–134]. Furthermore, repeated and prolonged stress leads to the shortening and debranching of dendrites in the hippocampus, a decrease in gray matter volume (GMV) in the anterior cingulate, hippocampus, and parahippocampal gyrus, and induces opposing effects on structural plasticity, such as dendritic atrophy and excitatory synapse loss in the hippocampus and prefrontal cortex [135–137].

Childhood stress not only affects the brain's structure but also alters the functional connectivity of neural networks [138]. Philip et al. [139] found that individuals who experienced early life stress showed a greater dissociation between the executive and default mode networks. Specifically, the left dorsolateral prefrontal cortex (DLPFC) exhibited increased local connectivity with the left middle frontal gyrus, while the right DLPFC showed negative connectivity with the inferior parietal lobule compared to the control group. Another rsfMRI study found that higher levels of early life stress are linked to reduced global connectivity and hub-like features in the left DLPFC [138]. Kim et al. [128] reported a positive coupling between the amygdala and the VLPFC in individuals from higher-income households in comparison to low-income households, with greater VLPFC activity resulting in decreased amygdala activation during emotion regulation.

Building upon the comprehensive analysis of rsMRI studies, rsEEG studies have sought to pinpoint the impact of income and stress on brain function. Generally, during development, there is a reduction in delta and theta Relative Power (RP), and a corresponding increase in alpha and beta RP with advancing age [53,58,59,66,68,69]. However, exposure to stressful environments and life events, which are commonly observed in children from low-income households, as previously discussed, has the potential to disrupt this developmental trajectory. Elevated physiological stress levels in mothers, typically gauged by hair cortisol concentration, is correlated with a relative increase in lower frequency power and a relative decrease in higher frequency power in their 6–12-month-old infants [62]. This alteration in the brain's maturational journey is often considered a sign of possible delay in cognitive development, working memory, learning, and attention [4,62,140–143]. The same results are observed in other rsEEG studies with older children [4,58,59], where children from low-income homes exhibit poorer performance on language tasks and distinct rsEEG patterns, including increased theta power and decreased alpha power. While these studies independently evaluated income and stress, they consistently yielded similar results, providing additional evidence supporting the assertion that stress functions as a mediator in the relationship between income and brain development. Moreover, among older adults, it has been observed that those who have experienced cumulative life stress exhibit elevations in delta power, which has been associated with the emergence of Alzheimer's disease and mild cognitive impairment, suggesting that income and stress have long lasting effects on brain development [38].

Researchers employing machine learning algorithms in an attempt to diagnose stress are contributing valuable insights into potential rsEEG characteristics that could effectively identify individuals exhibiting symptoms of stress [144,145]. Baumgartl et al. [144] accurately classified individuals with symptoms of high stress at a rate of 81.33% based on significantly higher power in lower frequencies, and significantly lower power in higher frequencies. In the analysis of source-level cortical rsEEG activity related to stress, Vanhollebeke et al. [146] found an increase in alpha power, indicative of decreased cortical activity in both the left and right precuneus and the right PCC, as well as an increase in functional connectivity between the left and right precuneus. This observed decrease in activity within the precuneus/PCC, a crucial component of the default mode network (DMN), may indicate a disruption in an individual's self-referential neural processes. Income-related, stress-dependent differences in rsEEG also have important implications for children's cognitive development, as researchers have found that increased rsEEG theta in toddlerhood [147] and during the school years (Schneider et al., under review), are associated with income status, and relates to lower scores on measures of IQ and working memory.

2. Understanding the Role of Cultural Context When Examining SES-Related Differences in Brain and Cognitive Development

To date, much of the research on SES-related differences in brain development draws conclusions from comparisons of two groups that differ markedly in one or more respects (e.g., SES, race, ethnicity, first language, bilingualism), [148]. This is problematic as important mechanisms through which SES might act on the developing brain to influence cognitive outcomes are often overlooked. Despite a number of recent studies demonstrating the mediating role of language input and physiological stress on brain and language development, there are a number of other important confounds which are related to SES that should be considered (see Figure 3). For example, culture, multilingualism, and household density may influence brain and cognitive development in unique ways that have not yet been examined [148]. Here, we outline confounds of SES that should be considered in future studies of SES and brain development. Consideration of these factors in future studies of the relationship between SES and brain development would enhance the scope of applicability and facilitate more informed policymaking to positively impact a broader demographic.

2.1. Consideration of Multigenerational Households

The 30-million-word gap has been a topic of debate in child development research since Hart and Risley first published their landmark study in 1995 [83]. They famously reported a difference in language exposure between children of different socioeconomic status (SES) homes by collecting and analyzing home speech samples of low-SES and high SES families. Specifically, they found that, by the age of 4, children from higher SES families heard, on average, 30 million more words than children from lower SES families [83]. Given that research has shown that the amount of language children hear impacts both brain structure and function [29,94,102], there has been an ever-growing interest in studying the home language environment.

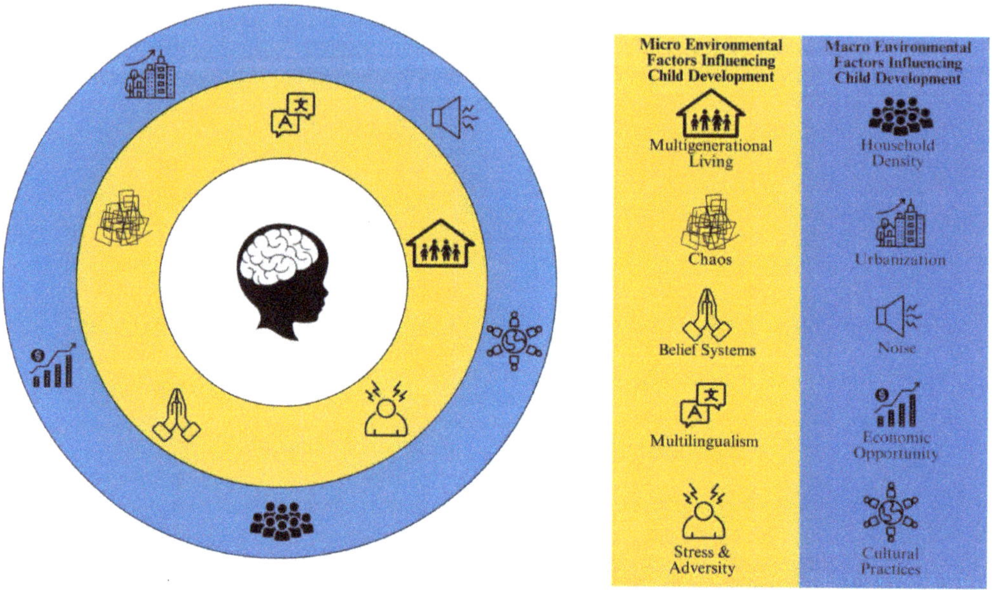

Figure 3. Micro and macro environmental factors that differentially influence child cognitive and brain development. Items highlighted in yellow represent micro-environmental factors that are more proximal to the child's development, while items highlighted in blue represent macro-environmental factors that are more distal to the child's development.

A critical issue with existing studies relating children's language environments to brain development is that they often fail to consider differences in cultural beliefs about multi-generational cohabitation. Households which include grandparents and/or extended family are common in collectivist societies throughout South America, Africa, and Asia [149], and are increasing in prevalence among developing countries [150]. Due to cultural and economic shifts, the number of multiple generational homes in the US has quadrupled in the last 50 years [151]. How this might influence the child's language development is still an area of debate, as multigenerational households can offer multiple communication partners but lead to reductions in one-on-one interactions. Research that expands the definition of language exposure to include both language input from a primary caregiver and language produced to and around the child by others (e.g., siblings, aunts/uncles, other primary caregivers, etc.) results in a vastly diminished word gap on the basis of SES while also taking into account socio-cultural differences in family structure [152–154]. Specifically, when considering all speech in a child's environment instead of just maternal or parental speech, children from different SES groups do not hear vastly different language quantities [152,153]. This is critical to consider in future studies of language input and brain development, as lower-SES families and racial and ethnic minority groups, including foreign-born, Asian, Black, and Hispanic Americans are more likely to live in larger, multigenerational households (also termed co-residence) [151,155]. Thus, children commonly examined in studies of SES-related differences in brain and cognitive development are more likely to experience environments with numerous sources of language input. The failure to consider how these variable sources of language input mediate the relationship between SES and brain development, therefore, severely impedes our ability to extrapolate a holistic perspective of child brain development.

2.2. Consideration of Caregiver Practices and Belief Systems

Given the importance of language input in brain development, recent research has examined how modifying the home language environment can promote brain development in children from lower SES households. Romeo et al. [156] revealed a positive link between improved turn-taking (following intervention) and cortical thickening in the left inferior frontal and supramarginal gyri. Notably, the supramarginal gyrus mediated the relationship between turn-taking changes and children's language development following intervention.

Despite the positive effects of these interventions, an unaddressed issue with much of this research is the lack of consideration of caregiver beliefs and practices about language input in the home environment. There is growing evidence of SES and race-based differences in parents' knowledge about general infant development and early language and literacy development [157–159]. For example, despite agreement between mid SES Black mothers and White mothers on a number of caregiving practices, low-SES Black mothers believe that their children's language develops naturally and does not require child-directed speech input [160]. Furthermore, Luo et al. [161] demonstrated that bilingual, Hispanic parents with higher levels of education and income were more likely than their lower SES counterparts to recognize the benefits of supporting language development in the home and school contexts. Although these studies focus primarily on caregiver beliefs and practices related to language input, their relevance for brain development should not be overlooked, as language input is consistently highlighted in studies of SES-related differences of brain development.

2.3. Consideration of Underrepresented Cultures

In an excellent review of the state of child development research, Draper et al. [162] highlight the lack of global representation in the field between 2006–2010. Fewer than 3% of participants in studies of child development were from Central and South America, Africa, Asia, or the Middle East, despite these contexts constituting 85% of the world's population [163]. This lack of representation is a shortcoming in existing research, as the environmental experiences of children from low-SES environments in the United States may differ drastically from those of children in non-WEIRD cultures. For example, although the influence of SES on child development is stable across countries [164], a meta-analysis of 39 low-income and middle-income countries found that the association between these variables systematically differs based on unique environmental experiences, such as country-level adult illiteracy, infant mortality, and food insecurity [165]. Similarly, in China, maternal education is the strongest predictor of children's educational attainment, however, this relationship is fully mediated by parenting practices and children's learning attitudes, which differ from Westernized parenting practices and child attitudes [166]. This lack of global representation severely limits our understanding of the role culture plays in brain and cognitive development.

As previously stated, language input mediates the pathway between SES and brain development [94], however, the exposure to language input substantially differs not just across households, but across cultures [167]. In Tseltal Mayan and Papua New Guinea cultures, children are rarely directly addressed by caregivers, but experience higher rates of overheard speech or speech that is not directed to them [168]. Similarly, children in rural farming communities hear significantly less input from caregivers than children in urban regions, where the bulk of extant research studies are conducted [169]. Importantly though, children in these global communities manage to extract the linguistic information they need to learn new words despite minimal directed speech [168]. Therefore, while westernized standards promote child directed speech, something commonly targeted in interventions aimed at improving children's language environments, a number of studies in other regions of the globe indicate that overheard speech is equally as beneficial for the development of children's language skills [170,171], and likely brain development.

However, to date, there are no studies examining how overheard speech mediates the relationship between SES and child brain development. Given that rsEEG is relatively low in cost and mobile, it offers an opportunity to explore how higher levels of overheard speech relate to variability in brain development across a range of cultures.

2.4. Consideration of Bilingual Populations

There are more bilingual and multilingual children around the world who grow up learning and using two or more languages than there are monolingual children [172,173]. Reported rates of bilingualism in places such as Europe (67%), Canada (55%), India (25%), and the United States (20%) indicate that bilingualism is both prevalent and growing [174]. Rates of bilingualism in the United States are even higher among children (26% of 5–17-year-olds nationwide), with certain regions reporting nearly 50% of the child population being bilingual (44% in California [175]) and 49% in Texas [176]. Despite the dominance of bi- and multi-lingualism in global and local contexts, the influence of bilingualism on child brain development remains poorly understood in relation to what is known about monolingual populations [172]. In fact, many studies of brain and language development in the United States have historically excluded bilingual participants.

Monolingual and bilingual children experience different language environments, which, in turn, may result in changes in the development of cortical networks underlying language and cognitive development. Unlike monolingual families, bilingual families differ in which languages are spoken by whom, and specifically when each language is spoken to the child [177]. For example, in a study of Spanish–English bilingual mothers, the effect of maternal education on child language abilities was only observed in the language in which the mother received her highest level of education [97]. Additionally, research has shown that language input from older siblings, but not from parents, is associated with stronger lexical, grammatical, and narrative abilities in English among young bilingual speakers [178,179]. The language environment of bilingual children has been shown to be highly dynamic as different individuals join or leave the household [180]. These findings are important as they highlight variability in children's language environments that are unaccounted for when conducting research in monolingual populations only.

It is especially important to consider variability in bilingual children's environments, as research has shown that bilingualism can serve as an important neuromodulator [181–183]. Specifically, variability in bilingual language experience modulates neural adaptability during development [181,183,184], finding that bilingualism permanently shifts underlying brain mechanisms, leading to long-term structural and functional changes [185–189]. This suggests that life experiences related to bilingualism can ultimately induce brain changes, such as volumetric changes, differences in spatial distribution and functional connectivity, as well as more distributed cortical activation, which converges in frontal regions [182,185,187,188,190]. Patterns of structural differences appear to depend on whether two languages were acquired simultaneously from birth, or sequentially before age five [191], suggesting that timing of bilingualism interacts with brain development [192]. Thus, bilingual children's early language experiences not only differ from monolingual children and result in different trajectories for brain development, but the age at which they experience bilingual language input results in variability in neural circuitry as well. Despite these associations, little to no research has directly measured the quantity and quality of language input by bilingual caregivers within each language, and how these differences influence the trajectory of children's brain development, limiting our ability to generalize existing findings to the increasingly dominant bilingual population.

2.5. Current Demographic Shifts: Household Density, Urbanization, Chaos, and Noise

Stress is also a key mechanism linking children's exposure to poverty-related adversity and brain development. Chronic stressors related to urbanization include higher household density, noise, and household chaos, which alter the physiologic response to stress, leading to potentially teratogenic effects of stress-related hormones on the developing brain and to a range of adverse outcomes in the development of executive function, socio-emotional, language, and the brain [193–201]. Understanding how the factors associated with urbanization influence caregiver and child stress and, in turn, impacting child brain development, is critical as the world is currently experiencing a wave of urbanization and population growth. Individuals living in more urban regions (based on measures of population density and duration of residency) exhibit differences in brain volume, cortical surface area, and brain network connectivity in the medial prefrontal cortex and cerebellum, related to depressive symptoms and difficulties perspective-taking. Importantly, the effects of living in a more urban area on brain development are the greatest in the cerebellum during childhood, and in the prefrontal cortex from childhood to adolescence [202], suggesting that urban environments alter different aspects of the brain across the course of development. Understanding how chronic stressors associated with urbanization, such as higher household density, noise, and household chaos, influence brain and language development is vital for addressing differences in these domains on the basis of SES.

Household density is often measured as the number of people in a household divided by number of rooms (or bedrooms) and is more common among lower income households, with the available number of rooms per household member being significantly lower in lower income neighborhoods [122,203]. Household density is often viewed negatively, with past research indicating that household density is negatively correlated with language outcomes in children [204–207]. However, this deficit-based perspective regarding the relationship between household density and child language development lacks cultural responsivity. One key issue is that most studies of household density fail to differentiate adults from children in the household. For example, when using traditional ratios of number of people to number of bedrooms in the home, household density is negatively associated with language outcomes. Importantly though, when the family make up is calculated as adult to child ratio, there are positive outcomes for children having more adults in the home [122]. Furthermore, Sperry, Sperry, and Miller [152] revealed that when speech overheard by children from other adults in the home was included in calculations of the children's language input, there was no significant relationship between vocabulary and SES. Similar work related to brain development has not yet been conducted, but taking a more nuanced approach to household density in the future may provide greater insights into how individual members of the household might support or detract from a child's language environment.

Another stressor related to urbanization is the presence of noise in both the child's home and community environment. Noise can mask the speech signal of interest, making it difficult for children to access meaningful language input. This is particularly concerning as research has shown that children raised in noisy households face an increased risk of experiencing language, reading, and academic difficulties [208–210], as well as deficits in cognitive abilities that underlie language processing, such as working memory and attention [211,212]. Noisy environments are more common in high density households and can pose a risk to children's cognitive and brain development. Simon et al. [213] reported that children exposed to excessive levels of noise exhibited reduced cortical thickness in the left IFG, a region that is crucial for language processing. Furthermore, a study of 8–12-year-olds exposed to school road-traffic noise reported enhanced connectivity in the subcortical auditory pathway, suggesting that prolonged noise exposure may accelerate the neural maturation of the auditory pathway [214]. These findings add to a growing literature that explores the extent to which environmental noise is associated with neurobiological development underlying language and cognitive outcomes in children. Despite evidence of a negative association between noise and child brain and language development, children exposed to intense environmental noise may initially show an *increased* ability to tune out auditory distractors [215], making them better at learning from the language input they are provided. Thus, while noise in the environment is thought to typically have a negative effect on children's brain, vocabulary, and reading development, it is possible that noise exposure may initially offer an advantage to young children. Given that reducing noise in high density households may be difficult, it would be beneficial for future research to take a strengths-based approach to understanding which aspects of environmental noise may promote child brain and language development.

Lastly, urbanization may result in increased levels of chaos, a common measure of chronic stress, however, the measures utilized to assess chaos in the home may also be susceptible to examiner bias. Chaos is most commonly measured based on caregiver self-report using the Confusion, Hubbub and Order Scale (CHAOS) or by examiner observation using the Home Observation for Measurement (HOME) of the Environment. These two measures capture unique and different aspects of the environment that should be considered in context. The same environment can be viewed as chaotic to one person and a warm, active, lively household to another. In a study examining the psychometric properties of the HOME, there was substantial variability in its ecological validity across diverse cultural contexts [216]. Similarly, the CHAOS includes items with language that may be stigmatizing and off-putting to families from diverse backgrounds (e.g., "it's a real zoo in our house", "there is often a fuss going on at our home") [217]. Making assumptions without considering the larger context which influences behavior may reinforce stigmatization among communities with different cultural values or life experiences [217].

Urbanization may result in individuals being exposed to multiple stressors simultaneously, making it challenging to quantify the impact of a specific environmental factor on child brain development [218]. In summary, the multifaceted impact of urbanization related to household density, environmental noise, and chaos can influence language, cognitive, and brain development through various pathways, including overcrowding and caregiver stress. Recognizing and understanding these intricate relationships is essential to develop effective strategies to support language development in low-SES household settings.

2.6. Considerations for Texture and Style of Hair in Neuroimaging Techniques

One factor that may substantially influence the generalizability of rsEEG research is rooted in the technique by which this data is collected. EEG data collection requires access to the hair and scalp, indicating that the practical usage of EEG is affected by physical factors such as the texture and style of hair, as well as sociocultural factors relevant to having one's hair touched by someone else [219]. These physical factors have historically, and currently, resulted in the exclusion of groups of people, particularly Black communities [220–226]. As such, most EEG research studies have been dominated by individuals with privileged identities, particularly White, college-educated people [222,223,227,228]. However, due to long-standing systems of racial oppression, the representation of samples is often not reported, and findings are generalized as if they were drawn from representative samples [229]. These same systems of racial oppression and structural racism have resulted in a higher distribution of Black families living in lower-income environments [230,231], indicating that rsEEG studies examining socioeconomic differences may not be representative of individuals from diverse racial groups. To promote diversity in the participants recruited for EEG studies, researchers should consider engaging Black researchers in both the development and execution of EEG research, as this has been shown to result in a more positive experience for the participant while also collecting useful data [232–235]. Providing hair care products that are conducive to Black hair may also promote inclusivity and make the cleanup process less burdensome for participants [233,236]. Although new organizations such as Black in Neuro (blackinneuro.com) and SPARK Society (sparksociety.org) are conducting foundational work to raise awareness about systemic racism in neuroscience, researchers can also consider ways by which their EEG methods can be revised to promote diversity, equity, and inclusion.

3. Conclusions

SES is known to influence brain development through accumulated environmental experiences. Specifically, there appear to be two dominant pathways which account for this association: (1) maternal education results in differences in the quality and quantity of language input children hear, having downstream effects on the development of language-related brain regions and (2) differences in income alter physiological stress in the home, in turn impacting regions of the brain that are critical for self-regulation and working memory. While relatively underutilized in studies of neuroimaging, we provide evidence that rsEEG can uncover important differences in brain function that are subserved by these socioeconomic differences in environmental experiences. Furthermore, it is a cost effective and mobile technology that can capture brain development across a variety of households and cultures. This is critical in providing a more comprehensive understanding of the relationship between SES and brain development, one that includes the broader cultural context related to factors which may interact with SES or exert unique influences on brain development that are often overlooked, such bilingualism, household density, parenting practices, environmental noise, and chaos. Considering the broader cultural context in SES-related studies of brain development will advance the field of developmental cognitive neuroscience and provide researchers with a more generalizable, holistic model of child brain development.

Author Contributions: J.M.S.: Conceptualization, Visualization, Writing—original draft, Writing—review & editing. M.H.B.: Writing—original draft, Writing—review & editing. M.J.M.: Conceptualization, Visualization, Writing—original draft, Writing—review & editing. All authors were involved in the writing of the original draft, and subsequent revisions. All authors have read and agreed to the published version of the manuscript.

Funding: This research was funded by the National Science Foundation, Division of Behavioral and Cognitive Sciences, to M.J.M. (ID #2317734) and J.M.S. (ID # 2141007).

Conflicts of Interest: The authors declare no conflict of interest.

References

1. Farah, M.J. The Neuroscience of Socioeconomic Status: Correlates, Causes, and Consequences. *Neuron* **2017**, *96*, 56–71. [CrossRef] [PubMed]
2. Duncan, G.J.; Magnuson, K. Socioeconomic Status and Cognitive Functioning: Moving from Correlation to Causation. *WIREs Cogn. Sci.* **2012**, *3*, 377–386. [CrossRef] [PubMed]
3. Harmony, T.; Alvarez, A.; Pascual, R.; Ramos, A.; Marosi, E.; Díaz De León, A.E.; Valdés, P.; Becker, J. EEG Maturation on Children with Different Economic and Psychosocial Characteristics. *Int. J. Neurosci.* **1988**, *41*, 103–113. [CrossRef] [PubMed]
4. Maguire, M.J.; Schneider, J.M. Socioeconomic Status Related Differences in Resting State EEG Activity Correspond to Differences in Vocabulary and Working Memory in Grade School. *Brain Cogn.* **2019**, *137*, 103619. [CrossRef]
5. Noble, K.G.; Houston, S.M.; Kan, E.; Sowell, E.R. Neural Correlates of Socioeconomic Status in the Developing Human Brain. *Dev. Sci.* **2012**, *15*, 516–527. [CrossRef] [PubMed]
6. Noble, K.G.; Houston, S.M.; Brito, N.H.; Bartsch, H.; Kan, E.; Kuperman, J.M.; Akshoomoff, N.; Amaral, D.G.; Bloss, C.S.; Libiger, O.; et al. Family Income, Parental Education and Brain Structure in Children and Adolescents. *Nat. Neurosci.* **2015**, *18*, 773–778. [CrossRef] [PubMed]
7. Olson, L.; Chen, B.; Fishman, I. Neural Correlates of Socioeconomic Status in Early Childhood: A Systematic Review of the Literature. *Child Neuropsychol.* **2021**, *27*, 390–423. [CrossRef] [PubMed]
8. Otero, G.A. Poverty, Cultural Disadvantage and Brain Development: A Study of Pre-School Children in Mexico. *Electroencephalogr. Clin. Neurophysiol.* **1997**, *102*, 512–516. [CrossRef] [PubMed]
9. Duncan, G.J.; Yeung, W.J.; Brooks-Gunn, J.; Smith, J.R. How Much Does Childhood Poverty Affect the Life Chances of Children? *Am. Sociol. Rev.* **1998**, *63*, 406–423. [CrossRef]
10. Cohen, S.; Janicki-Deverts, D.; Chen, E.; Matthews, K.A. Childhood Socioeconomic Status and Adult Health. *Ann. N. Y. Acad. Sci.* **2010**, *1186*, 37–55. [CrossRef]
11. Evans, G.W.; Cassells, R.C. Childhood Poverty, Cumulative Risk Exposure, and Mental Health in Emerging Adults. *Clin. Psychol. Sci.* **2014**, *2*, 287–296. [CrossRef] [PubMed]
12. Schnack, H.G.; van Haren, N.E.M.; Brouwer, R.M.; Evans, A.; Durston, S.; Boomsma, D.I.; Kahn, R.S.; Hulshoff Pol, H.E. Changes in Thickness and Surface Area of the Human Cortex and Their Relationship with Intelligence. *Cereb. Cortex* **2015**, *25*, 1608–1617. [CrossRef] [PubMed]
13. Sowell, E.R.; Peterson, B.S.; Thompson, P.M.; Welcome, S.E.; Henkenius, A.L.; Toga, A.W. Mapping Cortical Change across the Human Life Span. *Nat. Neurosci.* **2003**, *6*, 309–315. [CrossRef] [PubMed]
14. Tamnes, C.K.; Herting, M.M.; Goddings, A.-L.; Meuwese, R.; Blakemore, S.-J.; Dahl, R.E.; Güroğlu, B.; Raznahan, A.; Sowell, E.R.; Crone, E.A.; et al. Development of the Cerebral Cortex across Adolescence: A Multisample Study of Inter-Related Longitudinal Changes in Cortical Volume, Surface Area, and Thickness. *J. Neurosci.* **2017**, *37*, 3402–3412. [CrossRef] [PubMed]

15. Brouwer, R.M.; Klein, M.; Grasby, K.L.; Schnack, H.G.; Jahanshad, N.; Teeuw, J.; Thomopoulos, S.I.; Sprooten, E.; Franz, C.E.; Gogtay, N.; et al. Genetic Variants Associated with Longitudinal Changes in Brain Structure across the Lifespan. *Nat. Neurosci.* **2022**, *25*, 421–432. [CrossRef] [PubMed]
16. Bethlehem, R.A.I.; Seidlitz, J.; White, S.R.; Vogel, J.W.; Anderson, K.M.; Adamson, C.; Adler, S.; Alexopoulos, G.S.; Anagnostou, E.; Areces-Gonzalez, A.; et al. Brain Charts for the Human Lifespan. *Nature* **2022**, *604*, 525–533. [CrossRef]
17. Grady, C. The Cognitive Neuroscience of Ageing. *Nat. Rev. Neurosci.* **2012**, *13*, 491–505. [CrossRef]
18. Luna, B. Developmental Changes in Cognitive Control through Adolescence. In *Advances in Child Development and Behavior*; Bauer, P., Ed.; JAI: Greenwich, CT, USA, 2009; Volume 37, pp. 233–278.
19. Hackman, D.A.; Farah, M.J. Socioeconomic Status and the Developing Brain. *Trends Cogn. Sci.* **2009**, *13*, 65–73. [CrossRef]
20. Mackey, A.P.; Finn, A.S.; Leonard, J.A.; Jacoby-Senghor, D.S.; West, M.R.; Gabrieli, C.F.O.; Gabrieli, J.D.E. Neuroanatomical Correlates of the Income-Achievement Gap. *Psychol. Sci.* **2015**, *26*, 925–933. [CrossRef]
21. Hanson, J.L.; Hair, N.; Shen, D.G.; Shi, F.; Gilmore, J.H.; Wolfe, B.L.; Pollak, S.D. Family Poverty Affects the Rate of Human Infant Brain Growth. *PLoS ONE* **2013**, *8*, e80954. [CrossRef]
22. Jednoróg, K.; Altarelli, I.; Monzalvo, K.; Fluss, J.; Dubois, J.; Billard, C.; Dehaene-Lambertz, G.; Ramus, F. The Influence of Socioeconomic Status on Children's Brain Structure. *PLoS ONE* **2012**, *7*, e42486. [CrossRef] [PubMed]
23. Lawson, G.M.; Duda, J.T.; Avants, B.B.; Wu, J.; Farah, M.J. Associations between Children's Socioeconomic Status and Prefrontal Cortical Thickness. *Dev. Sci.* **2013**, *16*, 641–652. [CrossRef]
24. Kim, D.-J.; Davis, E.P.; Sandman, C.A.; Glynn, L.; Sporns, O.; O'Donnell, B.F.; Hetrick, W.P. Childhood Poverty and the Organization of Structural Brain Connectome. *NeuroImage* **2019**, *184*, 409–416. [CrossRef]
25. Moriguchi, Y.; Shinohara, I. Socioeconomic Disparity in Prefrontal Development during Early Childhood. *Sci. Rep.* **2019**, *9*, 2585. [CrossRef] [PubMed]
26. Rakesh, D.; Whittle, S. Socioeconomic Status and the Developing Brain—A Systematic Review of Neuroimaging Findings in Youth. *Neurosci. Biobehav. Rev.* **2021**, *130*, 379–407. [CrossRef]
27. Sheridan, M.A.; Sarsour, K.; Jutte, D.; D'Esposito, M.; Boyce, W.T. The Impact of Social Disparity on Prefrontal Function in Childhood. *PLoS ONE* **2012**, *7*, e35744. [CrossRef] [PubMed]
28. Su, M.; Li, P.; Zhou, W.; Shu, H. Effects of Socioeconomic Status in Predicting Reading Outcomes for Children: The Mediation of Spoken Language Network. *Brain Cogn.* **2021**, *147*, 105655. [CrossRef]
29. Romeo, R.R.; Leonard, J.A.; Robinson, S.T.; West, M.R.; Mackey, A.P.; Rowe, M.L.; Gabrieli, J.D.E. Beyond the 30-Million-Word Gap: Children's Conversational Exposure Is Associated with Language-Related Brain Function. *Psychol. Sci.* **2018**, *29*, 700–710. [CrossRef] [PubMed]
30. King, L.S.; Camacho, M.C.; Montez, D.F.; Humphreys, K.L.; Gotlib, I.H. Naturalistic Language Input Is Associated with Resting-State Functional Connectivity in Infancy. *J. Neurosci.* **2021**, *41*, 424–434. [CrossRef]
31. McDermott, C.L.; Seidlitz, J.; Nadig, A.; Liu, S.; Clasen, L.S.; Blumenthal, J.D.; Reardon, P.K.; Lalonde, F.; Greenstein, D.; Patel, R.; et al. Longitudinally Mapping Childhood Socioeconomic Status Associations with Cortical and Subcortical Morphology. *J. Neurosci.* **2019**, *39*, 1365–1373. [CrossRef]
32. Fox, M.D.; Raichle, M.E. Spontaneous Fluctuations in Brain Activity Observed with Functional Magnetic Resonance Imaging. *Nat. Rev. Neurosci.* **2007**, *8*, 700–711. [CrossRef] [PubMed]
33. De Luca, M.; Beckmann, C.F.; De Stefano, N.; Matthews, P.M.; Smith, S.M. fMRI Resting State Networks Define Distinct Modes of Long-Distance Interactions in the Human Brain. *NeuroImage* **2006**, *29*, 1359–1367. [CrossRef] [PubMed]
34. Damoiseaux, J.S.; Rombouts, S.A.R.B.; Barkhof, F.; Scheltens, P.; Stam, C.J.; Smith, S.M.; Beckmann, C.F. Consistent Resting-State Networks across Healthy Subjects. *Proc. Natl. Acad. Sci. USA* **2006**, *103*, 13848–13853. [CrossRef] [PubMed]
35. Mantini, D.; Perrucci, M.G.; Del Gratta, C.; Romani, G.L.; Corbetta, M. Electrophysiological Signatures of Resting State Networks in the Human Brain. *Proc. Natl. Acad. Sci. USA* **2007**, *104*, 13170–13175. [CrossRef] [PubMed]
36. Leopold, D.A.; Murayama, Y.; Logothetis, N.K. Very Slow Activity Fluctuations in Monkey Visual Cortex: Implications for Functional Brain Imaging. *Cereb. Cortex* **2003**, *13*, 422–433. [CrossRef] [PubMed]
37. Bice, K.; Yamasaki, B.L.; Prat, C.S. Bilingual Language Experience Shapes Resting-State Brain Rhythms. *Neurobiol. Lang.* **2020**, *1*, 288–318. [CrossRef] [PubMed]
38. Marshall, A.C.; Cooper, N.R. The Association between High Levels of Cumulative Life Stress and Aberrant Resting State EEG Dynamics in Old Age. *Biol. Psychol.* **2017**, *127*, 64–73. [CrossRef] [PubMed]
39. Gellci, K.; Marusak, H.A.; Peters, C.; Elrahal, F.; Iadipaolo, A.S.; Rabinak, C.A. Community and Household-Level Socioeconomic Disadvantage and Functional Organization of the Salience and Emotion Network in Children and Adolescents. *NeuroImage* **2019**, *184*, 729–740. [CrossRef] [PubMed]
40. Gao, W.; Alcauter, S.; Elton, A.; Hernandez-Castillo, C.R.; Smith, J.K.; Ramirez, J.; Lin, W. Functional Network Development During the First Year: Relative Sequence and Socioeconomic Correlations. *Cereb. Cortex* **2015**, *25*, 2919–2928. [CrossRef]
41. Barch, D.; Pagliaccio, D.; Belden, A.; Harms, M.P.; Gaffrey, M.; Sylvester, C.M.; Tillman, R.; Luby, J. Effect of Hippocampal and Amygdala Connectivity on the Relationship between Preschool Poverty and School-Age Depression. *Am. J. Psychiatry* **2016**, *173*, 625–634. [CrossRef]
42. Tooley, U.A.; Mackey, A.P.; Ciric, R.; Ruparel, K.; Moore, T.M.; Gur, R.C.; Gur, R.E.; Satterthwaite, T.D.; Bassett, D.S. Associations between Neighborhood SES and Functional Brain Network Development. *Cereb. Cortex* **2020**, *30*, 1–19. [CrossRef] [PubMed]

43. Tooley, U.A.; Bassett, D.S.; Mackey, A.P. Environmental Influences on the Pace of Brain Development. *Nat. Rev. Neurosci.* **2021**, *22*, 372–384. [CrossRef] [PubMed]
44. Buzsáki, G. *Rhythms of the Brain*; Oxford University Press: Oxford, UK, 2006; ISBN 978-0-19-986371-6.
45. Dustman, R.E.; Shearer, D.E.; Emmerson, R.Y. Life Span Changes in EEG Spectral Amplitude, Amplitude Variability and Mean Frequency. *Clin. Neurophysiol.* **1999**, *110*, 1399–1409. [CrossRef] [PubMed]
46. Lüchinger, R.; Michels, L.; Martin, E.; Brandeis, D. Brain State Regulation during Normal Development: Intrinsic Activity Fluctuations in Simultaneous EEG–fMRI. *NeuroImage* **2012**, *60*, 1426–1439. [CrossRef] [PubMed]
47. Sakai, J. Functional Near-Infrared Spectroscopy Reveals Brain Activity on the Move. *Proc. Natl. Acad. Sci. USA* **2022**, *119*, e2208729119. [CrossRef] [PubMed]
48. Whitford, T.J.; Rennie, C.J.; Grieve, S.M.; Clark, C.R.; Gordon, E.; Williams, L.M. Brain Maturation in Adolescence: Concurrent Changes in Neuroanatomy and Neurophysiology. *Hum. Brain Mapp.* **2007**, *28*, 228–237. [CrossRef] [PubMed]
49. Lüchinger, R.; Michels, L.; Martin, E.; Brandeis, D. EEG–BOLD Correlations during (Post-)Adolescent Brain Maturation. *NeuroImage* **2011**, *56*, 1493–1505. [CrossRef] [PubMed]
50. Benninger, C.; Matthis, P.; Scheffner, D. EEG Development of Healthy Boys and Girls. Results of a Longitudinal Study. *Electroencephalogr. Clin. Neurophysiol.* **1984**, *57*, 1–12. [CrossRef]
51. Clarke, A.R.; Barry, R.J.; McCarthy, R.; Selikowitz, M. Age and Sex Effects in the EEG: Development of the Normal Child. *Clin. Neurophysiol.* **2001**, *112*, 806–814. [CrossRef]
52. Cragg, L.; Kovacevic, N.; McIntosh, A.R.; Poulsen, C.; Martinu, K.; Leonard, G.; Paus, T. Maturation of EEG Power Spectra in Early Adolescence: A Longitudinal Study. *Dev. Sci.* **2011**, *14*, 935–943. [CrossRef]
53. Harmony, T.; Marosi, E.; Díaz de León, A.E.; Becker, J.; Fernández, T. Effect of Sex, Psychosocial Disadvantages and Biological Risk Factors on EEG Maturation. *Electroencephalogr. Clin. Neurophysiol.* **1990**, *75*, 482–491. [CrossRef]
54. Matthis, P.; Scheffner, D.; Benninger, C.; Lipinski, C.; Stolzis, L. Changes in the Background Activity of the Electroencephalogram According to Age. *Electroencephalogr. Clin. Neurophysiol.* **1980**, *49*, 626–635. [CrossRef]
55. Somsen, R.J.M.; van't Klooster, B.J.; van der Molen, M.W.; van Leeuwen, H.M.P.; Licht, R. Growth Spurts in Brain Maturation during Middle Childhood as Indexed by EEG Power Spectra. *Biol. Psychol.* **1997**, *44*, 187–209. [CrossRef]
56. Tröndle, M.; Popov, T.; Dziemian, S.; Langer, N. Decomposing the Role of Alpha Oscillations during Brain Maturation. *eLife* **2022**, *11*, e77571. [CrossRef]
57. Wilkinson, C.L.; Pierce, L.J.; Sideridis, G.; Wade, M.; Nelson, C.A. Associations between EEG Trajectories, Family Income, and Cognitive Abilities over the First Two Years of Life. *Dev. Cogn. Neurosci.* **2023**, *61*, 101260. [CrossRef]
58. Otero, G.A. Eeg Spectral Analysis in Children with Sociocultural Handicaps. *Int. J. Neurosci.* **1994**, *79*, 213–220. [CrossRef]
59. Otero, G.A.; Pliego-Rivero, F.B.; Fernández, T.; Ricardo, J. EEG Development in Children with Sociocultural Disadvantages: A Follow-up Study. *Clin. Neurophysiol.* **2003**, *114*, 1918–1925. [CrossRef]
60. Cantiani, C.; Piazza, C.; Mornati, G.; Molteni, M.; Riva, V. Oscillatory Gamma Activity Mediates the Pathway from Socioeconomic Status to Language Acquisition in Infancy. *Infant Behav. Dev.* **2019**, *57*, 101384. [CrossRef]
61. Tomalski, P.; Moore, D.G.; Ribeiro, H.; Axelsson, E.L.; Murphy, E.; Karmiloff-Smith, A.; Johnson, M.H.; Kushnerenko, E. Socioeconomic Status and Functional Brain Development—Associations in Early Infancy. *Dev. Sci.* **2013**, *16*, 676–687. [CrossRef]
62. Troller-Renfree, S.V.; Brito, N.H.; Desai, P.M.; Leon-Santos, A.G.; Wiltshire, C.A.; Motton, S.N.; Meyer, J.S.; Isler, J.; Fifer, W.P.; Noble, K.G. Infants of Mothers with Higher Physiological Stress Show Alterations in Brain Function. *Dev. Sci.* **2020**, *23*, e12976. [CrossRef]
63. Barry, R.J.; Clarke, A.R.; Hajos, M.; McCarthy, R.; Selikowitz, M.; Dupuy, F.E. Resting-State EEG Gamma Activity in Children with Attention-Deficit/Hyperactivity Disorder. *Clin. Neurophysiol.* **2010**, *121*, 1871–1877. [CrossRef]
64. Behboudi, M.H.; Castro, S.; Chalamalasetty, P.; Maguire, M.J. Development of Gamma Oscillation during Sentence Processing in Early Adolescence: Insights into the Maturation of Semantic Processing. *Brain Sci.* **2023**, *13*, 1639. [CrossRef]
65. Benasich, A.A.; Gou, Z.; Choudhury, N.; Harris, K.D. Early Cognitive and Language Skills Are Linked to Resting Frontal Gamma Power across the First 3 Years. *Behav. Brain Res.* **2008**, *195*, 215–222. [CrossRef]
66. Anderson, A.J.; Perone, S. Developmental Change in the Resting State Electroencephalogram: Insights into Cognition and the Brain. *Brain Cogn.* **2018**, *126*, 40–52. [CrossRef]
67. Brito, N.H.; Fifer, W.P.; Myers, M.M.; Elliott, A.J.; Noble, K.G. Associations among Family Socioeconomic Status, EEG Power at Birth, and Cognitive Skills during Infancy. *Dev. Cogn. Neurosci.* **2016**, *19*, 144–151. [CrossRef]
68. Marshall, P.J.; Fox, N.A.; Group, B.C. A Comparison of the Electroencephalogram between Institutionalized and Community Children in Romania. *J. Cogn. Neurosci.* **2004**, *16*, 1327–1338. [CrossRef]
69. Uhlhaas, P.J.; Singer, W. Abnormal Neural Oscillations and Synchrony in Schizophrenia. *Nat. Rev. Neurosci.* **2010**, *11*, 100–113. [CrossRef]
70. Callaghan, B.L.; Tottenham, N. The Stress Acceleration Hypothesis: Effects of Early-Life Adversity on Emotion Circuits and Behavior. *Curr. Opin. Behav. Sci.* **2016**, *7*, 76–81. [CrossRef]
71. Leonard, J.A.; Romeo, R.R.; Park, A.T.; Takada, M.E.; Robinson, S.T.; Grotzinger, H.; Last, B.S.; Finn, A.S.; Gabrieli, J.D.E.; Mackey, A.P. Associations between Cortical Thickness and Reasoning Differ by Socioeconomic Status in Development. *Dev. Cogn. Neurosci.* **2019**, *36*, 100641. [CrossRef]

72. Park, A.T.; Leonard, J.A.; Saxler, P.K.; Cyr, A.B.; Gabrieli, J.D.E.; Mackey, A.P. Amygdala–Medial Prefrontal Cortex Connectivity Relates to Stress and Mental Health in Early Childhood. *Soc. Cogn. Affect. Neurosci.* **2018**, *13*, 430–439. [CrossRef]
73. Qin, S.; Hermans, E.J.; van Marle, H.J.F.; Luo, J.; Fernández, G. Acute Psychological Stress Reduces Working Memory-Related Activity in the Dorsolateral Prefrontal Cortex. *Biol. Psychiatry* **2009**, *66*, 25–32. [CrossRef]
74. Noble, K.G.; Norman, M.F.; Farah, M.J. Neurocognitive Correlates of Socioeconomic Status in Kindergarten Children. *Dev. Sci.* **2005**, *8*, 74–87. [CrossRef]
75. Farah, M.J.; Shera, D.M.; Savage, J.H.; Betancourt, L.; Giannetta, J.M.; Brodsky, N.L.; Malmud, E.K.; Hurt, H. Childhood Poverty: Specific Associations with Neurocognitive Development. *Brain Res.* **2006**, *1110*, 166–174. [CrossRef]
76. Noble, K.G.; McCandliss, B.D.; Farah, M.J. Socioeconomic Gradients Predict Individual Differences in Neurocognitive Abilities. *Dev. Sci.* **2007**, *10*, 464–480. [CrossRef]
77. U.S. Bureau of Labor Statistics and U.S. Census Bureau (2023). HINC-01. Selected Characteristics of Households by Total Money Income. Current Population Survey (CPS) Annual Social and Economic (ASEC) Supplement. Available online: https://www.census.gov/data/tables/time-series/demo/income-poverty/cps-hinc/hinc-01.html (accessed on 6 April 2024).
78. Khanam, R.; Nghiem, S. Family Income and Child Cognitive and Noncognitive Development in Australia: Does Money Matter? *Demography* **2016**, *53*, 597–621. [CrossRef]
79. Mistry, R.; Biesanz, J.; Taylor, L.; Burchinal, M.; Cox, M. Family income and its relation to preschool children's adjustment for families in the NICHD Study of Early Child Care. *Dev. Psychol.* **2004**, *40*, 727–745. [CrossRef]
80. Guo, G.; Harris, K.M. The mechanisms mediating the effects of poverty on children's intellectual development. *Demography* **2000**, *37*, 431–447. [CrossRef]
81. Erola, J.; Jalonen, S.; Lehti, H. Parental education, class and income over early life course and children's achievement. *Res. Soc. Stratif. Mobil.* **2016**, *44*, 33–43. [CrossRef]
82. Lean, R.E.; Paul, R.A.; Smyser, C.D.; Rogers, C.E. Maternal intelligence quotient (IQ) predicts IQ and language in very preterm children at age 5 years. *J. Child Psychol. Psychiatry* **2018**, *59*, 150–159. [CrossRef]
83. Hart, B.; Risley, T.R.; Kirby, J.R. Meaningful Differences in the Everyday Experience of Young American Children. *Can. J. Educ.* **1997**, *22*, 323.
84. Bradley, R.H.; Whiteside, L.; Caldwell, B.M.; Casey, P.H.; Kelleher, K.; Pope, S.; Swanson, M.; Barrett, K.; Cross, D. Maternal IQ, the home environment, and child IQ in low birthweight, premature children. *Int. J. Behav. Dev.* **1993**, *16*, 61–74. [CrossRef]
85. Ronfani, L.; Vecchi Brumatti, L.; Mariuz, M.; Tognin, V.; Bin, M.; Ferluga, V.; Knowles, A.; Montico, M.; Barbone, F. The complex interaction between home environment, socioeconomic status, maternal IQ and early child neurocognitive development: A multivariate analysis of data collected in a newborn cohort study. *PLoS ONE* **2015**, *10*, e0127052. [CrossRef] [PubMed]
86. Rowe, M.L. A Longitudinal Investigation of the Role of Quantity and Quality of Child-Directed Speech in Vocabulary Development. *Child Dev.* **2012**, *83*, 1762–1774. [CrossRef]
87. Huttenlocher, J.; Vasilyeva, M.; Cymerman, E.; Levine, S. Language Input and Child Syntax. *Cogn. Psychol.* **2002**, *45*, 337–374. [CrossRef]
88. Huttenlocher, J.; Vasilyeva, M.; Waterfall, H.R.; Vevea, J.L.; Hedges, L.V. The Varieties of Speech to Young Children. *Dev. Psychol.* **2007**, *43*, 1062–1083. [CrossRef]
89. Huttenlocher, J.; Waterfall, H.; Vasilyeva, M.; Vevea, J.; Hedges, L.V. Sources of Variability in Children's Language Growth. *Cogn. Psychol.* **2010**, *61*, 343–365. [CrossRef] [PubMed]
90. Ferjan Ramírez, N.; Lytle, S.R.; Kuhl, P.K. Parent Coaching Increases Conversational Turns and Advances Infant Language Development. *Proc. Natl. Acad. Sci. USA* **2020**, *117*, 3484–3491. [CrossRef]
91. Hirsh-Pasek, K.; Adamson, L.B.; Bakeman, R.; Owen, M.T.; Golinkoff, R.M.; Pace, A.; Yust, P.K.S.; Suma, K. The Contribution of Early Communication Quality to Low-Income Children's Language Success. *Psychol. Sci.* **2015**, *26*, 1071–1083. [CrossRef]
92. Rowe, M.L. Child-Directed Speech: Relation to Socioeconomic Status, Knowledge of Child Development and Child Vocabulary Skill. *J. Child Lang.* **2008**, *35*, 185–205. [CrossRef]
93. Schwab, J.F.; Lew-Williams, C. Language Learning, Socioeconomic Status, and Child-Directed Speech. *WIREs Cogn. Sci.* **2016**, *7*, 264–275. [CrossRef]
94. Romeo, R.R.; Segaran, J.; Leonard, J.A.; Robinson, S.T.; West, M.R.; Mackey, A.P.; Yendiki, A.; Rowe, M.L.; Gabrieli, J.D.E. Language Exposure Relates to Structural Neural Connectivity in Childhood. *J. Neurosci.* **2018**, *38*, 7870–7877. [CrossRef] [PubMed]
95. Hoff, E.; Burridge, A.; Ribot, K.M.; Giguere, D. Language Specificity in the Relation of Maternal Education to Bilingual Children's Vocabulary Growth. *Dev. Psychol.* **2018**, *54*, 1011–1019. [CrossRef] [PubMed]
96. Hoff-Ginsberg, E. Mother-Child Conversation in Different Social Classes and Communicative Settings. *Child Dev.* **1991**, *62*, 782–796. [CrossRef] [PubMed]
97. Hoff, E.; Core, C.; Place, S.; Rumiche, R.; Señor, M.; Parra, M. Dual Language Exposure and Early Bilingual Development. *J. Child Lang.* **2012**, *39*, 1–27. [CrossRef] [PubMed]
98. Schneider, J.M.; McIlvain, G.; Johnson, C.L. Mechanical Properties of the Developing Brain Are Associated with Language Input and Vocabulary Outcome. *Dev. Neuropsychol.* **2022**, *47*, 258–272. [CrossRef] [PubMed]
99. Raizada, R.D.S.; Richards, T.L.; Meltzoff, A.; Kuhl, P.K. Socioeconomic Status Predicts Hemispheric Specialisation of the Left Inferior Frontal Gyrus in Young Children. *NeuroImage* **2008**, *40*, 1392–1401. [CrossRef] [PubMed]

100. Brito, N.H.; Troller-Renfree, S.V.; Leon-Santos, A.; Isler, J.R.; Fifer, W.P.; Noble, K.G. Associations among the Home Language Environment and Neural Activity during Infancy. *Dev. Cogn. Neurosci.* **2020**, *43*, 100780. [CrossRef] [PubMed]
101. Pierce, L.J.; Reilly, E.; Nelson, C.A. Associations between Maternal Stress, Early Language Behaviors, and Infant Electroencephalography during the First Year of Life. *J. Child Lang.* **2021**, *48*, 737–764. [CrossRef] [PubMed]
102. Merz, E.C.; Wiltshire, C.A.; Noble, K.G. Socioeconomic Inequality and the Developing Brain: Spotlight on Language and Executive Function. *Child Dev. Perspect.* **2019**, *13*, 15–20. [CrossRef]
103. Rosen, M.L.; Sheridan, M.A.; Sambrook, K.A.; Meltzoff, A.N.; McLaughlin, K.A. Socioeconomic Disparities in Academic Achievement: A Multi-Modal Investigation of Neural Mechanisms in Children and Adolescents. *NeuroImage* **2018**, *173*, 298–310. [CrossRef]
104. Noble, K.G.; Giebler, M.A. The Neuroscience of Socioeconomic Inequality. *Curr. Opin. Behav. Sci.* **2020**, *36*, 23–28. [CrossRef] [PubMed]
105. NIMH Data Archive—Data Dictionary: Data Structure. Available online: https://nda.nih.gov/data_structure.html?short_name=inr01 (accessed on 18 January 2024).
106. Merz, E.C.; Desai, P.M.; Maskus, E.A.; Melvin, S.A.; Rehman, R.; Torres, S.D.; Meyer, J.; He, X.; Noble, K.G. Socioeconomic Disparities in Chronic Physiologic Stress Are Associated with Brain Structure in Children. *Biol. Psychiatry* **2019**, *86*, 921–929. [CrossRef] [PubMed]
107. Troller-Renfree, S.V.; Costanzo, M.A.; Duncan, G.J.; Magnuson, K.; Gennetian, L.A.; Yoshikawa, H.; Halpern-Meekin, S.; Fox, N.A.; Noble, K.G. The Impact of a Poverty Reduction Intervention on Infant Brain Activity. *Proc. Natl. Acad. Sci. USA* **2022**, *119*, e2115649119. [CrossRef] [PubMed]
108. Brooks-Gunn, J.; Duncan, G.J. The Effects of Poverty on Children. *Future Child.* **1997**, *7*, 55–71. [CrossRef] [PubMed]
109. McDonald, C.R.; Weckman, A.M.; Wright, J.K.; Conroy, A.L.; Kain, K.C. Developmental Origins of Disease Highlight the Immediate Need for Expanded Access to Comprehensive Prenatal Care. *Front. Public Health* **2022**, *10*, 1021901. [CrossRef] [PubMed]
110. Lautarescu, A.; Craig, M.C.; Glover, V. Chapter Two—Prenatal Stress: Effects on Fetal and Child Brain Development. In *International Review of Neurobiology*; Clow, A., Smyth, N., Eds.; Stress and Brain Health: Across the Life Course; Academic Press: New York, NY, USA, 2020; Volume 150, pp. 17–40.
111. Conger, R.D.; Wallace, L.E.; Sun, Y.; Simons, R.L.; McLoyd, V.C.; Brody, G.H. Economic Pressure in African American Families: A Replication and Extension of the Family Stress Model. *Dev. Psychol.* **2002**, *38*, 179–193. [CrossRef] [PubMed]
112. Conger, R.D.; Elder, G.H. *Families in Troubled Times: Adapting to Change in Rural America. Social Institutions and Social Change*; Aldine de Gruyter: Hawthorne, NY, USA, 1994; ISBN 0-202-30487-6/978-0-202-30488-5.
113. Hobkirk, A.L.; Krebs, N.M.; Muscat, J.E. Income as a Moderator of Psychological Stress and Nicotine Dependence among Adult Smokers. *Addict. Behav.* **2018**, *84*, 215–223. [CrossRef]
114. Luby, J.; Belden, A.; Botteron, K.; Marrus, N.; Harms, M.P.; Babb, C.; Nishino, T.; Barch, D. The Effects of Poverty on Childhood Brain Development: The Mediating Effect of Caregiving and Stressful Life Events. *JAMA Pediatr.* **2013**, *167*, 1135–1142. [CrossRef] [PubMed]
115. Schulz, A.J.; Israel, B.A.; Zenk, S.N.; Parker, E.A.; Lichtenstein, R.; Shellman-Weir, S.; Ab, L.K. Psychosocial Stress and Social Support as Mediators of Relationships between Income, Length of Residence and Depressive Symptoms among African American Women on Detroit's Eastside. *Soc. Sci. Med.* **2006**, *62*, 510–522. [CrossRef]
116. Attar, B.K.; Guerra, N.G.; Tolan, P.H. Neighborhood Disadvantage, Stressful Life Events and Adjustments in Urban Elementary-School Children. *J. Clin. Child Psychol.* **1994**, *23*, 391–400. [CrossRef]
117. Britt, C.L. Crime and Unemployment among Youths in the United States, 1958–1990. *Am. J. Econ. Sociol.* **1994**, *53*, 99–109. [CrossRef]
118. Evans, G.W. The Environment of Childhood Poverty. *Am. Psychol.* **2004**, *59*, 77–92. [CrossRef]
119. Santiago, C.D.; Wadsworth, M.E.; Stump, J. Socioeconomic Status, Neighborhood Disadvantage, and Poverty-Related Stress: Prospective Effects on Psychological Syndromes among Diverse Low-Income Families. *J. Econ. Psychol.* **2011**, *32*, 218–230. [CrossRef]
120. Evans, G.W.; Wells, N.M.; Moch, A. Housing and Mental Health: A Review of the Evidence and a Methodological and Conceptual Critique. *J. Soc. Issues* **2003**, *59*, 475–500. [CrossRef]
121. Evans, J.; Benefield, P. Systematic Reviews of Educational Research: Does the Medical Model Fit? *Br. Educ. Res. J.* **2001**, *27*, 527–541. [CrossRef]
122. Poudel, S.; Denicola-Prechtl, K.; Nelson, J.A.; Behboudi, M.H.; Benitez-Barrera, C.; Castro, S.; Maguire, M.J. Rethinking Household Size and Children's Language Environment. *Dev. Psychol.* **2024**, *60*, 159–169. [CrossRef]
123. Benner, A.D.; Kim, S.Y. Understanding Chinese American Adolescents' Developmental Outcomes: Insights from the Family Stress Model. *J. Res. Adolesc.* **2010**, *20*, 1–12. [CrossRef] [PubMed]
124. Linver, M.R.; Brooks-Gunn, J.; Kohen, D.E. Family Processes as Pathways from Income to Young Children's Development. *Dev. Psychol.* **2002**, *38*, 719–734. [CrossRef] [PubMed]
125. Neppl, T.K.; Senia, J.M.; Donnellan, M.B. The Effects of Economic Hardship: Testing the Family Stress Model over Time. *J. Fam. Psychol.* **2016**, *30*, 12–21. [CrossRef]

126. Lupien, S.J.; Maheu, F.; Tu, M.; Fiocco, A.; Schramek, T.E. The Effects of Stress and Stress Hormones on Human Cognition: Implications for the Field of Brain and Cognition. *Brain Cogn.* **2007**, *65*, 209–237. [CrossRef]
127. Shonkoff, J.P.; Garner, A.S.; Committee on Psychosocial Aspects of Child and Family Health, Committee on Early Childhood, Adoption, and Dependent Care, and Section on Developmental and Behavioral Pediatrics; Siegel, B.S.; Dobbins, M.I.; Earls, M.F.; Garner, A.S.; McGuinn, L.; Pascoe, J.; Wood, D.L. The Lifelong Effects of Early Childhood Adversity and Toxic Stress. *Pediatrics* **2012**, *129*, e232–e246. [CrossRef]
128. Kim, P.; Evans, G.W.; Angstadt, M.; Ho, S.S.; Sripada, C.S.; Swain, J.E.; Liberzon, I.; Phan, K.L. Effects of Childhood Poverty and Chronic Stress on Emotion Regulatory Brain Function in Adulthood. *Proc. Natl. Acad. Sci. USA* **2013**, *110*, 18442–18447. [CrossRef] [PubMed]
129. McEwen, B.S. Stress, Sex, and Neural Adaptation to a Changing Environment: Mechanisms of Neuronal Remodeling. *Ann. N. Y. Acad. Sci.* **2010**, *1204*, 38–59. [CrossRef] [PubMed]
130. Wager, T.D.; Sylvester, C.-Y.C.; Lacey, S.C.; Nee, D.E.; Franklin, M.; Jonides, J. Common and Unique Components of Response Inhibition Revealed by fMRI. *NeuroImage* **2005**, *27*, 323–340. [CrossRef] [PubMed]
131. Hanson, J.L.; Chandra, A.; Wolfe, B.L.; Pollak, S.D. Association between Income and the Hippocampus. *PLoS ONE* **2011**, *6*, e18712. [CrossRef]
132. Biegler, R.; McGregor, A.; Krebs, J.R.; Healy, S.D. A Larger Hippocampus Is Associated with Longer-Lasting Spatial Memory. *Proc. Natl. Acad. Sci. USA* **2001**, *98*, 6941–6944. [CrossRef]
133. Gianaros, P.J.; Jennings, J.R.; Sheu, L.K.; Greer, P.J.; Kuller, L.H.; Matthews, K.A. Prospective Reports of Chronic Life Stress Predict Decreased Grey Matter Volume in the Hippocampus. *NeuroImage* **2007**, *35*, 795–803. [CrossRef]
134. Piccolo, L.R.; Noble, K.G.; Pediatric Imaging, Neurocognition, and Genetics Study. Perceived Stress Is Associated with Smaller Hippocampal Volume in Adolescence. *Psychophysiology* **2018**, *55*, e13025. [CrossRef] [PubMed]
135. McEwen, B.S. Effects of Adverse Experiences for Brain Structure and Function. *Biol. Psychiatry* **2000**, *48*, 721–731. [CrossRef]
136. Papagni, S.A.; Benetti, S.; Arulanantham, S.; McCrory, E.; McGuire, P.; Mechelli, A. Effects of Stressful Life Events on Human Brain Structure: A Longitudinal Voxel-Based Morphometry Study. *Stress* **2011**, *14*, 227–232. [CrossRef]
137. Radley, J.J.; Morrison, J.H. Repeated Stress and Structural Plasticity in the Brain. *Ageing Res. Rev.* **2005**, *4*, 271–287. [CrossRef] [PubMed]
138. Cisler, J.M.; James, G.A.; Tripathi, S.; Mletzko, T.; Heim, C.; Hu, X.P.; Mayberg, H.S.; Nemeroff, C.B.; Kilts, C.D. Differential Functional Connectivity within an Emotion Regulation Neural Network among Individuals Resilient and Susceptible to the Depressogenic Effects of Early Life Stress. *Psychol. Med.* **2013**, *43*, 507–518. [CrossRef]
139. Philip, N.S.; Valentine, T.R.; Sweet, L.H.; Tyrka, A.R.; Price, L.H.; Carpenter, L.L. Early Life Stress Impacts Dorsolateral Prefrontal Cortex Functional Connectivity in Healthy Adults: Informing Future Studies of Antidepressant Treatments. *J. Psychiatr. Res.* **2014**, *52*, 63–69. [CrossRef]
140. Corning, W.C.; Steffy, R.A.; Anderson, E.; Bowers, P. EEG "Maturational Lag" Profiles: Follow-up Analyses. *J. Abnorm. Child Psychol.* **1986**, *14*, 235–249. [CrossRef] [PubMed]
141. Harmony, T.; Hinojosa, G.; Marosi, E.; Becker, J.; Rodriguez, M.; Reyes, A.; Rocha, C. Correlation between Eeg Spectral Parameters and an Educational Evaluation. *Int. J. Neurosci.* **1990**, *54*, 147–155. [CrossRef]
142. McLaughlin, K.A.; Fox, N.A.; Zeanah, C.H.; Sheridan, M.A.; Marshall, P.; Nelson, C.A. Delayed Maturation in Brain Electrical Activity Partially Explains the Association between Early Environmental Deprivation and Symptoms of Attention-Deficit/Hyperactivity Disorder. *Biol. Psychiatry* **2010**, *68*, 329–336. [CrossRef] [PubMed]
143. Pierce, L.J.; Thompson, B.L.; Gharib, A.; Schlueter, L.; Reilly, E.; Valdes, V.; Roberts, S.; Conroy, K.; Levitt, P.; Nelson, C.A. Association of Perceived Maternal Stress during the Perinatal Period with Electroencephalography Patterns in 2-Month-Old Infants. *JAMA Pediatr.* **2019**, *173*, 561–570. [CrossRef]
144. Baumgartl, H.; Fezer, E.; Buettner, R. Two-Level Classification of Chronic Stress Using Machine Learning on Resting-State EEG Recordings. In Proceedings of the 25th Americas Conference on Information Systems, Cancún, Mexico, 12–16 August 2020.
145. Saeed, S.M.U.; Anwar, S.M.; Majid, M.; Bhatti, A.M. Psychological Stress Measurement Using Low Cost Single Channel EEG Headset. In Proceedings of the 2015 IEEE International Symposium on Signal Processing and Information Technology (ISSPIT), Abu Dhabi, United Arab Emirates, 7–10 December 2015; pp. 581–585.
146. Vanhollebeke, G.; Kappen, M.; De Raedt, R.; Baeken, C.; van Mierlo, P.; Vanderhasselt, M.-A. Effects of Acute Psychosocial Stress on Source Level EEG Power and Functional Connectivity Measures. *Sci. Rep.* **2023**, *13*, 8807. [CrossRef]
147. Tan, E.; Tang, A.; Debnath, R.; Humphreys, K.L.; Zeanah, C.H.; Nelson, C.A.; Fox, N.A. Resting Brain Activity in Early Childhood Predicts IQ at 18 Years. *Dev. Cogn. Neurosci.* **2023**, *63*, 101287. [CrossRef]
148. Ellwood-Lowe, M.E.; Sacchet, M.D.; Gotlib, I.H. The Application of Neuroimaging to Social Inequity and Language Disparity: A Cautionary Examination. *Dev. Cogn. Neurosci.* **2016**, *22*, 1–8. [CrossRef]
149. Bester, S.; Malan-Van Rooyen, M. Emotional Development, Effects of Parenting and Family Structure On. In *International Encyclopedia of the Social & Behavioral Sciences*, 2nd ed.; Wright, J.D., Ed.; Elsevier: Oxford, UK, 2015; pp. 438–444, ISBN 978-0-08-097087-5.
150. Ruggles, S.; Heggeness, M. Intergenerational Coresidence in Developing Countries. *Popul. Dev. Rev.* **2008**, *34*, 253–281. [CrossRef]

151. Cohn, D.; Horowitz, J.M.; Minkin, R.; Fry, R.; Hurst, K. The Demographics of Multigenerational Households. *Pew Research Center's Social & Demographic Trends Project*. 2022. Available online: https://www.pewresearch.org/social-trends/2022/03/24/the-demographics-of-multigenerational-households/ (accessed on 16 April 2024).
152. Sperry, D.E.; Sperry, L.L.; Miller, P.J. Reexamining the Verbal Environments of Children from Different Socioeconomic Backgrounds. *Child Dev.* **2019**, *90*, 1303–1318. [CrossRef] [PubMed]
153. Dailey, S.; Bergelson, E. Language Input to Infants of Different Socioeconomic Statuses: A Quantitative Meta-Analysis. *Dev. Sci.* **2022**, *25*, e13192. [CrossRef]
154. Gilkerson, J.; Richards, J.A.; Warren, S.F.; Montgomery, J.K.; Greenwood, C.R.; Kimbrough, O.D.; Hansen, J.H.L.; Paul, T.D. Mapping the Early Language Environment Using All-Day Recordings and Automated Analysis. *Am. J. Speech-Lang. Pathol.* **2017**, *26*, 248–265. [CrossRef] [PubMed]
155. Keene, J.R.; Batson, C.D. Under One Roof: A Review of Research on Intergenerational Coresidence and Multigenerational Households in the United States. *Sociol. Compass* **2010**, *4*, 642–657. [CrossRef]
156. Romeo, R.R.; Leonard, J.A.; Grotzinger, H.M.; Robinson, S.T.; Takada, M.E.; Mackey, A.P.; Scherer, E.; Rowe, M.L.; West, M.R.; Gabrieli, J.D.E. Neuroplasticity Associated with Changes in Conversational Turn-Taking Following a Family-Based Intervention. *Dev. Cogn. Neurosci.* **2021**, *49*, 100967. [CrossRef]
157. Rowe, M.L.; Denmark, N.; Harden, B.J.; Stapleton, L.M. The Role of Parent Education and Parenting Knowledge in Children's Language and Literacy Skills among White, Black, and Latino Families. *Infant Child Dev.* **2016**, *25*, 198–220. [CrossRef]
158. Gonzalez, J.E.; Acosta, S.; Davis, H.; Pollard-Durodola, S.; Saenz, L.; Soares, D.; Resendez, N.; Zhu, L. Latino Maternal Literacy Beliefs and Practices Mediating Socioeconomic Status and Maternal Education Effects in Predicting Child Receptive Vocabulary. *Early Educ. Dev.* **2017**, *28*, 78–95. [CrossRef]
159. Suskind, D.L.; Leung, C.Y.Y.; Webber, R.J.; Hundertmark, A.C.; Leffel, K.R.; Fuenmayor Rivas, I.E.; Grobman, W.A. Educating Parents about Infant Language Development: A Randomized Controlled Trial. *Clin. Pediatr.* **2018**, *57*, 945–953. [CrossRef]
160. Hammer, C.S.; Weiss, A.L. African American Mothers' Views of Their Infants' Language Development and Language-Learning Environment. *Am. J. Speech-Lang. Pathol.* **2000**, *9*, 126–140. [CrossRef]
161. Luo, R.; Song, L.; Villacis, C.; Santiago-Bonilla, G. Parental Beliefs and Knowledge, Children's Home Language Experiences, and School Readiness: The Dual Language Perspective. *Front. Psychol.* **2021**, *12*, 661208. [CrossRef] [PubMed]
162. Draper, C.E.; Barnett, L.M.; Cook, C.J.; Cuartas, J.A.; Howard, S.J.; McCoy, D.C.; Merkley, R.; Molano, A.; Maldonado-Carreño, C.; Obradović, J.; et al. Publishing Child Development Research from around the World: An Unfair Playing Field Resulting in Most of the World's Child Population under-Represented in Research. *Infant Child Dev.* **2023**, *32*, e2375. [CrossRef]
163. Nielsen, M.; Haun, D.; Kärtner, J.; Legare, C.H. The Persistent Sampling Bias in Developmental Psychology: A Call to Action. *J. Exp. Child Psychol.* **2017**, *162*, 31–38. [CrossRef] [PubMed]
164. Jackson, M.I.; Kiernan, K.; McLanahan, S. Maternal education, changing family circumstances, and children's skill development in the United States and UK. *ANN. Am. Acad. Polit. Soc. Sci.* **2017**, *674*, 59–84. [CrossRef] [PubMed]
165. Jeong, J.; Kim, R.; Subramanian, S.V. How consistent are associations between maternal and paternal education and child growth and development outcomes across 39 low-income and middle-income countries? *J. Epidemiol. Community Health* **2018**, *72*, 434–441. [CrossRef]
166. Qi, D.; Wu, Y. Family's social economic status and child educational outcomes in China: The mediating effects of parenting practices and children's learning attitudes. *Child. Youth Serv. Rev.* **2020**, *118*, 105387. [CrossRef]
167. Rowe, M.L.; Weisleder, A. Language Development in Context. *Annu. Rev. Dev. Psychol.* **2020**, *2*, 201–223. [CrossRef]
168. Casillas, M.; Brown, P.; Levinson, S.C. Early Language Experience in a Tseltal Mayan Village. *Child Dev.* **2020**, *91*, 1819–1835. [CrossRef]
169. Cristia, A. A Systematic Review Suggests Marked Differences in the Prevalence of Infant-Directed Vocalization across Groups of Populations. *Dev. Sci.* **2023**, *26*, e13265. [CrossRef]
170. Cristia, A.; Dupoux, E.; Gurven, M.; Stieglitz, J. Child-Directed Speech Is Infrequent in a Forager-Farmer Population: A Time Allocation Study. *Child Dev.* **2019**, *90*, 759–773. [CrossRef]
171. Shneidman, L.A.; Goldin-Meadow, S. Language Input and Acquisition in a Mayan Village: How Important Is Directed Speech? *Dev. Sci.* **2012**, *15*, 659–673. [CrossRef] [PubMed]
172. Byers-Heinlein, K.; Esposito, A.G.; Winsler, A.; Marian, V.; Castro, D.C.; Luk, G. The Case for Measuring and Reporting Bilingualism in Developmental Research. *Collabra Psychol.* **2019**, *5*, 37. [CrossRef] [PubMed]
173. Tucker, G.R. A Global Perspective on Bilingualism and Bilingual Education. In *Georgetown University Round Table on Languages and Linguistics 1999*; Georgetown University Press: Washington, DC, USA, 2001.
174. Surrain, S.; Luk, G. Describing Bilinguals: A Systematic Review of Labels and Descriptions Used in the Literature between 2005–2015. *Biling. Lang. Cogn.* **2019**, *22*, 401–415. [CrossRef]
175. KIDS COUNT Data Center. Children Who Speak a Language Other than English at Home. Available online: https://datacenter.aecf.org/data/tables/81-children-who-speak-a-language-other-than-english-at-home (accessed on 2 April 2024).
176. Katsiaficas, M.P.; O'Toole, A. Caitlin Dual Language Learners: A National Demographic and Policy Profile. Available online: https://www.migrationpolicy.org/research/dual-language-learners-national-demographic-and-policy-profile (accessed on 3 April 2024).

177. Espinosa, L.M.; LaForett, D.R.; Burchinal, M.; Winsler, A.; Tien, H.-C.; Peisner-Feinberg, E.S.; Castro, D.C. Child Care Experiences among Dual Language Learners in the United States: Analyses of the Early Childhood Longitudinal Study–Birth Cohort. *AERA Open* **2017**, *3*, 2332858417699380. [CrossRef]
178. Rojas, R.; Iglesias, A.; Bunta, F.; Goldstein, B.; Goldenberg, C.; Reese, L. Interlocutor Differential Effects on the Expressive Language Skills of Spanish-Speaking English Learners. *Int. J. Speech-Lang. Pathol.* **2016**, *18*, 166–177. [CrossRef] [PubMed]
179. Duncan, T.S.; Paradis, J. Home Language Environment and Children's Second Language Acquisition: The Special Status of Input from Older Siblings. *J. Child Lang.* **2020**, *47*, 982–1005. [CrossRef]
180. Verdon, S.; McLeod, S.; Winsler, A. Language Maintenance and Loss in a Population Study of Young Australian Children. *Early Child. Res. Q.* **2014**, *29*, 168–181. [CrossRef]
181. DeLuca, V.; Rothman, J.; Bialystok, E.; Pliatsikas, C. Duration and Extent of Bilingual Experience Modulate Neurocognitive Outcomes. *NeuroImage* **2020**, *204*, 116222. [CrossRef]
182. Li, L.; Abutalebi, J.; Zou, L.; Yan, X.; Liu, L.; Feng, X.; Wang, R.; Guo, T.; Ding, G. Bilingualism Alters Brain Functional Connectivity between "Control" Regions and "Language" Regions: Evidence from Bimodal Bilinguals. *Neuropsychologia* **2015**, *71*, 236–247. [CrossRef]
183. Pliatsikas, C.; DeLuca, V.; Voits, T. The Many Shades of Bilingualism: Language Experiences Modulate Adaptations in Brain Structure. *Lang. Learn.* **2020**, *70*, 133–149. [CrossRef]
184. Dash, T.; Berroir, P.; Joanette, Y.; Ansaldo, A.I. Alerting, Orienting, and Executive Control: The Effect of Bilingualism and Age on the Subcomponents of Attention. *Front. Neurol.* **2019**, *10*, 483266. [CrossRef] [PubMed]
185. Berken, J.A.; Chai, X.; Chen, J.-K.; Gracco, V.L.; Klein, D. Effects of Early and Late Bilingualism on Resting-State Functional Connectivity. *J. Neurosci.* **2016**, *36*, 1165–1172. [CrossRef] [PubMed]
186. Grady, C.L.; Luk, G.; Craik, F.I.M.; Bialystok, E. Brain Network Activity in Monolingual and Bilingual Older Adults. *Neuropsychologia* **2015**, *66*, 170–181. [CrossRef] [PubMed]
187. Gullifer, J.W.; Chai, X.J.; Whitford, V.; Pivneva, I.; Baum, S.; Klein, D.; Titone, D. Bilingual Experience and Resting-State Brain Connectivity: Impacts of L2 Age of Acquisition and Social Diversity of Language Use on Control Networks. *Neuropsychologia* **2018**, *117*, 123–134. [CrossRef] [PubMed]
188. Sun, X.; Li, L.; Ding, G.; Wang, R.; Li, P. Effects of Language Proficiency on Cognitive Control: Evidence from Resting-State Functional Connectivity. *Neuropsychologia* **2019**, *129*, 263–275. [CrossRef] [PubMed]
189. Yamasaki, B.L.; Stocco, A.; Liu, A.S.; Prat, C.S. Effects of Bilingual Language Experience on Basal Ganglia Computations: A Dynamic Causal Modeling Test of the Conditional Routing Model. *Brain Lang.* **2019**, *197*, 104665. [CrossRef] [PubMed]
190. Bialystok, E.; Craik, F.I.M.; Luk, G. Bilingualism: Consequences for Mind and Brain. *Trends Cogn. Sci.* **2012**, *16*, 240–250. [CrossRef] [PubMed]
191. Berken, J.A.; Gracco, V.L.; Chen, J.-K.; Watkins, K.E.; Baum, S.; Callahan, M.; Klein, D. Neural Activation in Speech Production and Reading Aloud in Native and Non-Native Languages. *NeuroImage* **2015**, *112*, 208–217. [CrossRef] [PubMed]
192. Jasinska, K.K.; Petitto, L.A. How Age of Bilingual Exposure Can Change the Neural Systems for Language in the Developing Brain: A Functional near Infrared Spectroscopy Investigation of Syntactic Processing in Monolingual and Bilingual Children. *Dev. Cogn. Neurosci.* **2013**, *6*, 87–101. [CrossRef]
193. Lauharatanahirun, N.; Maciejewski, D.; Holmes, C.; Deater-Deckard, K.; Kim-Spoon, J.; King-Casas, B. Neural Correlates of Risk Processing among Adolescents: Influences of Parental Monitoring and Household Chaos. *Child Dev.* **2018**, *89*, 784–796. [CrossRef]
194. Andrews, J.L.; Ahmed, S.P.; Blakemore, S.-J. Navigating the Social Environment in Adolescence: The Role of Social Brain Development. *Biol. Psychiatry* **2021**, *89*, 109–118. [CrossRef]
195. Iwinski, S.; Donovan, S.M.; Fiese, B.; Bost, K. The Impact of Household Chaos and Dietary Intake on Executive Function in Young Children. *Nutrients* **2021**, *13*, 4442. [CrossRef]
196. Razza, R.A.; Martin, A.; Brooks-Gunn, J. The Implications of Early Attentional Regulation for School Success among Low-Income Children. *J. Appl. Dev. Psychol.* **2012**, *33*, 311–319. [CrossRef]
197. Blair, C.; Raver, C.C. Poverty, Stress, and Brain Development: New Directions for Prevention and Intervention. *Acad. Pediatr.* **2016**, *16*, S30–S36. [CrossRef]
198. Lecheile, B.M.; Spinrad, T.L.; Xu, X.; Lopez, J.; Eisenberg, N. Longitudinal Relations among Household Chaos, SES, and Effortful Control in the Prediction of Language Skills in Early Childhood. *Dev. Psychol.* **2020**, *56*, 727–738. [CrossRef] [PubMed]
199. Vernon-Feagans, L.; Willoughby, M.; Garrett-Peters, P. Predictors of Behavioral Regulation in Kindergarten: Household Chaos, Parenting, and Early Executive Functions. *Dev. Psychol.* **2016**, *52*, 430–441. [CrossRef]
200. Coley, R.L.; Lynch, A.D.; Kull, M. Early Exposure to Environmental Chaos and Children's Physical and Mental Health. *Early Child. Res. Q.* **2015**, *32*, 94–104. [CrossRef] [PubMed]
201. Deater-Deckard, K.; Sewell, M.D.; Petrill, S.A.; Thompson, L.A. Maternal Working Memory and Reactive Negativity in Parenting. *Psychol. Sci.* **2010**, *21*, 75–79. [CrossRef]
202. Xu, J.; Liu, X.; Li, Q.; Goldblatt, R.; Qin, W.; Liu, F.; Chu, C.; Luo, Q.; Ing, A.; Guo, L.; et al. Global Urbanicity Is Associated with Brain and Behaviour in Young People. *Nat. Hum. Behav.* **2022**, *6*, 279–293. [CrossRef] [PubMed]
203. Evans, G.W.; Gonnella, C.; Marcynyszyn, L.A.; Gentile, L.; Salpekar, N. The Role of Chaos in Poverty and Children's Socioemotional Adjustment. *Psychol. Sci.* **2005**, *16*, 560–565. [CrossRef] [PubMed]

204. Essen, J.; Fogelman, K.; Head, J. Children's Housing and Their Health and Physical Development. *Child Care Health Dev.* **1978**, *4*, 357–369. [CrossRef]
205. Evans, G.W.; Maxwell, L.E.; Hart, B. Parental Language and Verbal Responsiveness to Children in Crowded Homes. *Dev. Psychol.* **1999**, *35*, 1020–1023. [CrossRef]
206. Evans, G.W.; Saegert, S.; Harris, R. Residential Density and Psychological Health among Children in Low-Income Families. *Environ. Behav.* **2001**, *33*, 165–180. [CrossRef]
207. Havron, N.; Lovcevic, I.; Kee, M.Z.L.; Chen, H.; Chong, Y.S.; Daniel, M.; Broekman, B.F.P.; Tsuji, S. The Effect of Older Sibling, Postnatal Maternal Stress, and Household Factors on Language Development in Two- to Four-Year-Old Children. *Dev. Psychol.* **2022**, *58*, 2096–2113. [CrossRef]
208. Klatte, M.; Bergstroem, K.; Lachmann, T. Does Noise Affect Learning? A Short Review on Noise Effects on Cognitive Performance in Children. *Front. Psychol.* **2013**, *4*, 55965. [CrossRef] [PubMed]
209. Kujala, T.; Brattico, E. Detrimental Noise Effects on Brain's Speech Functions. *Biol. Psychol.* **2009**, *81*, 135–143. [CrossRef]
210. Maxwell, L.E.; Evans, G.W. The effects of noise on pre-school children's pre-reading skills. *J. Environ. Psychol.* **2000**, *20*, 91–97. [CrossRef]
211. Foraster, M.; Esnaola, M.; López-Vicente, M.; Rivas, I.; Álvarez-Pedrerol, M.; Persavento, C.; Sebastian-Galles, N.; Pujol, J.; Dadvand, P.; Sunyer, J. Exposure to Road Traffic Noise and Cognitive Development in Schoolchildren in Barcelona, Spain: A Population-Based Cohort Study. *PLoS Med.* **2022**, *19*, e1004001. [CrossRef] [PubMed]
212. Werchan, D.M.; Brandes-Aitken, A.; Brito, N.H. Signal in the Noise: Dimensions of Predictability in the Home Auditory Environment Are Associated with Neurobehavioral Measures of Early Infant Sustained Attention. *Dev. Psychobiol.* **2022**, *64*, e22325. [CrossRef]
213. Simon, K.R.; Merz, E.C.; He, X.; Noble, K.G. Environmental Noise, Brain Structure, and Language Development in Children. *Brain Lang.* **2022**, *229*, 105112. [CrossRef]
214. Martínez-Vilavella, G.; Pujol, J.; Blanco-Hinojo, L.; Deus, J.; Rivas, I.; Persavento, C.; Sunyer, J.; Foraster, M. The Effects of Exposure to Road Traffic Noise at School on Central Auditory Pathway Functional Connectivity. *Environ. Res.* **2023**, *226*, 115574. [CrossRef]
215. Cohen, S.; Krantz, D.S.; Evans, G.W.; Stokols, D.; Kelly, S. Aircraft Noise and Children: Longitudinal and Cross-Sectional Evidence on Adaptation to Noise and the Effectiveness of Noise Abatement. *J. Personal. Soc. Psychol.* **1981**, *40*, 331–345. [CrossRef]
216. Jones, P.C.; Pendergast, L.L.; Schaefer, B.A.; Rasheed, M.; Svensen, E.; Scharf, R.; Shrestha, R.; Maphula, A.; Roshan, R.; Rasmussen, Z.; et al. Measuring Home Environments across Cultures: Invariance of the HOME Scale across Eight International Sites from the MAL-ED Study. *J. Sch. Psychol.* **2017**, *64*, 109–127. [CrossRef] [PubMed]
217. Garcini, L.M.; Arredondo, M.M.; Berry, O.; Church, J.A.; Fryberg, S.; Thomason, M.E.; McLaughlin, K.A. Increasing Diversity in Developmental Cognitive Neuroscience: A Roadmap for Increasing Representation in Pediatric Neuroimaging Research. *Dev. Cogn. Neurosci.* **2022**, *58*, 101167. [CrossRef]
218. Polemiti, E.; Hese, S.; Schepanski, K.; Yuan, J.; Schumann, G.; environMENTAL Consortium. How Does the Macroenvironment Influence Brain and Behaviour—A Review of Current Status and Future Perspectives. *medRxiv* **2023**. [CrossRef]
219. Choy, T.; Baker, E.; Stavropoulos, K. Systemic Racism in EEG Research: Considerations and Potential Solutions. *Affect. Sci.* **2022**, *3*, 14–20. [CrossRef]
220. Ricard, J.A.; Parker, T.C.; Dhamala, E.; Kwasa, J.; Allsop, A.; Holmes, A.J. Confronting Racially Exclusionary Practices in the Acquisition and Analyses of Neuroimaging Data. *Nat. Neurosci.* **2023**, *26*, 4–11. [CrossRef]
221. Louis, C.C.; Webster, C.T.; Gloe, L.M.; Moser, J.S. Hair Me out: Highlighting Systematic Exclusion in Psychophysiological Methods and Recommendations to Increase Inclusion. *Front. Hum. Neurosci.* **2022**, *16*, 1058953. [CrossRef] [PubMed]
222. Gatzke-Kopp, L.M. Diversity and Representation: Key Issues for Psychophysiological Science. *Psychophysiology* **2016**, *53*, 3–13. [CrossRef] [PubMed]
223. Roberts, S.O.; Bareket-Shavit, C.; Dollins, F.A.; Goldie, P.D.; Mortenson, E. Racial Inequality in Psychological Research: Trends of the Past and Recommendations for the Future. *Perspect. Psychol. Sci.* **2020**, *15*, 1295–1309. [CrossRef]
224. Fulvio, J.M.; Akinnola, I.; Postle, B.R. Gender (Im)Balance in Citation Practices in Cognitive Neuroscience. *J. Cogn. Neurosci.* **2021**, *33*, 3–7. [CrossRef]
225. Taylor, C.M.; Pritschet, L.; Jacobs, E.G. The Scientific Body of Knowledge—Whose Body Does It Serve? A Spotlight on Oral Contraceptives and Women's Health Factors in Neuroimaging. *Front. Neuroendocrinol.* **2021**, *60*, 100874. [CrossRef] [PubMed]
226. Goldfarb, M.G.; Brown, D.R. Diversifying Participation: The Rarity of Reporting Racial Demographics in Neuroimaging Research. *NeuroImage* **2022**, *254*, 119122. [CrossRef] [PubMed]
227. Fisher, J.A.; Kalbaugh, C.A. Challenging Assumptions About Minority Participation in US Clinical Research. *Am. J. Public Health* **2011**, *101*, 2217–2222. [CrossRef]
228. Wendler, D.; Kington, R.; Madans, J.; Wye, G.V.; Christ-Schmidt, H.; Pratt, L.A.; Brawley, O.W.; Gross, C.P.; Emanuel, E. Are Racial and Ethnic Minorities Less Willing to Participate in Health Research? *PLoS Med.* **2005**, *3*, e19. [CrossRef]
229. Webb, E.K.; Etter, J.A.; Kwasa, J.A. Addressing Racial and Phenotypic Bias in Human Neuroscience Methods. *Nat. Neurosci.* **2022**, *25*, 410–414. [CrossRef] [PubMed]
230. Long Shadows: The Black-White Gap in Multigenerational Poverty. Available online: https://www.brookings.edu/articles/long-shadows-the-black-white-gap-in-multigenerational-poverty/ (accessed on 3 April 2024).

231. Heard-Garris, N.; Boyd, R.; Kan, K.; Perez-Cardona, L.; Heard, N.J.; Johnson, T.J. Structuring Poverty: How Racism Shapes Child Poverty and Child and Adolescent Health. *Acad. Pediatr.* **2021**, *21*, S108–S116. [CrossRef]
232. Simmons, A.; Taylor, E.K.; Abdurokhmonova, G.; Romeo, R.R. *Developing Best Practices for Inclusion in Pediatric fNIRS Research: Equity for Participants with Afro-Textured Hair*; Flux: Developmental Cognitive Neuroscience Society: Santa Rosa, CA, USA, 2023.
233. Brown, L.; Rollock, D.; Foti, D. Conducting Electroencephalography with Black Individuals: Barriers, Recommendations, and Impact on Generalizability. *Policy Insights Behav. Brain Sci.* **2023**, *10*, 178–185. [CrossRef]
234. Marx, D.M.; Goff, P.A. Clearing the Air: The Effect of Experimenter Race on Target's Test Performance and Subjective Experience. *Br. J. Soc. Psychol.* **2005**, *44*, 645–657. [CrossRef]
235. Thorson, K.R.; Mendes, W.B.; West, T.V. Controlling the Uncontrolled: Are There Incidental Experimenter Effects on Physiologic Responding? *Psychophysiology* **2020**, *57*, e13500. [CrossRef]
236. Bradford, D.E.; DeFalco, A.; Perkins, E.R.; Carbajal, I.; Kwasa, J.; Goodman, F.R.; Jackson, F.; Richardson, L.N.S.; Woodley, N.; Neuberger, L.; et al. Whose Signals Are Being Amplified? Toward a More Equitable Clinical Psychophysiology. *Clin. Psychol. Sci.* **2024**, *12*, 237–252. [CrossRef]

brain sciences

MDPI

Article

Neonatal Physical Growth Predicts Electroencephalography Power in Rural South African Children

Amanda R. Tarullo [1,*], Denise Evans [2], Lezanie Coetzee [2], Diana C. Lopera-Perez [1], Shaina P. Brady [1], Laurel J. Gabard-Durnam [3], Günther Fink [4], Davidson H. Hamer [5,6], Aisha K. Yousafzai [7] and Peter C. Rockers [5]

1 Department of Psychological and Brain Sciences, College of Arts and Sciences, Boston University, Boston, MA 02215, USA; dlopera@bu.edu (D.C.L.-P.); spbrady@bu.edu (S.P.B.)
2 Health Economics and Epidemiology Research Office, Faculty of Health Sciences, University of the Witwatersrand, Johannesburg 2041, South Africa; devans@heroza.org (D.E.)
3 Department of Psychology, Northeastern University, Boston, MA 02115, USA; l.gabard-durnam@northeastern.edu
4 Swiss Tropical and Public Health Institute, 4123 Basel, Switzerland; guenther.fink@swisstph.ch
5 Department of Global Health, Boston University School of Public Health, Boston, MA 02118, USA; dhamer@bu.edu (D.H.H.); prockers@bu.edu (P.C.R.)
6 Section of Infectious Diseases, Department of Medicine, Boston University School of Medicine, Boston, MA 02118, USA
7 Department of Global Health and Population, Harvard T.H. Chan School of Public Health, Boston, MA 02120, USA; ayousafzai@hsph.harvard.edu
* Correspondence: atarullo@bu.edu

Abstract: Anthropometric measures at birth, indexing prenatal growth, are associated with later cognitive development. Children in low- and middle-income countries (LMIC) are at elevated risk for impaired prenatal and early postnatal growth and enduring cognitive deficits. However, the associations of neonatal physical growth with neural activity are not well-characterized in LMIC contexts, given the dearth of early childhood neuroimaging research in these settings. The current study examined birth length, weight, and head circumference as predictors of EEG relative power over the first three years of life in rural Limpopo Province, South Africa, controlling for postnatal growth and socioeconomic status (SES). A larger head circumference at birth predicted lower relative gamma power, lower right hemisphere relative beta power, and higher relative alpha and theta power. A greater birth length also predicted lower relative gamma power. There were interactions with timepoints such that the associations of birth head circumference and length with EEG power were most pronounced at the 7-month assessment and were attenuated at the 17- and 36-month assessments. The results identify birth head circumference and length as specific predictors of infant neural activity within an under-resourced context.

Keywords: EEG; socioeconomic status; growth; head circumference; poverty; infancy; low- and middle-income countries; global south; early adversity

1. Introduction

Fetal growth restriction and stunting in the early postnatal years place children at risk for impaired cognitive development [1,2]. Impaired fetal growth reflects epigenetic and environmental factors in the prenatal environment, such as chronic micro- and macronutrient deficiencies and maternal stress [3–6]. Slowed postnatal growth likewise is an indicator of chronic malnutrition [7,8]. In combination, impaired prenatal and postnatal growth contribute to stunting, i.e., low height for age. Globally, stunting is present in at least 165 million children under age 5, and wasting is occurring in at least 52 million children, as indexed by low weight for height [7]. In low- and middle-income countries (LMIC), children face numerous co-occurring early adversities [7], and a staggering one out of three young children fails to reach basic developmental milestones [9]. The processes of growth restriction that manifest as stunting also have profound effects on cognitive development [2,8], contributing to this epidemic of unrealized cognitive potential. However, the mechanistic pathways through which growth restriction impacts neural development in LMIC contexts are not well understood, given the scarcity of early childhood neuroimaging studies in these contexts where children are most at risk for poor neurocognitive outcomes. More work is needed to understand the role of normative variations in growth and of impaired growth for neural activity within LMIC contexts.

The implications of both fetal and postnatal growth for cognitive development are well established. In a large-scale study conducted in eight LMIC countries, growth faltering over the first two postnatal years was correlated with cognitive development, with small head circumference being the growth measure most strongly related to cognitive outcomes [2]. Both postnatal stunting (i.e., height for age) and being underweight have been associated with impaired cognitive, motor, and language development in Bangladeshi children [8,10].

In studies that disentangle the effects of prenatal and postnatal growth, it is evident that prenatal growth makes specific contributions to cognitive outcomes over and above postnatal growth. For example, in a study of Finnish children, birth head circumference, length, and weight were all significant predictors of cognitive outcomes, with slower postnatal growth accounting for additional variance in some cognitive domains [1]. In a large study of pregnant women in six countries, fetal cranial growth, determined through ultrasound, uniquely predicted neurocognitive outcomes at age two years [11]. Barker's fetal origins hypothesis posits that fetal growth conditions have far-reaching influences on lifelong health and well-being via fetal programming [12]. Thus, growth measures at birth can serve both as proxies for prenatal adversity and as predictors of neurodevelopmental trajectories.

The fetal and postnatal environments have irreparable impacts on brain volume, microstructure, neural connectivity, and neurotransmitter function [4,5,13,14]. Prenatal iron deficiency, for example, is linked to intrauterine growth restriction (IUGR) and low birth weight (LBW) and is associated with widespread and lasting impairments in neuronal growth, differentiation, and myelination [4]. Children who are small for their gestational age have been shown to have neonatal abnormalities in EEG activity, including spectral power, which persist into childhood [15].

EEG is recorded from the scalp, indexing neuronal oscillations. EEG spectral power can be partitioned into frequency bands indexing faster and slower neural oscillations, each of which have distinct functional correlates. In high-income countries (HIC), there are age-related changes in the spectral power distribution such that the proportion of spectral power in high-frequency bands increases and the proportion of power in lower-frequency bands decreases with development [16]. Relative EEG power—the proportion of total spectral power in a given frequency band—can therefore be interpreted as indexing neural maturation. These age-related changes are not universal, however, and even within HIC, socioeconomic risk has been associated with altered developmental trajectories of EEG power in the first two postnatal years [17]. EEG developmental trajectories are less well-characterized in LMIC contexts due to the paucity of EEG research in these settings.

There is emerging evidence linking postnatal growth to EEG power and functional connectivity. EEG gamma power, indexing high-frequency oscillations, has been correlated with head circumference in early infancy [13]. In a study in Bangladesh, the postnatal height-for-age z-score was inversely associated with EEG functional connectivity in the theta and beta bands in toddlers, though no such association was observed in an infant cohort [8]. However, the implications of prenatal growth for EEG activity in LMIC contexts are not well-characterized.

The current study leverages a subsample of a birth cohort of Black children in rural Limpopo Province, South Africa [18]. The aim is to test neonatal length, weight, and head circumference as predictors of spontaneous EEG relative power in 7-, 17-, and 36-month laboratory assessments. We statistically control for concurrent growth measures to disambiguate the contributions of restricted prenatal growth from slowed postnatal growth.

Further, we control for general socioeconomic status, indexed via household assets within this under-resourced context, to examine the specific role of physical growth in relation to neural activity within the context of general socioeconomic deprivation.

2. Materials and Methods

2.1. Participants

The sample for the current analyses was comprised of 258 Black children (137 girls) in the Greater Tzaneen subdistrict, Mopani district, Limpopo Province, South Africa. Participants were enrolled in a cluster-randomized controlled trial soon after birth, which took place in the Greater Tzaneen and Greater Giyani subdistricts [18]. The trial was aimed at improving child development through community health worker (CHW) home visits from birth through 24 months, with the endline conducted at 36 months. Caregivers were at least 18 years of age. The current analyses consist of a subsample who participated in laboratory-based assessments of neural function. Invitations to participate in these laboratory-based assessments were restricted to those who (a) lived in the Greater Tzaneen subdistrict, which was closer to the laboratory location in the town of Tzaneen; (b) were born full term (>37 gestational weeks) with a birthweight > 2500 g; and (c) did not have any diagnosed developmental disorder at the time of the first lab assessment at seven months. Participants from this subsample were included in the current analyses if they had useable EEG data from at least two of the three laboratory assessments. Sample characteristics are shown in Table 1.

Table 1. Sample demographics.

Variable	M (SD)
Household wealth (z-score)	0.208 (1.012)
Caregiver age at childbirth (years)	31.66 (9.755)
Caregiver education (years)	10.35 (2.42)
Child number of older siblings	1.268 (1.153)

2.2. Procedures

Participants were enrolled in the cluster-randomized controlled trial soon after birth, and a baseline assessment was conducted. Intervention CHWs delivered home-visit interventions from birth to 24 months using a job aid with information on child developmental milestones, health, and nutrition, and strategies to encourage developmentally appropriate play activities. Control CHWs provided the local standard of care on their home visits. The endline was delayed due to the COVID-19 pandemic and occurred at 36 months. The current analyses leverage growth measures obtained at birth, along with EEG and growth data collected from a subsample of participants at 7 months old, 17 months old, and 36 months old. The intervention group was unrelated to EEG relative power measures but was included as a covariate to ensure that treatment would have no impact on the models. Data were collected in a laboratory that was established in the town of Tzaneen. Assessors were extensively trained to collect the data before each of the three timepoints. To enhance the quality of the EEG data, an electromagnetic meter was used to ensure that the assessment rooms had low electrical noise. Study staff scheduled lab visits with caregivers via phone call. On the day of the appointment, participants were transported to and from the lab in a study vehicle. Most caregivers were receptive to the study and to having their infants participate in the EEG assessments. The average time to complete the full assessment was approximately 45 min.

2.3. Measures

2.3.1. Electrophysiological Recording and Analysis

At each of the three lab-based assessments—ages 7 months, 17 months, and 36 months—EEG was recorded using a 32-channel portable Geodesic EEG System 400 (by EGI/Philips

Neuro, Eugene, OR). As in many LMIC contexts, there were challenges to EEG data collection, particularly the prevalence of load shedding in South Africa, i.e., rolling blackouts due to electrical power shortages, which sometimes required the rescheduling of participants. Impedances were accepted if lower than 50 KOhm. Data were sampled at 500 Hz and referenced to the vertex (Cz). Data were recorded for six minutes while the child sat on the caregiver's lap with the lights dimmed in the room. This protocol was adopted because the children were unaccustomed to computer screens and in pilot testing appeared overstimulated and distressed by the rotating screensavers often used in baseline EEG data collection with infants and young children. Data were uploaded from laboratory computers to a secure file-sharing platform on a daily basis and regularly checked for data quality.

The raw files were first filtered using NetStation with a 50 Hz notch filter to remove electrical noise and a 70 Hz low pass filter. The data were exported to MATLAB and preprocessed using the Harvard Automated Processing Pipeline for EEG [19]. Data were segmented into two-second epochs. A wavelet-enhanced independent components analysis approach was used to correct for artifacts while retaining the entire length of the data file. Bad channels were replaced using spherical spline interpolation and re-referenced to Cz. Following independent components analysis, observations were excluded from analysis if they had greater than 15% bad channels, more than 5 interpolated channels, fewer than 40 epochs of usable data out of a possible 360 epochs, or median artifact probability greater than 33% for retained independent components.

EEG spectral power was then computed in MATLAB using the Batch Electroencephalography Automated Processing Platform, which is designed to streamline batch EEG processing and increase the reproducibility of results [20]. Power was partitioned into frequency bands, with gamma (30–48 Hz), beta (13–30 Hz), alpha (6–13 Hz), and theta (4–6 Hz) reported here. The delta (1–4 Hz) band was not examined as the infant EEG literature typically has less focus on this slow wave activity. Power was measured in $\mu V^2/Hz$. For each electrode site, these band-level power values were then divided by the total power from 1 to 48 Hz to yield a measure of relative power, i.e., the proportion of total power in that frequency band. The relative power values for all five regions (frontal; parietal; central; temporal; occipital) of both left and right hemispheres were calculated by averaging the appropriate channel measures for each region at each of the three timepoints.

2.3.2. Neonatal Growth Measures

After delivery, families were provided with a *Road to Health* booklet, a free health information booklet provided for all babies at the time of delivery by the South African Department of Health. At this time, clinical staff filled out three measures of growth at birth, which served as indices of prenatal growth: head circumference at birth (cm), birth length (cm), and birth weight (kg). These measures were later extracted from each participant's *Road to Health* booklet by research staff.

2.3.3. Covariates

Household asset information was collected during the baseline survey. Caregivers were asked to endorse how many of 20 household assets they owned, such as a radio, gas stove, or cellphone. These data were reduced using principal component analysis, generating a score that was then z-scored [21] to index relative SES, such that a family with the mean household assets for this sample would have a z-score of 0. This measure serves as an appropriate index of socioeconomic status (SES) for this sample. Concurrent anthropometric measures of growth were collected by research staff at each of the three laboratory assessments where EEG was collected, with head circumference (cm), length/height (cm), and weight (kg) at each timepoint.

The intervention group was unrelated to any of the EEG relative power and growth measures but was included as a covariate as a precaution, to ensure that treatment would have no impact on the models.

2.4. Data Analysis Plan

Hypotheses were tested using a series of repeated measures analyses of variance (RM-ANOVAs), separately for each frequency band. All analyses were conducted using R V4.3.1. [22,23]. RM-ANOVAs were used due to recordings occurring across three timepoints, as well as across multiple regions and two hemispheres. Data were missing at random. Prior to analysis, multiple imputation was used to address missing data (both neural variables and growth variables) with the *mice* package [24]. For all RM-ANOVAs, the relative power for that frequency band was the dependent repeated measure, and within-subjects factors were timepoint (7 months; 17 months; 36 months), region (frontal; parietal; central; temporal; occipital), and hemisphere (right; left). Greenhouse–Geisser corrections were used when the sphericity assumption was violated.

All neonatal growth measures (head circumference at birth; birth length; birth weight) were entered as between-subjects predictors in the RM-ANOVAs, as well as their interactions with timepoint, region, and hemisphere. Next, covariates were added to the models. The number of household assets (z-score), our measure of SES, was added to the models as a between-subjects covariate in order to explore the role of prenatal growth on later EEG power, above and beyond the effects of global sociodemographic risk. Child sex and intervention groups were also added to the RM-ANOVA models as between-subjects covariates.

The interactions of concurrent growth (head circumference; length/height; and birth weight) and timepoint were also added as covariates to account for growth and potential stunting concurrent to EEG recordings, and to parse out the specific effects of prenatal growth.

As EEG power can be affected by data quality, the interactions with the timepoint of the number of usable channels prior to imputation, the number of usable epochs, and the room in which the EEG was conducted were included as between-subjects covariates as well. Controlling for the room was appropriate due to variations in electrical noise levels across testing rooms.

This approach enabled us to test the main effects and interactions of neonatal growth measures in relation to EEG gamma (30–48 Hz), beta (13–30 Hz), alpha (6–13 Hz), and theta (4–6 Hz) power over and above the effects of global environmental risk (SES), child sex, postnatal growth over the first few years of life, and EEG data quality.

3. Results

3.1. Preliminary Analyses

Descriptive statistics for neonatal and postnatal growth measures are reported in Table 2, along with the World Health Organization's (WHO) standardized z-score for the sample mean (https://www.who.int/tools/child-growth-standards, accessed on 29 April 2024). If the sample mean is half a standard deviation below the WHO mean for age and child sex, for instance, the z-score would be −0.50. The table also reports the percentage of children for each age, child sex, and growth measure who have impaired growth, defined as greater than two standard deviations below WHO means. Impaired growth would be labeled as stunting for length/height, and underweight for weight. These comparisons to WHO standards are provided to contextualize growth in the current sample.

Descriptive statistics for EEG variables are reported in Table 3. A series of Pearson correlations were conducted to examine the relationship between growth measures at birth, concurrent growth measures at each EEG assessment time, as well as sex and household assets (SES), as shown in Table 4. Concurrent growth measures at the three assessment points were strongly associated with each other, as well as with sex and SES. SES was not related to measures of neonatal growth. Birth length and birth weight were associated with postnatal growth. Please refer to the Supplementary Materials for further details on the neural data, including the correlations of EEG relative power at each timepoint with growth measures, along with power spectra graphs and topographical maps at each timepoint.

Table 2. Growth measures in relation to WHO z-scores by child sex.

(a) Neonatal	Boys			Girls		
	M (SD)	WHO z-score	% > -2SD WHO	M (SD)	WHO z-score	% > -2SD WHO
Birth Head Circumference	34.31 cm (1.68 cm)	−0.12	4.13%	34.32 cm (1.52 cm)	0.37	2.90%
Birth Length	50.33 cm (2.42 cm)	0.20	6.61%	49.43 cm (2.65 cm)	0.20	10.87%
Birth Weight	3.18 kg (0.39 kg)	−0.35	0%	3.04 kg (0.39 kg)	−0.43	1.45%
(b) Postnatal	**Boys**			**Girls**		
	M (SD)	WHO z-score	% > -2SD WHO	M (SD)	WHO z-score	% > -2SD WHO
7-Month Head Circumference	44.69 cm (1.64 cm)	0.58	4.13%	43.45 cm (1.53 cm)	0.47	2.17%
7-Month Length	69.18 cm (3.13 cm)	0.01	4.96%	67.66 cm (2.66 cm)	0.20	2.17%
7-Month Weight	8.90 kg (1.20 kg)	0.65	3.31%	8.16 kg (1.04 kg)	0.54	2.17%
17-Month Head Circumference	46.67 cm (1.57 cm)	−0.40	9.10%	45.49 cm (1.51 cm)	−0.41	11.59%
17-Month Height	78.07 cm (3.61 cm)	−1.20	28.93%	76.70 cm (2.97 cm)	−1.04	13.77%
17-Month Weight	10.85 kg (1.42 kg)	0.10	5.79%	10.12 kg (1.24 kg)	0.08	1.45%
36-Month Head Circumference	48.60 cm (1.67 cm)	−0.61	6.61%	47.74 cm (1.55 cm)	−0.55	16.67%
36-Month Height	95.21 cm (4.85 cm)	−0.23	8.26%	94.39 cm (3.91 cm)	−0.17	2.17%
36-Month Weight	14.45 kg (1.87 kg)	0.03	0.83%	13.79 kg (1.47 kg)	−0.09	1.45%

Table 3. EEG descriptive statistics.

Variable	M (SD)
7mo rel gamma	0.26 (0.11)
7mo rel beta	0.29 (0.05)
7mo rel alpha	0.10 (0.05)
7mo rel theta	0.11 (0.04)
7mo usable 2-s epochs	130.77 (54.94)
17mo rel gamma	0.18 (0.04)
17mo rel beta	0.28 (0.04)
17mo rel alpha	0.14 (0.02)
17mo rel theta	0.12 (0.02)
17mo usable 2-s epochs	106.54 (58.24)
36mo rel gamma	0.17 (0.06)
36mo rel beta	0.27 (0.04)
36mo rel alpha	0.15 (0.02)
36mo rel theta	0.12 (0.02)
36mo usable 2-s epochs	105.74 (59.61)

Table 4. Descriptive statistics and correlations for growth and demographic variables.

Variable	*M*	*SD*	1.	2.	3.	4.	5.	6.	7.	8.	9.	10.	11.	12.	13.	14.
1. sex	-	-	-	-	-	-	-	-	-	-	-	-	-	-	-	-
2. household assets	0.21	1.01	−0.10	-	-	-	-	-	-	-	-	-	-	-	-	-
3. birth head circ.	34.31	1.61	−0.00	−0.03	-	-	-	-	-	-	-	-	-	-	-	-
4. birth length	49.93	2.61	−0.18 **	0.10	0.20 **	-	-	-	-	-	-	-	-	-	-	-
5. birth weight	3.12	0.41	−0.14 *	0.06	0.29 **	0.42 **	-	-	-	-	-	-	-	-	-	-
6. 7mo head circ.	44.03	1.70	−0.36 **	0.14 *	0.06	0.18 **	0.15 *		-	-	-	-	-	-	-	-
7. 7mo length	68.38	2.99	−0.25 **	0.20 **	0.05	0.23 **	0.30 **	0.40 **	-	-	-	-	-	-	-	-
8. 7mo weight	8.51	1.18	−0.31 **	0.18 **	0.03	0.21 **	0.25 **	0.62 **	0.59 **	-	-	-	-	-	-	-
9. 17mo head circ.	46.04	1.63	−0.36 **	0.16 **	0.06	0.14 *	0.10	0.63 **	0.27 **	0.46 **	-	-	-	-	-	-
10. 17mo length	77.33	3.35	−0.20 **	0.18 **	0.11	0.27 **	0.27 **	0.34 **	0.59 **	0.53 **	0.34 **	-	-	-	-	-
11. 17mo weight	10.46	1.38	−0.26 **	0.25 **	0.10	0.23 **	0.22 **	0.47 **	0.52 **	0.79 **	0.49 **	0.61 **	-	-	-	-
12. 36mo head circ.	48.14	1.59	−0.28 **	0.10	0.12	0.13 *	0.11	0.61 **	0.29 **	0.43 **	0.65 **	0.31 **	0.39 **	-	-	-
13. 36mo height	94.67	4.47	−0.11	0.21 **	0.02	0.22 **	0.21 **	0.28 **	0.56 **	0.44 **	0.30 **	0.60 **	0.55 **	0.40 **	-	-
14. 36mo weight	14.05	1.66	−0.19 **	0.15 *	0.03	0.25 **	0.23 **	0.44 **	0.48 **	0.65 **	0.48 **	0.48 **	0.74 **	0.57 **	0.73 **	-

* $p < 0.05$; ** $p < 0.01$.

3.2. *Relative Gamma Power and Neonatal Growth*

Model 1's prediction of relative gamma power was significant (F = 34.78, $p < 0.001$) and explained 40.07% of the variance in gamma. There was a main effect of birth head circumference on gamma (30–48 Hz): F(1, 227) = 8.24, $p = 0.004$, partial $\eta^2 = 0.035$. A main effect was also found for birth length: F(1, 227) = 6.91, $p = 0.009$, partial $\eta^2 = 0.029$. Children who were born with a larger head circumference and with greater length demonstrated decreased relative gamma power. For both of these measures of neonatal growth, there was a significant interaction with timepoint: F(2, 484) = 10.85, $p < 0.001$, partial $\eta^2 = 0.043$ and F(2, 484) = 8.8, $p < 0.001$, partial $\eta^2 = 0.035$, respectively. For birth head circumference, the association with gamma power was significant at seven months only, as shown in Figure 1 (F(1, 243) = 25.65, $p < 0.001$, $\eta^2 = 0.095$). The same pattern was seen in birth length, which also was significant only at the 7-month timepoint (F(1, 243) = 21.11, $p < 0.001$, partial $\eta^2 = 0.08$), as shown in Figure 2. No main effect or interactions of birth weight on gamma were found. Concurrent growth measures were unrelated to gamma power in this model, and household assets showed a non-significant trend: F(1, 227) = 3.24, $p = 0.073$, partial $\eta^2 = 0.014$.

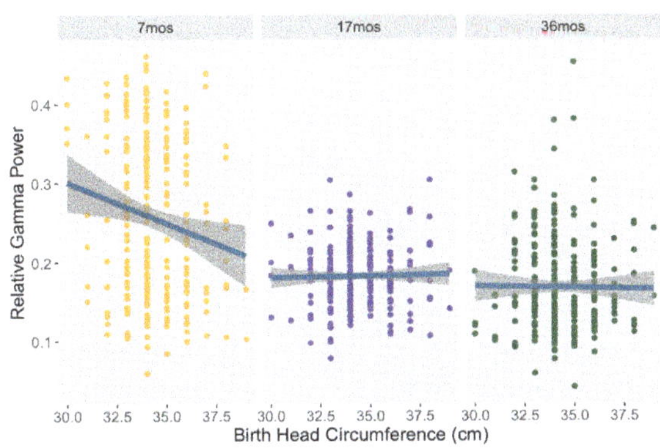

Figure 1. Relative gamma power as a function of birth head circumference across all three timepoints. Children born with a larger head circumference displayed lower relative gamma at seven months.

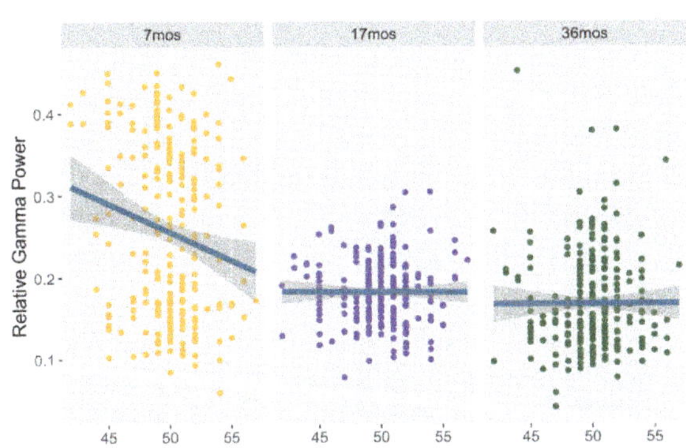

Figure 2. Relative gamma power as a function of birth length across all three timepoints. Children born with a longer birth length displayed lower relative gamma at seven months.

3.3. *Relative Beta Power and Neonatal Growth*

Model 2, predicting relative beta power, was significant (F = 8.07, $p < 0.001$) and explained 13.44% of the variance in gamma. There were no main effects of birth head circumference, birth length, or birth weight on beta (13–30 Hz). There was a significant interaction of head circumference at birth with hemisphere: F(1, 254) = 6.35, $p = 0.01$, partial $\eta^2 = 0.024$. The association of birth head circumference with beta power was significant only in the right hemisphere (F(1, 227) = 4.23, $p = 0.041$, $\eta^2 = 0.018$), such that larger head circumference at birth was associated with less relative beta power. This effect did not vary by timepoint. Household assets were not significantly related to beta power in this model, nor were concurrent head circumference or concurrent weight, but concurrent length/height was related ($F(1, 237) = 3.67$, $p = 0.013$, partial $\eta^2 = 0.006$), such that greater concurrent height was associated with less beta power.

3.4. Relative Alpha Power and Neonatal Growth

Model 3 examined growth predictors of relative alpha power ($F = 45.51$, $p < 0.001$) and explained 46.67% of the variance. A main effect was found of birth head circumference on relative alpha power (6–13 Hz): $F(1, 227) = 12.96$, $p < 0.001$, partial $\eta^2 = 0.054$. This effect was opposite to that of gamma, such that children with a larger head circumference at birth demonstrated higher relative alpha. This effect differed by timepoint ($F(2, 484) = 8.05$, $p < 0.001$, partial $\eta^2 = 0.032$), such that the relation of birth head circumference and alpha was only significant at the 7-month timepoint, as shown in Figure 3 ($F(1, 243) = 18.31$, $p < 0.001$, partial $\eta^2 = 0.07$). There were no interactions with hemisphere or region, indicating that the relationship between birth head circumference and gamma was consistent across the brain. Neither birth length nor birthweight was associated with alpha power. Concurrent growth measures were unrelated, but household assets were significantly related to alpha ($F(1, 227) = 4.51$, $p = 0.035$ partial $\eta^2 = 0.019$), such that children with higher household assets had more relative alpha power.

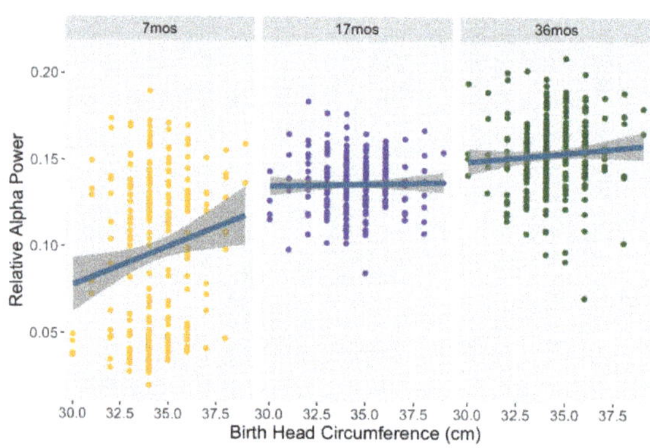

Figure 3. Relative alpha power as a function of birth head circumference across all three timepoints. At 7 months, children born with a larger head circumference displayed higher relative alpha power.

3.5. Relative Theta Power and Neonatal Growth

Model 4 assessed neonatal growth predictors of relative theta power ($F = 9.97$, $p < 0.001$), with 16.09% of the variance explained. There was a main effect of birth head circumference on relative theta power (4–6 Hz): $F(1, 227) = 4.54$, $p = 0.034$, partial $\eta^2 = 0.012$. Similar to relative alpha power, children with a larger head circumference at birth demonstrated higher relative theta. This effect also differed by timepoint: $F(2, 484) = 9.87$, $p < 0.001$, partial $\eta^2 = 0.039$. Similar to the other frequency bands, the relation of birth head circumference to theta was only significant at seven months and not for the latter two timepoints: $F(1, 243) = 12.21$, $p < 0.001$, partial $\eta^2 = 0.048$. As per Figure 4, children born with a larger head circumference displayed higher relative theta power at seven months. There were no interactions with the hemisphere or region. Concurrent growth measures, as well as household assets, were unrelated to relative theta power.

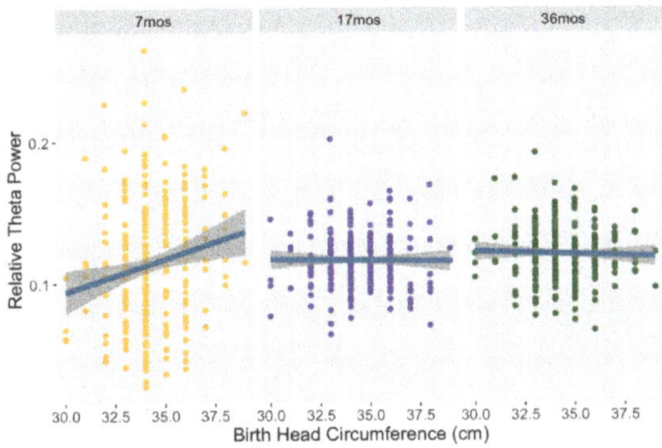

Figure 4. Relative theta power as a function of birth head circumference across all three timepoints. At 7 months, children born with a larger head circumference displayed higher relative theta power.

3.6. Follow-Up Analyses

Concurrent head circumference, height, and weight were not associated with EEG relative power in any of the three frequency bands, nor were there any interactions with region and hemisphere. Post-hoc partial correlations were conducted to examine the associations of neonatal and concurrent growth with household assets (SES). When controlling for birth head circumference, concurrent measures of head circumference were significantly related to household assets at seven months (r = 0.15, p = 0.018) and 17 months (r = 0.17, p = 0.007) but not at 36 months, parsing out the effects of postnatal brain growth. Concurrent height was positively associated with SES at all three timepoints (r = 0.19, p = 0.003; r = 0.16, p = 0.01; r = 0.19, p = 0.002), such that children with more household assets tended to be taller.

4. Discussion

The aim of the current study was to investigate the specific contributions of prenatal growth to neural activity within an LMIC context, above and beyond postnatal growth and general SES. We examined birth length, weight, and head circumference as predictors of neural activation at 7, 17, and 36 months among children in rural Limpopo Province, South Africa. While growth measures were within the normal range for the majority of the sample, there was an elevated incidence of impaired growth compared to WHO norms. There was a main effect of birth length on EEG relative gamma power, such that a greater birth length predicted lower relative gamma power. There were main effects of head circumference at birth on EEG relative power, such that a larger head circumference at birth predicted lower relative gamma and higher relative alpha and theta. There was also an interaction with the hemisphere such that a larger head circumference at birth predicted lower relative beta in the right hemisphere. Taken together, the results indicate that children who were longer at birth and had bigger heads, suggesting a more favorable fetal environment, went on to have an EEG pattern characterized by relatively fewer high-frequency oscillations and more slow-wave activity compared to their peers with a smaller birth length and head circumference. Timepoints interacted with the neonatal growth measures in predicting EEG relative power, such that birth length and head circumference were primarily related to 7-month EEG spectral power, and these associations were attenuated at the later assessments.

The findings are internally consistent, in that larger head circumference at birth predicted relatively fewer high-frequency oscillations (gamma) and more lower-frequency oscillations (alpha and theta). However, these results are the inverse of what one might expect if extrapolating from research in high-income countries, which shows a developmental shift towards more high-frequency activity and less slow-frequency activity [16]. From that perspective, the EEG pattern observed in the infants with better prenatal growth might be interpreted as less mature. It is important to consider the developmental context. In an environment characterized by widespread and persistent material deprivation, it is possible that rapid neural maturation is not advantageous. Rather, extending sensitive periods for brain development could be adaptive. It is plausible that the children who experienced relatively more favorable fetal conditions *within an overall context of significant adversity* were better able to adapt by slowing down postnatal neural maturation. The finding that birth head circumference and birth length were both mainly predictive of 7-month EEG relative power is consistent with this interpretation of a potential maturational lag. Within the 17- and 36-month timepoints, none of the neonatal growth measures predicted EEG activity. To test this speculative interpretation of the data, it would be of interest to extend follow-up into middle childhood to examine both cognitive and EEG trajectories. If the observed EEG pattern of relatively less high-frequency activity at 7 months is indeed adaptive, we would expect it to mediate pathways from neonatal growth to better cognitive outcomes.

The prior literature has also reported that the interplay of growth and EEG measures can vary across development, including in LMIC contexts. For instance, in Dhaka, Bangladesh, concurrent height-for-age z-score was related to lower EEG theta and beta functional connectivity at 24–36 months of age, but no such association was observed in an infant cohort [8]. The studies differ in LMIC context, type of EEG measure, and whether the focus is on neonatal or concurrent measures of growth. However, taken together, they suggest the need for a nuanced consideration of the interplay of physical growth and neural activity across infancy and early childhood.

A strength of the current study is that concurrent growth, child sex, and general SES, indexed by household assets, were included in the models as covariates. Thus, the results indicate a specific role of neonatal head circumference and length in predicting infant EEG activity. When both neonatal and concurrent growth are included in models, the concurrent growth variables effectively index slower postnatal growth, whereas in studies that do not include birth measurements, the concurrent growth variables are influenced by both prenatal and postnatal growth. In these current analyses, greater concurrent height, indicating faster postnatal growth, was associated with less beta power. There were no other associations of postnatal growth with EEG activity. Post-hoc analyses revealed that while postnatal growth was correlated with household assets in our sample, the neonatal measures of growth were independent of individual differences in household assets within this under-resourced sample. Further research is needed with detailed assessments during the prenatal period to examine what specific environmental factors influence fetal growth conditions and the neonatal growth measures in this context. This sample was enrolled in a cluster-randomized controlled trial aimed at enhancing developmental outcomes. Intervention group status was unrelated to relative power or growth measures and was covaried in the models. Nonetheless, this is a potential limitation.

Head circumference at birth was the most sensitive predictor of EEG relative power, exhibiting associations with all three frequency bands at 7 months. This is consistent with prior research on the importance of fetal cranial growth [11] and is noteworthy given the existing evidence that head circumference is particularly predictive of early cognitive outcomes [2]. Prior research indicates that not only is growth restriction in birth head circumference a risk factor for poorer cognitive outcomes, but also, among children with growth in the normal range, larger birth head circumference positively predicts cognitive achievement [25]. Further, a relatively small proportion of the current sample met the criteria for impaired growth, though the prevalence of impaired growth was elevated compared to WHO norms, particularly for head circumference and length/height. It should be noted that criteria for inclusion in the sample included full-term birth and a birthweight > 2500 g. These initial selection criteria likely reduced the incidence of impaired growth in the current sample, as impaired growth reflects both prenatal growth restriction and postnatal slowing of growth. Given that all participants in the current study were born at full term, smaller neonatal measures were indicative of slower fetal growth, and possibly intrauterine growth restriction in some cases, rather than prematurity. The pattern of associations between birth measures and subsequent neural activity could be different in infants born preterm.

The contributions of measures such as birthweight and postnatal growth could be different within a population experiencing a greater prevalence of stunting and wasting. That said, the finding that birth head circumference predicted relative EEG power even within a full-term sample with a modest incidence of impaired growth suggests that birth head circumference may be a particularly sensitive proxy for the effects of adverse fetal conditions on brain structure and function in LMIC contexts.

5. Conclusions

The current study assessed neural activity in relation to neonatal and postnatal growth in a sample of rural South African infants in a resource-poor context. Within this sample with a modestly elevated risk of impaired growth, a greater head circumference at birth predicted neural activity in infancy, specifically a lower proportion of high-frequency activity and an increased proportion of low-frequency activity. The results indicate the importance of considering both prenatal and postnatal growth indices and exploring the significance of these measures for neural activity within an under-resourced LMIC context.

Supplementary Materials: The following supporting information can be downloaded at: https://www.mdpi.com/article/10.3390/brainsci14060552/s1, Table S1: Correlations of Relative EEG Power with Growth; Table S2: Absolute Power in ln[μV^2/Hz]; Analyses of the associations of growth with absolute EEG power; Figure S1: Average topographic distribution plots across 7, 17, and 36 months for frequency bands Theta (4–6 Hz), Alpha (6–13 Hz), Beta (13–30 Hz) and Gamma (30–50 Hz) after artifact removal; Figure S2: Average power spectra after preprocessing for all electrodes across the scalp at 7, 17, and 36 months. Refs. [26–28] are cited in Supplementary Materials.

Author Contributions: Conceptualization, A.R.T., D.E., D.C.L.-P., S.P.B., G.F., D.H.H., A.K.Y. and P.C.R.; methodology, A.R.T., D.E., D.C.L.-P., S.P.B., L.J.G.-D., G.F., D.H.H., A.K.Y. and P.C.R.; software, L.J.G.-D.; validation, A.R.T., D.C.L.-P. and L.J.G.-D.; formal analysis, A.R.T., S.P.B., D.C.L.-P. and P.C.R.; data curation, A.R.T., L.C. and D.C.L.-P.; writing—original draft preparation, A.R.T. and S.P.B.; writing—review and editing, A.R.T., D.E., L.C., D.C.L.-P., S.P.B., L.J.G.-D., G.F., D.H.H., A.K.Y. and P.C.R.; visualization, S.P.B.; supervision, A.R.T., D.E., L.C., D.C.L.-P. and P.C.R.; project administration, A.R.T., D.E., L.C., D.C.L.-P. and P.C.R.; funding acquisition, A.R.T., D.E. and P.C.R. All authors have read and agreed to the published version of the manuscript.

Funding: This research was funded by the South African Medical Research Council (www.samrc.ac.za) through a grant to the Health Economics and Epidemiology Research Office at the University of the Witwatersrand (received by D.E.).

Institutional Review Board Statement: The study was conducted in accordance with the Declaration of Helsinki and approved by the Human Research Ethics Committee at the University of the Witwatersrand (#M160251, #M180229, #M210570) and by the Institutional Review Board at Boston University (#H-37065).

Informed Consent Statement: Informed consent was obtained for all subjects involved in the study, with parents providing informed consent on behalf of their children.

Data Availability Statement: Data were deposited in the Dryad repository: https://doi.org/10.5061/dryad.qnk98sfm2 (accessed on 13 April 2023).

Acknowledgments: We thank Jukka Leppänen for his contributions to establishing the laboratory and cohort for this study. We thank Vongani Maluleke, Patricia Leshabana, and Eliud Nkuna for their invaluable assistance with data collection.

Conflicts of Interest: The authors declare no conflicts of interest. The funders had no role in the design of the study; in the collection, analyses, or interpretation of data; in the writing of the manuscript; or in the decision to publish the results.

References

1. Heinonen, K.; Räikkönen, K.; Pesonen, A.-K.; Kajantie, E.; Andersson, S.; Eriksson, J.G.; Niemelä, A.; Vartia, T.; Peltola, J.; Lano, A. Prenatal and Postnatal Growth and Cognitive Abilities at 56 Months of Age: A Longitudinal Study of Infants Born at Term. *Pediatrics* **2008**, *121*, e1325–e1333. [CrossRef] [PubMed]
2. Scharf, R.J.; Rogawski, E.T.; Murray-Kolb, L.E.; Maphula, A.; Svensen, E.; Tofail, F.; Rasheed, M.; Abreu, C.; Vasquez, A.O.; Shrestha, R.; et al. Early childhood growth and cognitive outcomes: Findings from the MAL-ED study. *Matern. Child Nutr.* **2018**, *14*, e12584. [CrossRef] [PubMed]
3. Camerota, M.; Bollen, K.A. Birth Weight, Birth Length, and Gestational Age as Indicators of Favorable Fetal Growth Conditions in a US Sample. *PLoS ONE* **2016**, *11*, e0153800. [CrossRef] [PubMed]
4. Georgieff, M.K. Iron deficiency in pregnancy. *Am. J. Obstet. Gynecol.* **2020**, *223*, 516–524. [CrossRef]
5. Lautarescu, A.; Craig, M.C.; Glover, V. Prenatal stress: Effects on fetal and child brain development. *Int. Rev. Neurobiol.* **2020**, *150*, 17–40. [CrossRef] [PubMed]
6. Lindsay, K.L.; Buss, C.; Wadhwa, P.D.; Entringer, S. The Interplay Between Nutrition and Stress in Pregnancy: Implications for Fetal Programming of Brain Development. *Biol. Psychiatry* **2019**, *85*, 135–149. [CrossRef]
7. Black, R.E.; Victora, C.G.; Walker, S.P.; Bhutta, Z.A.; Christian, P.; de Onis, M.; Ezzati, M.; Grantham-McGregor, S.; Katz, J.; Martorell, R.; et al. Maternal and child undernutrition and overweight in low-income and middle-income countries. *Lancet* **2013**, *382*, 427–451. [CrossRef]
8. Xie, W.; Jensen, S.K.G.; Wade, M.; Kumar, S.; Westerlund, A.; Kakon, S.H.; Haque, R.; Petri, W.A.; Nelson, C.A. Growth faltering is associated with altered brain functional connectivity and cognitive outcomes in urban Bangladeshi children exposed to early adversity. *BMC Med.* **2019**, *17*, 199. [CrossRef]
9. McCoy, D.C.; Peet, E.D.; Ezzati, M.; Danaei, G.; Black, M.M.; Sudfeld, C.R.; Fawzi, W.; Fink, G. Early childhood developmental status in low- and middle-income countries: National, regional, and global prevalence es-timates using predictive modeling. *PLoS Med.* **2016**, *13*, e1002034. [CrossRef]
10. Nahar, B.; Hossain, M.; Mahfuz, M.; Islam, M.M.; Hossain, I.; Murray-Kolb, L.E.; Seidman, J.C.; Ahmed, T. Early childhood development and stunting: Findings from the MAL-ED birth cohort study in Bangladesh. *Matern. Child Nutr.* **2020**, *16*, e12864. [CrossRef]
11. Villar, J.; Gunier, R.B.; Tshivuila-Matala, C.O.O.; Rauch, S.A.; Nosten, F.; Ochieng, R.; Restrepo-Méndez, M.C.; McGready, R.; Barros, F.C.; Fernandes, M.; et al. Fetal cranial growth trajectories are associated with growth and neurodevelopment at 2 years of age: INTERBIO-21st Fetal Study. *Nat. Med.* **2021**, *27*, 647–652. [CrossRef] [PubMed]
12. Barker, D.J.P. Fetal origins of coronary heart disease. *BMJ* **1995**, *311*, 171–174. [CrossRef] [PubMed]
13. Pivik, R.; Andres, A.; Tennal, K.B.; Gu, Y.; Downs, H.; Bellando, B.J.; Jarratt, K.; Cleves, M.A.; Badger, T.M. Resting gamma power during the postnatal critical period for GABAergic system development is modulated by infant diet and sex. *Int. J. Psychophysiol.* **2019**, *135*, 73–94. [CrossRef] [PubMed]
14. Miguel, P.M.; Pereira, L.O.; Silveira, P.P.; Meaney, M.J. Early environmental influences on the development of children's brain structure and function. *Dev. Med. Child Neurol.* **2019**, *61*, 1127–1133. [CrossRef] [PubMed]
15. Stevenson, N.J.; Lai, M.M.; Starkman, H.E.; Colditz, P.B.; Wixey, J.A. Electroencephalographic studies in growth-restricted and small-for-gestational-age neonates. *Pediatr. Res.* **2022**, *92*, 1527–1534. [CrossRef] [PubMed]
16. Wang, X.-J.; Lee, J.J.; Schmit, B.D.; Bellet, J.; Chen, C.-Y.; Hafed, Z.M.; Hoseini, M.S.; Pobst, J.; Clawson, W.; Shew, W.; et al. Neurophysiological and Computational Principles of Cortical Rhythms in Cognition. *Physiol. Rev.* **2010**, *90*, 1195–1268. [CrossRef] [PubMed]
17. Wilkinson, C.L.; Pierce, L.J.; Sideridis, G.; Wade, M.; Nelson, C.A. Associations between EEG trajectories, family income, and cognitive abilities over the first two years of life. *Dev. Cogn. Neurosci.* **2023**, *61*, 101260. [CrossRef] [PubMed]
18. Rockers, P.C.; Leppänen, J.M.; Tarullo, A.; Coetzee, L.; Fink, G.; Hamer, D.H.; Yousafzai, A.K.; Evans, D. Evaluation of a community health worker home visit intervention to improve child development in South Africa: A cluster-randomized controlled trial. *PLoS Med.* **2023**, *20*, e1004222. [CrossRef] [PubMed]
19. Gabard-Durnam, L.J.; Leal, A.S.M.; Wilkinson, C.L.; Levin, A.R. The Harvard Automated Processing Pipeline for Electroencephalography (HAPPE): Standardized processing software for developmental and high-artifact data. *Front. Neurosci.* **2018**, *12*, 316496. [CrossRef]
20. Levin, A.R.; Leal, A.S.M.; Gabard-Durnam, L.J.; O'leary, H.M. BEAPP: The batch electroencephalography automated processing platform. *Front. Neurosci.* **2018**, *12*, 316258. [CrossRef]
21. Vyas, S.; Kumaranayake, L. Constructing socio-economic status indices: How to use principal components analysis. *Health Policy Plan.* **2006**, *21*, 459–468. [CrossRef] [PubMed]
22. R Core Team. *R: A Language and Environment for Statistical Computing*; R Foundation for Statistical Computing: Vienna, Austria, 2013.
23. RStudio Team. *RStudio: Integrated Development Environment for R*; R Foundation for Statistical Computing: Vienna, Austria, 2021.
24. van Buuren, S.; Groothuis-Oudshoorn, K. mice: Multivariate Imputation by Chained Equations in R. *J. Stat. Softw.* **2011**, *45*, 1–67. [CrossRef]
25. Bach, C.C.; Henriksen, T.B.; Larsen, R.T.; Aagaard, K.; Matthiesen, N.B. Head circumference at birth and school performance: A nationwide cohort study of 536,921 children. *Pediatr. Res.* **2020**, *87*, 1112–1118. [CrossRef] [PubMed]

26. Clarke, A.R.; Barry, R.J.; McCarthy, R.; Selikowitz, M. Age and sex effects in the EEG: Development of the normal child. *Clin. Neurophysiol.* **2001**, *112*, 806–814. [CrossRef]
27. Pierce, L.J.; Thompson, B.L.; Gharib, A.; Schlueter, L.; Reilly, E.; Valdes, V.; Roberts, S.; Conroy, K.; Levitt, P.; Nelson, C.A. Association of Perceived Maternal Stress During the Perinatal Period with Electroencephalography Patterns in 2-Month-Old Infants. *JAMA Pediatr.* **2019**, *173*, 561. [CrossRef]
28. Troller-Renfree, S.V.; Brito, N.H.; Desai, P.M.; Leon-Santos, A.G.; Wiltshire, C.A.; Motton, S.N.; Meyer, J.S.; Isler, J.; Fifer, W.P.; Noble, K.G. Infants of mothers with higher physiological stress show alterations in brain function. *Dev. Sci.* **2020**, *23*, e12976. [CrossRef]

brain sciences

MDPI

Review

From Infancy to Childhood: A Comprehensive Review of Event- and Task-Related Brain Oscillations

Esra Ünsal [1,2], Rümeysa Duygun [1,2], İrem Yemeniciler [1,2,3], Elifnur Bingöl [1,2,3], Ömer Ceran [4] and Bahar Güntekin [2,3,*]

[1] Department of Neuroscience, Graduate School of Health Sciences, Istanbul Medipol University, 34810 Istanbul, Turkey; esra.unsal@std.medipol.edu.tr (E.Ü.); rumeysa.duygun@std.medipol.edu.tr (R.D.); irem.yemeniciler@std.medipol.edu.tr (İ.Y.); elifnur.bingol@std.medipol.edu.tr (E.B.)

[2] Neuroscience Research Center, Research Institute for Health Sciences and Technologies (SABITA), Istanbul Medipol University, 34810 Istanbul, Turkey

[3] Department of Biophysics, School of Medicine, Istanbul Medipol University, 34810 Istanbul, Turkey

[4] Department of Pediatrics, School of Medicine, Istanbul Medipol University, 34810 Istanbul, Turkey; oceran@medipol.edu.tr

* Correspondence: bguntekin@medipol.edu.tr

Abstract: Brain development from infancy through childhood involves complex structural and functional changes influenced by both internal and external factors. This review provides a comprehensive analysis of event and task-related brain oscillations, focusing on developmental changes across different frequency bands, including delta, theta, alpha, beta, and gamma. Electroencephalography (EEG) studies highlight that these oscillations serve as functional building blocks for sensory and cognitive processes, with significant variations observed across different developmental stages. Delta oscillations, primarily associated with deep sleep and early cognitive demands, gradually diminish as children age. Theta rhythms, crucial for attention and memory, display a distinct pattern in early childhood, evolving with cognitive maturation. Alpha oscillations, reflecting thalamocortical interactions and cognitive performance, increase in complexity with age. Beta rhythms, linked to active thinking and problem-solving, show developmental differences in motor and cognitive tasks. Gamma oscillations, associated with higher cognitive functions, exhibit notable changes in response to sensory stimuli and cognitive tasks. This review underscores the importance of understanding oscillatory dynamics to elucidate brain development and its implications for sensory and cognitive processing in childhood. The findings provide a foundation for future research on developmental neuroscience and potential clinical applications.

Keywords: infant; childhood; electroencephalography; event-related oscillations; task-related oscillations

1. Introduction

Neuron migration and differentiation, followed by the formation of abundant synaptic connections and specialized neural networks, are some of the fundamental events that occur during the developmental process [1–3]. This process is the most intensive within layer IV of the cortex, which is known as a major recipient layer of thalamocortical projection. However, the process of address selection [4] by thalamocortical input promotes elaboration of the cortical circuitry. The elaboration of thalamocortical circuitry and interaction with sensory input strengthens sensory-driven activity, further shaping fine connectivity within the cortical columns [5]. MRI studies have provided compelling insights into the structural changes occurring in the brain during childhood and adolescence. The maturation of the cortex parallels cognitive milestones in human development. Regions supporting primary functions, such as motor and sensory systems, mature earliest, while temporal–parietal areas associated with basic language skills and spatial attention develop subsequently.

Higher-order integration regions, such as the prefrontal and lateral temporal cortices, are among the last to mature. The physiological basis for this maturation is supported by mechanisms of experience-dependent synapse formation and dendritic arborization [6–9]. As growth and development progress, the electroencephalography (EEG) exhibits a trend towards continuity, although slow activity transients are present, indicating functional immaturity [10,11].

The EEG is used to record the electrophysiological activity generated by neural networks and is thus extremely sensitive to changes in their organization throughout developmental processes, as well as in cases of pathological disturbances. Moreover, these generated brain oscillations are considered the "functional building blocks" of sensory and cognitive processes [12]. Hence, EEG brain oscillations of different age and disease groups demonstrate varying characteristics based on the maturity and integrity of their sensory and cognitive processing systems [13,14]. Erol Başar conceptualized how the brain generates responses to external and internal stimuli and its dynamic response mechanisms through the framework of susceptibility rules. Within this framework, it is articulated that the brain responds at different frequencies depending on the types of stimuli and cognitive processes. Oscillatory activity can be modulated by the type of stimulus, whether it is visual, auditory, or tactile. Similarly, the oscillatory activity obtained varies depending on the type of cognitive function, such as attention, memory, or problem-solving [15].

The most common hypothesis proposed to explain the decrease in the absolute spectral power during development refers to the reduction in gray matter associated with maturation [16–18]. In fact, a parallel developmental trajectory between the reduction in cortical thickness and the decrease in spectral power with age has been observed [16]. It has been suggested that the increase from low to high frequencies in the brain oscillations and the increase in the alpha frequency band from childhood to adulthood would be due to the increase in the speed of neural communication, basically produced by the increase in the speed of action potentials due to myelination and/or an increase in the axon diameter [19–22]. Considering that brain plasticity, which forms the basis of the ability to adapt to environmental conditions, is at its peak in early childhood, it becomes more essential to follow neuronal development processes in childhood [23,24]. EEG brain oscillations are produced randomly by various groups of neural generators oscillating at different frequencies. However, by delivering sensory stimulation, these generators are combined and act together harmoniously. Synchronized activities lead to evoked and induced rhythms. These combined waves undergo filtration to generate event-related brain oscillations spanning a frequency spectrum from 0.5 Hz to 100 Hz, encompassing delta through gamma frequencies. This entails the segmentation of EEG frequencies into distinct bands, including delta (0.5–3.5 Hz), theta (3.5–7 Hz), alpha (8–13 Hz), beta (18–30 Hz), and gamma (30–70 Hz) oscillations [25]. Alpha frequency is in the range of 8–13 Hz, which reflects the essential physiological characteristics of the brain. It occurs in sensory, motor, cognitive, and memory tasks. Furthermore, the differences in the evolutionary and developmental emergence of the alpha frequency are noteworthy in the studies [26–29]. In a longitudinal investigation focusing on infants' event-related responses to auditory stimuli, analysis of EEG data showed that infants exhibit a stimulus-evoked neural synchrony pattern similar to adults, characterized by a 1/f distribution. Furthermore, the introduction of the stimulus triggers simultaneous increases in theta, beta, and gamma power [30].

The frequency of brain oscillations generally exhibits a negative correlation with their amplitude, indicating that the amplitude decreases as frequency increases. The amplitude of oscillations is considered to be proportional to the number of synchronously active neural elements, with slower oscillating cell populations composed of more neurons than faster oscillating cells [31,32]. In this context, it is expressed that changes in power related to events (event-related synchronization—ERS and desynchronization—ERD) reflect alterations in the activity of local interactions between principal neurons and interneurons that control the frequency components of the ongoing EEG [31]. The heightened short-term alpha enhancement observed in participants while anticipating a specific sensory stimulus serves as a strong example of power changes related to the event [33]. Oscillations related to events are divided into two categories based on their relationship with stimuli: induced and evoked. Evoked oscillations occur after stimulation and are phase-locked to the stimulus. In contrast, induced oscillations arise without phase-locking to the stimulus [34]. Theta and alpha oscillations play a significant role in attention, memory, and learning processes. Theta oscillations are generally associated with memory and learning, while alpha oscillations are linked to the direction of attention and visual processing. Beta oscillations are related to motor control and task-focused attention, whereas gamma oscillations play a critical role in the integration of cognitive processes. Therefore, the analysis of brain oscillations is an important tool for assessing and understanding individuals' cognitive abilities [35].

This review aims to identify developmental changes in the characteristics of event-related and task-related oscillations. To this end, oscillatory responses in delta, theta, alpha, beta, and gamma frequency bands will be examined across different developmental periods, starting from infancy to childhood. In this context, examining childhood oscillations will contribute to providing an updated framework for cognitive developmental research by emphasizing the functional relevance of frequency-specific event-related oscillations and networks

2. Materials and Methods

The literature search was performed using Medline (PubMed) and Science Direct databases. Keywords included brain oscillation, event-related, and children (or similar words; infants, toddlers, task-related) [36]. The first literature search was conducted in August 2022, with updated searches in August 2023 and October 2023. This article aims to understand the development of event/task-related brain oscillations without any pathological conditions. The articles that comprise the diagnostic groups or cases were excluded.

Eligible studies were identified based on the following criteria:

(a) Inclusion of event- or task-related oscillations in children, toddlers, or infants.
(b) Recording of EEG activity while participants performed a task (i.e., not resting state EEG).
(c) Examination of EEG power, phase, or other time-frequency analyses in response to task events, such as event-related increases or decreases.
(d) No case studies or review articles.
(e) Written in English.

Articles were imported to the open-source software Rayyan, where duplicate articles were excluded. Ouzzanni and his team developed Rayyan (http://rayyan.qcri.org, accessed on 2 July 2024), a free web and mobile application that accelerates the initial screening of abstracts and titles through a semi-automated process while maintaining a high level of usability [37]. To minimize the risk of bias in the included studies, the first and second authors independently screened titles and abstracts of all studies for eligibility and marked studies as "exclude", "maybe", or "include". All studies marked as "include" or "maybe" by either author were carried forward to the full-text screening, performed independently by the same two authors. Finally, 86 studies meeting all eligibility criteria were selected for inclusion. Theta oscillations are classified as childhood's most studied frequency bands (Figures 1 and 2).

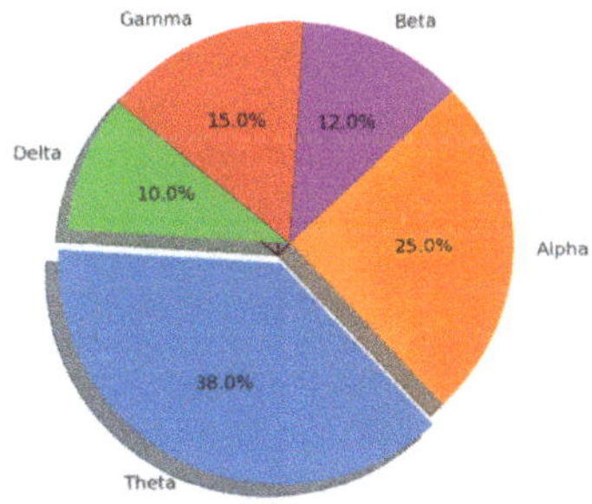

Figure 1. The distribution of event-related EEG studies based on frequencies.

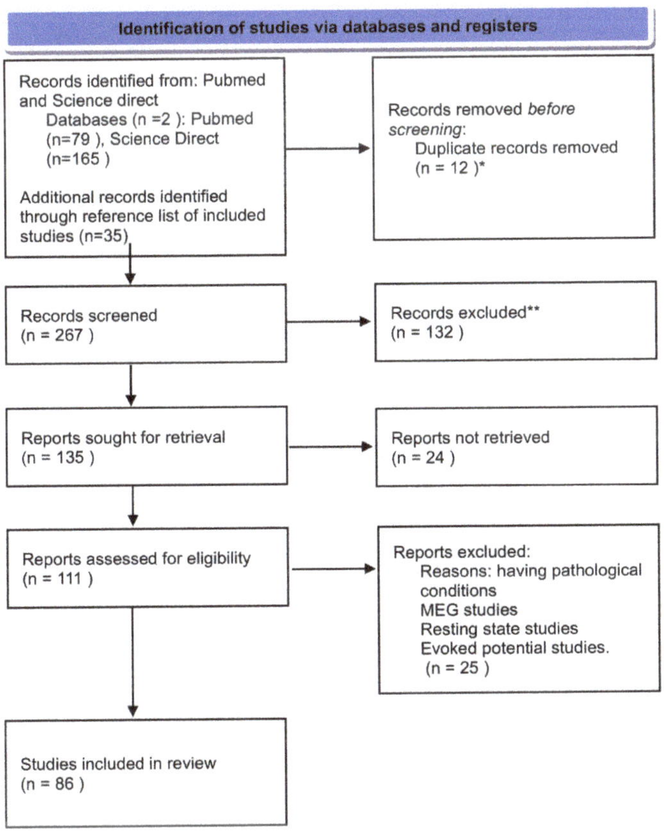

Figure 2. Flow diagram displaying the screening, selection and review process. * The automation tool (Rayyan) was used, indicates how many records were excluded by automation tools. ** Automation tool was used, indicates how many records were excluded by a human.

3. Brain Oscillation from Infancy to Childhood

3.1. Delta Oscillations: The Foundation of Early Neural Development

As previously acknowledged, delta oscillations in children are scarce. Consequently, there is a need to provide a condensed overview of delta oscillations based on the most extensively researched topics and concepts. The slow wave rhythm, regarding the delta frequency, is predominantly seen during the deepest phases of sleep and significantly diminishes in the wakeful states of adults [38]. The presence of delta activity during wakefulness is often interpreted as an indicator of brain injury or pathological states stemming from neurological harm or psychiatric disorders [39]. In children, delta EEG activity exceeds that of adults due to the ongoing process of cortical maturation, and it is often more pronounced in children who have learning disabilities [40,41]. The study of language and language acquisition in children is a regular focus. The early stages of life are particularly critical, marked by numerous neurological changes that facilitate the acquisition of new abilities, with language being a prominent one. Language acquisition is recognized as involving the processing of both semantic and phonological information, thereby imposing cognitive demands. In our previous review article, which discussed evoked and event-related delta responses, it was observed that heightened delta oscillatory responses in frontal and central brain regions were associated with increased cognitive demands in various tasks [42]. Conversely, heightened delta responses in temporal, parietal, and occipital regions were linked to the perception of facial expressions and the recognition of faces in adults. While a substantial body of research in adults has established connections between event-related delta oscillations and various cognitive, perceptual, or emotional processes, the available studies investigating these oscillatory responses in children are limited [43–45]. A developmental trajectory observed in neuroimaging studies highlights changes in the connectivity between the amygdala and the ventromedial prefrontal cortex (vmPFC), a network essential for processing and updating emotional stimuli. This trajectory is characterized by an inverse pattern, transitioning from positive connectivity in childhood to negative connectivity during adolescence and continuing into adulthood [46].

In a study investigating different age groups, 10-year-old children showed bilaterally distributed delta responses to judging whether written word pairs matched their semantic, phonological, or orthographic features, whereas adults had higher left-lateralized delta responses. This difference was interpreted as the incomplete maturation of the left lateralization of the children's linguistic network [41]. It is known from fMRI studies that by age three months, the infant cortex is already structured into several regions of functional importance for speech processing, for example, Broca's area and anterior temporal regions [47], while the neural mechanisms underlying synchronization with the auditory envelope during speech listening are actively debated in the adult literature [48]. In a study comparing filtered visual evoked potentials of 3-year-old children and adults, children's delta responses were similar to adults [49]. However, the post-stimulus latencies of their responses varied greatly. Furthermore, when perceiving ambiguous motion patterns, 10-year-old children had lower inter-trial coherence in their delta responses than adults [50].

Research indicates that in infants, cortical signals in the delta and theta frequency bands align with the rhythmic patterns observed in the speech envelope. Evidence suggests that in the first year of life, delta band tracking of the speech envelope is significantly more robust compared to theta band tracking, with the strongest delta tracking observed at four months. The power spectral density (PSD) of the delta signal produced in response to the stimulus maintains a consistent level from the age of 4 months, while the power in the theta band shows progressive development from 4 to 11 months [51]. It has been suggested that the delta power associated with the speech envelope could serve as a basis for the attentional and grouping mechanisms essential for the perceptual structuring of the speech signal, as evidenced by research conducted on adults [51,52].

Slow-frequency oscillations are fundamentally involved in the operation of widespread networks, whereas fast-frequency oscillations are primarily associated with the functioning of more localized networks [53]. Lower frequency bands, including delta and theta, might reflect the activity of interconnected neuronal groups and could play a role in facilitating information exchange across different regions of the brain [54,55]. In Nanova et al.'s (2011) investigation into event-related responses to auditory stimuli, an elevation in the delta, theta, and slow alpha power across all electrodes was noted in girls during the first 300 milliseconds following the stimulus. A deeper exploration of the data uncovered that in older girls, the observed alterations in delta frequency amplitude were underpinned by event-related phase coherence. The observation of gender-specific developmental differences in delta responses is considered additional evidence of the participation of extensive neural networks in auditory processing. The outcomes of the article reveal that the quickened functional maturation of processing networks is selective to particular frequencies and cannot solely be explained by a broad enhancement of neural connectivity [56]. Delta–beta coupling analyses are employed to indicate regulation capabilities in resting state recordings of infants, yet intra-individual fluctuations are observed within very short time spans. In interpretations made in this context, it is suggested that improvements in regulation abilities among children might be indicated by increased delta–beta coupling scores [57,58].

A review of the literature reveals that delta oscillations have been examined in contexts such as early language acquisition and stimulus processing, often in conjunction with slower frequencies like theta or low alpha. While delta oscillations suggest broad cortical activity, it is important to recognize that they do not solely capture cortical communication in early development. It is suggested that future research should explore the integration of delta oscillations with faster frequencies.

3.2. Theta Oscillations: Bridging Perception and Cognition in Early Years

The enhancement of theta rhythms in EEG has been linked to attention and anticipation processes between associative cortex areas and the hippocampal-frontal-parietal regions, leading to the identification of the 'Theta-Response System" [59,60]. Additionally, the theta rhythm is recognized as a phylogenetically ancient mechanism crucial for neuronal organization within the brain, playing a pivotal role in learning [61–63]. From six months, infants start to direct their attention towards environmental stimuli. Köster et al. (2019) observed that visually entrained theta oscillations intensified in response to unexpected stimuli, indicating disrupted infant expectations. They utilized a novel experimental approach to manipulate oscillations in infants, offering fresh insights. Understanding the impact of stimuli on oscillations in children is vital for precise comprehension of brain dynamics [63].

Xie et al. (2018) indicated that sustained attention in infants is linked with theta (2–6 Hz) synchronization and alpha (6–9 Hz) desynchronization. During periods of prolonged attention, theta synchronization is evident across the frontal pole, temporal, and parietal electrodes, signifying enhanced allocation of attention [64]. Orekhova and Stroganova's 1999 study, which examined the development of theta frequency in infants aged 8–11 months, discovered a theta frequency response in the frontal region of 8-month-old infants within the 3.6–6.0 Hz range across three distinct attentional states [65]. However, it was observed that while this response contributes to prolonged attention spans in the 9th and 10th months, the theta frequency response diminishes with age. This reduction in theta frequency response is believed to be a result of the decreasing need for activity to sustain attention as the brain matures and attentional capabilities improve [65,66]. In addition, the study examined the relationship between medial frontal theta power and inhibitory control in a large developmental sample and defined that non-phase locked medial frontal theta power plays a significant role in inhibitory control and task performance among children and adolescents [67].

Meyer et al. (2022) identified that infants displayed enhanced attentional engagement, as shown by increased theta power in the frontal area when they were presented with action demonstrations involving movements of varying amplitude, compared to demonstrations with consistent or high amplitude movements. Infants who exhibited increased theta power in response to the variable amplitude movements were more adept at successfully carrying out actions and engaging with novel objects relevant to the task at hand. This study supports the notion that infant-directed actions play a critical role in amplifying infants' attentional capacities, learning processes, and exploratory activities [68].

Even though the resting state EEG studies have mentioned that the activation of theta with age is decreasing, there is a study which was conducted in 8-month-old babies where the working memory task creates a theta frequency response in a large part of the cortex [69,70]. In children, it is observed that the theta response is elicited in the medial frontal region, and the theta power decreases with age in the entire topography [71,72]. In addition, it is thought that the theta increase in the frontal regions in children occurs due to its relationship with executive functions [70,73,74].

In attention-demanding tasks, the increase in frontal theta oscillations helps individuals focus their attention on the target task while preventing distraction. Theta oscillations are associated with the processing of information and the activation of information relevant to the current task. Additionally, frontal theta activity establishes a connection between attention and memory, promoting better processing and retention of information [75]. As a result, bursts in frontal theta activity, which are facilitated by the activation of the protease calpain, increase myelination and connectivity [76]. In their 2018 study, Conejero et al. found that toddlers displayed a higher theta power at frontal electrode site when making errors compared to when they were correct. They propose that frontal theta activity plays a crucial role in modifying white matter fibers, potentially as a key factor in fostering the development of ideal structural links within the executive attention network [77]. In their 2021 research, Köster et al. explored the impact of prediction errors on brain oscillations in 9-month-old infants, finding heightened activity within the 4–5 Hz power range during unexpected events. The study demonstrated a significant rise in theta activity specifically associated with processing prediction errors in the infant brain. The authors theorize that besides encompassing all known information about theta oscillations, the theta rhythm might facilitate a computational strategy that condenses real-time events into a quicker neural timeframe. Furthermore, this could potentially enhance cognitive processes by allowing them to operate ahead of actual time and assist in incorporating new events into established networks [78].

The findings reveal that theta power increases in the frontomedial region when children are engaged in a task and increases further during the cognitive engagement. Theta power is also stronger over left fronto-temporal sites for language-related demands compared to action-related demands. The results suggest that theta oscillations in the frontomedial and anterior cingulate cortex reflect top-down control processes in 4-year-olds. Theta oscillations have the potential to serve as a neural marker of top-down control in early childhood and can provide insights into aspects of early development that are not directly reflected in overt behavior [79]. In the research involving infants exposed to Spanish and English syllables, Ortiz-Mantilla et al. proposed that an elevation in delta/theta power and phase coherence might support the auditory discrimination of syllabic content in 6-month-old infants [80].

As it is termed working memory, it depends on the persistent firing of neurons. There is converging evidence that this long term potention has an oscillatory character and that the frequencies involved are in the theta (4–8 Hz) and gamma (30–100 Hz) ranges [81,82]. In an animal study examining memory performance and theta oscillations, it was noted that theta responses were recorded during the encoding and retrieval phases, while gamma responses became more prevalent as the decision-making duration increased [35]. Further information about the necessity of the oscillations comes from the examination of the firing of individual cells while the short-term memory processes drive the information into long-term memory. In many brain regions, the neurons' firing rate is very low, and cells do not fire during every cycle of the theta cycle. However, the firing is nevertheless organized by theta oscillations, as demonstrated by the fact that firing occurs selectively at certain phases of the theta cycle [83]. Superior performance on an infant working memory task has been associated with baseline-to-task increases in 6–9 Hz brain electrical activity at frontal sites during the stages of short-term encoding, storage, and retrieval [84]. Infants who successfully displayed ordered recall showed a pattern of increasing EEG from baseline to task at anterior temporal scalp locations, whereas infants showing no ordered recall displayed no changes in EEG from baseline to the task. The data in the study are the first to provide evidence to trace the continuous electrophysiological change in infant declarative memory processing throughout the creation of new event memories and long-term retrieval of events [85].

Although the medial prefrontal cortex (mPFC) is known as a neural generator of frontal theta oscillations, its potential maturational effects are not clearly understood. Moreover, mPFC develops significantly during early childhood [86]. There is an fMRI study which was conducted with 3-month-old infants. They have explained the hippocampal mechanism for the statistical learning paradigm and the activity of frontal sides during the task. Despite the hippocampus doubling anatomical volume across infancy, learning-related functional activity bore no relationship to age. They suggested that the hippocampus is recruited for statistical learning at the youngest ages in their sample. Outside of the hippocampus, they observed the involvement of the mPFC in statistical learning [87]. This is extremely surprising, given the dramatic changes in frontal lobe anatomy over development [88]. Consistent with recent research suggesting that the frontal cortex plays an important role in infant cognition [89,90]. In adults, mPFC strongly interacts with the hippocampus during memory formation, facilitating encoding based on related past experiences (i.e., schemas) to promote memory integration [91].

The study examined how proactive and reactive processing modes impact sensory and cognitive information processing in children aged 7–10 years. The findings indicated that the proactive processing mode was associated with increased theta activity before the stimulus and a reduction in the temporal synchronization of theta/alpha oscillations related to the event within the initial 300 ms following the stimulus. The outcomes suggest that holding internal task representations in working memory activates oscillatory networks, which in turn can influence the processing of incoming sensory data [92]. In research involving a large sample size of 239 children and adolescents, it was discovered that delta, theta, and alpha oscillations collectively contribute to the attentional engagement and focus necessary for working memory tasks. Within this synergistic relationship, slower frequencies, specifically delta and theta, play a crucial role in maintaining the active state of the working memory trace [93].

In a study investigating the impact of gender differences on working memory and auditory processing, it was observed that only the slower frequency bands (delta, theta, and slow alpha) showed an increase in phase-locking among girls, with no such increase noted in the faster (fast alpha) bands. It is thought that the rapid maturation of processing networks is selective for specific frequencies and cannot be attributed merely to a widespread strengthening of neural connections that would boost synchronization across all oscillatory networks [56]. Research investigating source memory identified elevated theta activity in 6–8-year-old children during a memory task, with no significant variation observed between conditions (fact and source recall). The increase in activity from baseline to task, recorded via frontal, temporal, and parietal electrodes, is speculated to indicate hippocampal cortical oscillations, which might signify episodic memory involvement [94].

Pre-verbal children have top-down mechanisms related to language encoding and attentional prediction of speech [95,96]. Another remarkable situation is the emergence of the theta response to emotional stimuli in infants [73,97]. A study examined the theta response of infants and children and compared them in three different attention conditions. The conclusion is the theta response increases in both infants and children in attention conditions. While theta frequency was observed at 3.6–5.6 Hz in the posterior temporal, occipital–parietal–temporal regions of infants under attention conditions. Theta frequencies in the frontal region of children are observed to be in the range of 4–7.6 Hz. One of the important results of the study is that the theta response, which occurs under attention conditions, increases at a greater topography in infants compared to children [73]. Cuevas et al. (2022) analyzed variations in EEG power across three frequency ranges (3–5 Hz, 6–9 Hz, 10–12 Hz) during a verbal recall activity in children aged two. The study highlighted that theta (3–5 Hz) and low alpha (6–9 Hz) rhythms provide significant insights into toddler memory processes. They found that as performance improved, the power values in these theta and alpha bands also increased, enabling the differentiation of toddlers according to their performance levels [98].

In a study focusing on 5-month-old infants observing changes in behavior and physiology in response to facial expressions, an increase in theta activity within the 4.8 Hz range was noted across most anterior and right posterior temporal regions of the scalp when infants were exposed to a blank face [99]. These findings suggest that infants are responsive to variations in adult facial expressions and indicate the possibility of distinct neurophysiological processes underpinning the infant's gaze towards smiling versus neutral faces. Moreover, the presence of right temporal theta activity could reflect interactions between temporal cortical areas and the hippocampus. Given that hippocampal theta activity relies on cholinergic inputs from the nucleus of the solitary tract (the sensory center of the visceromotor component of the social engagement system), it may facilitate the cholinergic effects on the temporal cortex's role in directing anticipatory attention towards the emotional content of faces [99–101]. In research examining speech perception through the integration of auditory and visual stimuli, the synchronization between theta phase and gamma amplitude was characterized as the interplay between top-down and bottom-up mechanisms in speech processing. The study found that the right hemisphere exhibits greater theta–gamma coupling under conditions where stimuli match, whereas the left hemisphere shows enhanced theta–gamma coupling when stimuli do not match. This distribution across hemispheres suggests a rapid global processing approach in the right hemisphere for congruent conditions and a more meticulous, slower processing method in the left hemisphere for dealing with incongruent situations [102].

The functional roles of theta oscillations have been elucidated in a variety of contexts, ranging from information processing to working memory capabilities, as detailed earlier. Table 1 comprehensively summarizes the technical specifications for all the articles reviewed in this study. The spatial distribution of event-related theta oscillations during childhood seems crucial for cognitive functions. Frontal theta activity is linked with attention processes and engaging in memory tasks, whereas temporal theta activity is associated with the processing of facial expressions and social engagement. Furthermore, with maturation, theta frequencies can be distinctly differentiated from those in the alpha range. For this reason, studying both theta and alpha frequency ranges concurrently in children is believed to be beneficial for gaining a deeper understanding.

Table 1. Summary of key articles reviewed.

	Development Stages and Focused Oscillations	Modality and Paradigms	Inference
Musacchia et al. (2015)	211 infant recording sessions with 98.6% of data retention 57 participants Theta, beta, and gamma oscillations	The auditory stimuli in free-field. Stimuli are presented in two blocks that include a fundamental frequency of either 800 or 1200 Hz and 15 harmonics (6 dB roll-off per octave).	Time-frequency analyses of the EEG data that are collected from infants' data show that stimulus evoked 1/f pattern neural synchrony like adults. The stimulus onset shows synchronous bursts of theta, beta, and gamma power in both the right and left auditory regions of the brain. Suggestion about the creation of a model to understand the integrity of early auditory processing mechanisms.
Güntekin et al. (2023)	40 participants (children and young adults with rates of 1 in 2) Alpha oscillations	The day-night stroop task. Event-related spectral perturbation (ERSP) and event-related phase-locking analyses were performed.	Different alpha dynamics in children. Shifting of alpha power from posterior to anterior with age. The posterior sides of the brain are involved in the inhibitory processes. Increased alpha power in early time windows, decreased event-related desynchronization (ERD) in later time windows.

Table 1. *Cont.*

	Development Stages and Focused Oscillations	Modality and Paradigms	Inference
Gómez et al. (2023)	239 participants (6–29 years) Delta, theta, and alpha oscillations	The Delayed-Match-to-Sample Test. ERSP in the range of 1–25 Hz.	Delta, theta (7–8 Hz) and alpha-2 (8.5–10 Hz) oscillations work for attentional engagement and focusing for working memory. Beta and low-frequency components work together for encoding, while alpha sharpens this process. Low-frequency component works to keep the WM trace active. Anticipatory attention is characterized by an increase in alpha and beta in late adolescents and adults. Behavioral responses which are represented with reaction time have the most consistent relationship with alpha-2.
Wang et al. (2014)	17 children (mean age 10 years, 5 months) Theta and gamma oscillations	Speech perception paradigm consists of the match and mismatch conditions with auditory and visual stimulus. Theta–gamma coupling analysis.	The coupling of theta phase and gamma amplitude demonstrates the interaction of top-down and bottom-up processes in speech perception. Right hemisphere has higher theta–gamma coupling in the match conditions, and the left hemisphere has stronger theta–gamma coupling in the mismatch conditions. This hemispheric pattern is interpreted as a fast global processing strategy and a slow detailed strategy to adapt to congruent and incongruent conditions.
Ortiz-Mantilla et al. (2013)	28 infants (6-month-olds) Theta and gamma band	English and Spanish syllable contrasts. Single-trial time frequency analysis. Temporal spectral evaluation (TSE). Intertrial phase locking (ITPL).	Increased delta/theta power and phase alignment could subserve auditory discrimination of syllabic information in 6-month-old infants. Preferential processing of specific native features are represented by frontal gamma power. Six-month-old infants can distinguish phoneme information across languages.
Maróti et al. (2019)	22 children (mean age 6.5 years) Beta and gamma frequency range	Auditory stimulus includes the tones at the beta and gamma range in isochronous sequences. Steady-state evoked potentials.	Neural responses which are related with the entrainment associated with the different frequency oscillations do not develop at the same level.
Musacchia et al. (2017)	49 infants A longitudinal study at the fourth and seventh months Theta and gamma oscillations	Active and passive auditory exposure. Go/no-go task. Passive oddball paradigm. Three different types of paired acoustic stimuli. TSE and ITPL.	A representation of maturation from 4 to 7 months, which is increased spead of event-related source response and decreased theta phase stability. Differentiation of theta and gamma oscillatory dynamics between active and passive auditory exposure protocols. Acoustic experience appears to promote more mature oscillatory and lateralized patterns for the auditory system.

Table 1. *Cont.*

	Development Stages and Focused Oscillations	Modality and Paradigms	Inference
Colomer et al. (2023)	17 participants (9-month-olds) Theta and beta	Action observation task with familiar and novel conditions. Action execution task to engage the infants actively. Time–frequency analyses (4–20 Hz). Inter-Site Phase Clustering (ISPC) to perform phase-based connectivity analysis.	Infants' brain has greater ISPC with age. It is suggested that the global connectivity at the whole brain might be related to maturation. Greater visual–motor coherence is shown in infants with better motor skills. That could have downstream consequences on how infants encode and learn others' actions.
Hao et al. (2020)	16 children (mean age 7.7) and 13 adults Mu and beta rhythms	Motor-related experiment. Event-related spectral pertubations (ERSPs). Inter-trial coherence (ITC).	The power of mu rhythm (8–14 Hz) decreases significantly during movement, and it returned to baseline after movement for both children and adults. Motor related mu supression develops early in children. Beta power (15–30 Hz) has different dynamics between children and adults. Movement induces beta ERD in both, but post-movement beta power rebound (PMBR) is observed only in adults. PMBR might be associated with a more prolonged motor development process.
Adam et al. (2020)	87 children (52 preschool and 35 school children) Theta frequency range	Heart and flower task to evaluate cognitive control process. Behavioral results (response time and accuracy). Time–frequency analysis. Midfrontal theta peak latency.	Midfrontal theta power is interpreted as a generic indicator of cognitive control process that supports both inhibitory and flexibility functions. MFT might be accepted as a general neural mechanism to coordinate cognitive process across the development.
Cantiani et al. (2022)	A longitudinal study at seventh and ninth months 25 infants (mean age 8 months)	Three different rhythmic stimuli (African, Tabla, and speech). Steady-state evoked potentials (SS-EPs).	Low frequency cortical tracking of the stimulus envelope is already measurable at 4 months of age. There are no hemispheric differences in the magnitude of SS-EPs. There is left-lateralized SS-EPs at higher frequencies for speech stimulus. Infants can entrain like adults, but there is difference between frequency ranges for same stimulus.
Noordt et al. (2022)	432 participants (224 females, 208 males) Theta	Go/no-go task. Raven's progressive matrices to measure reasoning ability. Threshold free cluster enhancement and permutation testing to analyze spatial and temporal dynamics of theta power. Phase evaluation. Latent class analysis.	Greater non-phase locked theta power during response inhibition is associated with more efficient response control. Developmental changes in frontally mediated inhibitory control processes are represented by a greater non-phase locked theta power and a shift to less phase locked activity.

Table 1. *Cont.*

	Development Stages and Focused Oscillations	Modality and Paradigms	Inference
Antognini et al. (2019)	20 toddlers (18 months), 27 toddlers (24 months) Mu	Three types of stimuli-related actions (auditory, visual and play material).	The lateralization of mu suppression towards the left hemisphere in response to action verbs in terms of similarities with language processing. Very strong occipital alpha suppression for action observation differs between age groups, whereas the lateralization of central activity is towards right hemisphere. Although the neural processing of action verbs and pseudoverbs is related to the left central cluster, the sensorimotor activity is shown just for action verbs.
Köster et al. (2021)	36 infants (9-month-olds) Theta	The violation of expectation paradigm to evaluate neural proccesing of prediction error. Time–frequency analysis and event-related potentails	All scalp electrodes has increased activity in 4–5 Hz power in unexpected events. Theta substantially increased for the processing of prediction errors in the infant brain. Theta rhythms are located in the parietal regions in infants. Theta serves as a facilitation mehanism to integration of new information into existing networks.
Patzwald et al. (2020)	38 infants (18-month-olds) Mu	Action observation with verbal cues. Mid-latency ERP 300–800 ms. Central mu power (6–9 Hz).	Reduced mu power is shown for congruent to incongruent stimuli. Central electrode sides (C3–C4) might evaluate as a primary channel of interest to study mu rhythm (6–9 Hz). Occipital channels are unavailable because of the widespread posterior alpha effect.
Bryant et al. (2019)	21 children (mean age 4.89) Mu and beta	Two tool sets and plastic aquatic animal toys. Active and observational training sessions to handle the toys. Spectral analysis of signals. Time-locked ERD.	Central–parietal desynchronization during the action observation and execution represents neural mirroring, and 3- to 6-year-old children exhibit neural mirroring within mu ryhthm. Greater occipital alpha (8–10 Hz) ERD is found during perception of the active training task. This suggests that occipital alpha may be facilitating the processing of attended stimuli while suppressing attention to irrelevant information.
Angelini et al. (2023)	17 infants (9-month-olds) Theta (4–5 Hz) and alpha (6–8 Hz)	Face-to-face live paradigm with a triadic social interaction. Time–frequency analysis.	Theta power increases in response to an unexpected event that is interpreted as a marker of information prediction about social behaviors. Higher alpha power is detected in infants in relation to the inhibition of interfering stimuli needed to sustain internally controlled attention. The live paradigm favors the expectation of a usual event that could have an enhanced effect on alpha power.

Table 1. *Cont.*

	Development Stages and Focused Oscillations	Modality and Paradigms	Inference
Ramos-Escobar et al. (2021)	60 children (mean age: 9 years and 5 months) Syllable and word frequencies in ROIs	Audio-visual artificial language streams. Frequency tagging analysis of spectral power. Weighted Phase-Lag Index.	Frontal and parieto-occipital ROIs at target frequencies are detected with long-distance EEG phase synchronization. Word segmentation and meaning mapping might be modulated by attentional mechanisms to multimodal word learning.
Weiss et al. (2018)	80 children (6 to 8 years old) Mu and alpha	Tactile stimulus. Flanker task. ERSP.	Anticipatory ERD of the mu rhythm is related to executive function skills, whereas it is not associated with the accuracy of children's behavioral responses to tactile stimulation.
Mikhailova et al. (2021)	36 children (17–41 months) Mu	Observation of actions (fake, real, and performing the action). Bayley test for speech comprehension. Time–frequency analysis.	Observing a real action situation led to greater power desynchronization in the μ range in the medial frontal and medial parietal loci compared to observing a fake action situation. Children with higher levels of speech comprehension showed greater EEG power desynchronization in the μ range when observing a real action in the frontal and parietal loci of the left hemisphere.
Cantiani et al. (2019)	56 participants Typically developing (15 F/17 M) Age: 6 months and 20 months of age (longitudinal) High familial risk for Language and Learning Impairment (LLI) (11 F/13 M) Age: 6 months and 20 months of age (longitudinal) Delta/theta (2–12 Hz) Gamma (30–80 Hz)	Non-speech double oddball paradigm. Cluster permutation testing: Spectral power (TSE) Phase coherence (ITPL). Tests: BSID-III Language Development Survey (LDS) Communicative Development Inventories (CDIs).	Higher bilateral theta activation in auditory cortices in response to deviant stimuli in both groups Infants at risk for LLI: reduced gamma activity in left auditory cortex and higher gamma activity in right auditory cortex
Phelps et al. (2022)	48 participants 24 bilinguals (8 F/16 M) Age: (M = 9.3 years, SD = 1.83) 24 monolinguals (11 F/13 M) Age: (M = 9.6 years, SD = 1.48) Low-frequency neural oscillations	Dichotic listening task. Multivariate temporal response function (mTRF). Speech envelopes.	Different EEG patterns in monolingual and bilingual children during encoding Weaker differentiation of linguistic distractors in bilingual children
Köster et al. (2020)	42 participants (26 F/16 M) Age: 10-month-olds 36 participants (15 F/21 M) Age: 20-month-olds 2–14 Hz	Action demonstration videos and generalization test, followed by imitation test. Spectral perturbation analysis.	20-month-old infants: increased 7–10 Hz activity at C3-C4 during action observation; desynchronization in 3–6 Hz peaked at posterior electrodes (P3, P4, Pz, P7, P8, O1, O2)

Table 1. *Cont.*

	Development Stages and Focused Oscillations	Modality and Paradigms	Inference
Ossmy et al. (2021)	22 participants (11 F/11 M) Preschoolers Age: 3.09 to 5.49 years 22 participants (14 F/8 M) Adults Age: 19.37 to 26.40 years 6–9 Hz 8–13 Hz (Mu)	Videos of efficient and inefficient displays of tool use (multi-step actions). Eye-tracking, pupil dilation, EEG, and machine learning. Nonparametric cluster analysis. Event-related spectral perturbation (ERSP).	Increased event-related desynchronization in the mu frequency band at sensorimotor regions after action observation. Differential EEG responses to efficient versus inefficient initial grips observed only in adults.
Anderson et al. (2021)	55 participants (31 F/24 M) Age: 6–12 months Alpha (6–9 Hz)	EEG data: baseline (1–2 min), 90 s of mother/infant play Behavioral data: parent/infant play, infant temperament. Fast Fourier Transform (FFT). Tests: The Infant Behavior Questionnaire (IBQ-R).	Significant positive correlation between baseline frontal alpha asymmetry (FAA) and object exploration Only the cuddliness subscale (in IBQ-R) was associated with a leftward shift in FAA.
Georgieva et al. (2020)	14 participants (8 F/6 M) Age: (M = 338.85 days) Delta (1–3 Hz) Theta (3–6 Hz) Beta (∼15 Hz)	EEG data: resting state, spontaneous motion. Fast Fourier Transform (FFT). Cluster-Based Permutation Test.	Movement EEG scalp topology was similar to resting state EEG: high delta/theta power at posterior regions and high beta power at orbitofrontal regions. Upper limb movements characterized by increased beta power and generated more widespread artifacts. Decreased theta and alpha power at central in all motion types.
Ortiz-Mantilla et al. (2022)	100 participants (47 F/53 M) Age: 4 months (longitudinal visits at 7, 9, 12, and 18 months) P1 Theta (frequency clusters in 2–6 Hz range)	Auditory go/no-go task (target stimulus is paired with a reward video). Intertrial Phase Synchrony.	Lower theta phase synchrony at 7 months is linked to better expressive language at 12 and 18 months and better receptive language at 9 months.
Buzzell et al. (2020)	136 participants 68 receiving care as usual (CAUG) (35 F/33 M) 68 receiving foster care (FCG) (34 F/34 M) Age: (M age = 21.6 months) Mediofrontal theta (4–8 Hz) oscillations	Go/no-go task. Behavioral assessment of risk for psychopathology (HBQ). Theta power was calculated as the total power measure weighted by the average power time-frequency PCA loadings.	Higher error-related mediofrontal theta power in FCG compared to CAUG.
Rayson et al. (2019)	23 participants (10 F/13 M, Age: 6.5 months (M = 200.91 days, SD = 5.86) 24 infants (13 F/11 M), Age: 9.5 months (M = 292.92 days, SD = 7.88) Alpha	Preferential looking stimuli. Gaze following stimuli. Event-related spectral perturbation analysis.	At both ages, alpha ERD was higher in the congruent condition than the incongruent condition. The higher the alpha ERD in the congruent condition, the more preference toward congruent stimuli (9.5-month-old infants).

Table 1. *Cont.*

	Development Stages and Focused Oscillations	Modality and Paradigms	Inference
Haartsen et al. (2020)	73 participants (38 F/35 M, Age: (M = 302 days, SD = 13) EEG data for the second session consisted of 64 infants (9 drop-outs) Alpha oscillations Graph theory metrics (normalized weighted clustering coefficient, normalized weighted path length, and small-worldness index)	Presentation of naturalistic dynamic videos and moving toys (60 s each)—2 sessions with one-week interval. Phase lag index (PLI). Debiased weighted PLI (dbWPLI) measured using Fourier coefficients.	Intra-class correlations (ICCs) for whole brain connectivity were higher than ICCs for the other graph metrics.
Hoehl et al. (2014)	24 participants (9 F/15 M) Age: 8 months and 29 days Alpha oscillations	EEG recording during eye-contact and no eye-contact conditions. Event-related spectral perturbation.	Infants responded with alpha desynchronization to eye-contact condition. In no eye-contact condition no alpha synchronization or desynchronization effect was observed.
Conejero et al. (2018)	66 participants 52 toddlers (26 F/26 M) Age: (M = 16.75 months; SD = 0.67) 14 adults (13 F/1 M) Age: (M = 21.93; SD = 2.34) Error-related negativity (ERN) Theta (4–7 Hz)	27 different animal head/body combinations (conceptual error condition). ERP. Time–frequency (power).	Higher fronto-central negativity for incorrect trials compared to correct trials in all participants. Significantly greater increase in theta power in the error condition compared to the correct condition in children of highly educated parents. SES significantly contributed to the amplitude of the ERN.
Jones et al. (2015)	168 participants 88 infants (39 F/49 M) Age: 6 months 80 infants (40 F/40 M) Age: 12-month Theta (3 to 6 Hz) Alpha (6 to 9 Hz)	Two movies of 1-min duration: naturalistic social and non-social FFT.	Socially selective theta responses showed increased power and topographical extent in both Live Action and Movie formats between 6 and 12 months. Theta power has the potential to be a more sensitive measure of social brain development in the first year of life.
Thorpe et al. (2016)	20 adults (11 F/9 M) Age: 18–21 years old, 47 participants (21 F/26 M, 9 drop-outs) Age: 45–68 months 50 participants (3 drop-outs) Age: 12 months Lower alpha band (8–10 Hz) Upper alpha band (10–12 Hz) Upper mu band (10–13 Hz) Lower mu band (8–10 Hz)	Grasp execution task. Power spectral density (PSD). Event-related desynchronization.	Both alpha bands' topographies peak over occipital–parietal electrodes in all subjects. Strong peak over frontal channels in both alpha bands in adults; no evidence of such a frontal peak in younger subject groups.

Table 1. *Cont.*

	Development Stages and Focused Oscillations	Modality and Paradigms	Inference
Friedrich et al. (2017)	107 participants (47 F/60 M) Age: 6–8 months 12–15 Hz	Learning phase and memory test (primed and unprimed word pairs were tested) ERP, FFT (late negativity, N400, spectral power of sleep spindle activity).	Evidence that infants as young as 6 months old can form semantic categories in long-term memory and associate them with word forms. The formation of perceptually based categories occur in the earlier stages of the sleep (NREM sleep stage 2).
Cuevas et al. (2011)	20 healthy, full-term infants (longitudinally seen from 5 months to 10 months of age) 6–9 Hz	EEG and ECG recordings: Baseline A-not-B task EEG power Coherence.	Baseline: (a) Power increased with age in all electrode sites except for the frontal pole and lateral frontal; (b) coherence increased between frontal pole/medial frontal, medial frontal/lateral frontal, medial frontal/medial parietal, and medial frontal/occipital; (c) heart rate decreased with age. Task: (a) evidence for working memory processing related increase in EEG power in 5-month-old infants; (b) only 5–7-month-old infants showed more localized task-related changes in EEG power in medial parietal region; (c) HR showed no task related changes.
Meyer et al. (2011)	7 participants (2 F/5 M) Age: 3-year-olds 17–21 Hz (beta)	Joint button-pressing game with three conditions: (1) joint action, (2) joint action observation, (3) individual action. Time–frequency representations. Discrete Fourier Transform (DFT)	Higher decrease in beta power in children during joint action observation when they were observing their own joint action partner. Enhanced motor system activity was observed at C3 electrode and was associated with fewer errors during task.
Swingler et al. (2017)	388 participants (199 F/189 M) Age: at 5 months and 10 months of age (longitudinal) 6–9 Hz	EEG Data: Baseline Visual attention tasks at 5- and 10-month follow-ups. Behavioral data: Attention (glove puppet presentation). Maternal behavior (child-mother interaction). Medial frontal EEG power (F3/F4) DFT—natural logarithmic normalization.	Increased EEG power difference at 10 months at the right frontal medial region was linked with better task performance. Higher positivity during mother–infant interactions promote better attentional engagement.
Cuevas et al. (2012)	122 participants (62 F/60 M) Age: 23 months–28 months (longitudinal) Theta (3–5 Hz) Alpha (6–9 Hz) Beta (10–12 Hz)	EEG Data: Baseline Verbal recall task Power—Discrete Fourier Transform (DFT).	Differences in theta and alpha responses during encoding and retrieval. Differences in theta, alpha, and beta responses in high and low performers.

Table 1. *Cont.*

	Development Stages and Focused Oscillations	Modality and Paradigms	Inference
Endedijk et al. (2017)	29 participants (19 F/10 M) Age: (M = 52.48 months, SD = 1.94) Mu (7–12 Hz) Beta (16–20 Hz)	The color-naming task. Action observation task Cooperation task (double-tube task). The entrainment task.	4-year-old children who exhibited greater motor system engagement during action observation, indicated by reduced beta-power, demonstrated higher success in early peer cooperation.
Liao et al. (2015)	21 participants: 11 toddlers (7 F/4 M) Age: (M = 41 months, SD = 4) 10 mothers' age: (M:35.5) Mu (7–9 Hz) Beta (15–18 Hz)	Component power spectral density (PSD). Component event-related spectral perturbations (ERSPs). IC clustering.	Age-matched head models indicated that mu sources were located in both the left and right sensorimotor cortex. When children observed their parents' actions, there was a noticeable suppression of power in the mu (7–9 Hz) and beta (15–18 Hz) frequency ranges. Remarkably, the children's mu suppression during action observation mirrored adult patterns in all observed properties.
Belalov et al. (2014)	91 participants: 41 orphans (14 F/27 M) Age: (M = 36 months, SD = 2) 50 children raised in a family (19 F/31 M) Age: (M = 35 months, SD = 3) Theta (4–6 Hz) Alpha (7–10 Hz) Beta (11–29 Hz) Gamma (30–45 Hz)	EEG was recorded in three situations: Spontaneous EEG (eyes open): 60 s; while the child listened to a short poem: 20 s; while the child heard meaningless speech: 20 s. Spectral power density (SPD). Tests: The Language Scale of the Bayley Scales of Infant and Toddler Development III (BSID-III).	The Bayley assessment showed that children raised in orphanages had significant delays in language development compared to those raised in families. During the speech stimulus, alpha oscillations become desynchronized, while theta, beta, and gamma oscillations show increased synchronization. Children raised in families showed notable increases in gamma rhythm SPDs across 13 leads in both hemispheres. Conversely, orphans exhibited this increase in only 8 locations, predominantly localized in the left hemisphere.
Schneider et al. (2016)	36 participants: 18 children (9 F/9 M) Age: (M = 10.94, SD = 0.94) 18 adults (9 F/9 M) Age: (M = 24.41, SD = 4.37) ERP (P600, N400) Theta (4–8 Hz) Beta (12–30 Hz)	Grammaticality judgment task. Event-related spectral perturbations (ERSPs). Fourier power spectra density (PSD).	Unlike adults, children exhibited a distinct pattern with a noticeable N400 effect and did not show a decrease in beta or theta activity in response to grammatical violations.
Orekhova et al. (2001)	60 participants Age: (M = 37.3 rd week, SD = 1.96) Alpha (6.4–10 Hz)	EEG: Attention to an object in the visual field peek-a-boo game. Behavioral data: the total time of anticipatory attention was calculated across all 10 trials.	Infants who displayed longer periods of anticipatory attention had higher absolute spectral amplitude across a wide frequency range in both experimental tasks. Alpha synchronization in the posterior parietal cortex actively inhibits certain networks involved in maintaining attention to the peripheral visual field, rather than representing a passive or idle state.

Table 1. *Cont.*

	Development Stages and Focused Oscillations	Modality and Paradigms	Inference
Orekhova et al. (2006)	47 participants: 28 infants (9 F/19 M) Age: (M = 9.2, SD = 1.3) 19 children (8 F/11 M) Age: (M = 5 years 5 months, SD = 359 days) Theta (3.6–5.6 Hz) Mu rhythms (6.0–8.8 Hz)	EEG was recorded during baseline (visual attention) and two test conditions: exploration of toys (manipulation), and attention to 'social' stimulation (speech). Event-related spectral power.	Theta responses increase in both infants and children during attention-requiring test conditions, with infants showing a broader topographical increase compared to children. (a) Theta frequency was observed at 3.6–5.6 Hz in the posterior temporal region and occipital–parietal–temporal regions of infants under attention conditions. (b) Theta frequency in the frontal region of children is observed to be in the range of 4–7.6 Hz.
Bell et al. (2007)	50 participants (22 F/28 M) Age: (8 months, when the 25 children were 4.5 years of age, they joined the research again) 6–9 Hz Alpha (8–13 Hz)	Spontaneous EEG. Discrete Fourier transform (DFT) Coherence. Infant working memory/inhibitory control task. The Day-Night Stroop-like task. The Yes-No task.	(a) In infancy, working memory was associated with widespread changes in EEG power and coherence across the entire scalp, indicating broad cortical involvement. (b) By early childhood, these associations were more localized, focusing primarily on medial frontal areas for EEG power and involving specific pairs like medial frontal/posterior temporal, and medial frontal/occipital for EEG coherence.
Blankenship et al. (2018)	242 participants (121 F/121 M) Age: (M = 6.64 years, SD = 0.44) Theta (4–7 Hz) Alpha (8–13 Hz)	Episodic memory (retrieval) tasks. Working memory tasks (A backward digit span (BDS) task). Tests: Woodcock–Johnson (WJ) III The Peabody Picture Vocabulary Test IV.	EEG alpha power in frontal regions and theta power in temporal regions during the working memory task were used to statistically predict both math and reading abilities.
Güntekin et al. (2020)	34 participants: 16 children Age: (M = 6.69 years, SD = 0.48) 18 young adults Age (M = 21.32 years, SD = 3.28) Theta (4–7 Hz) Alpha (8–13 Hz)	The Boston Naming Test (BNT). Free and Cued Selective Reminding Test (FCSRT). Tests: Wechsler Intelligence Scale for Children, Fourth Edition (WISC-IV).	Frontocentral theta and alpha phase-locking play a crucial role in brain maturation and the achievement of successful memory performance. Young adults had higher theta and alpha phase-locking than children over the frontal and central locations. Children exhibited heightened theta phase-locking and increased left alpha power in response to remembered objects compared to forgotten objects. The children had higher parietal–occipital alpha phase-locking than the young adults.
Rajan et al. (2021)	29 participants (19 F/10) Age: (M = 6.10 years, SD = 0.26) Theta (4–7 Hz)	Memory binding task. Discrete Fourier transform (DFT). Natural log (ln) transform.	Theta rhythms play a significant role in memory-binding processes during middle childhood.

Table 1. *Cont.*

	Development Stages and Focused Oscillations	Modality and Paradigms	Inference
Stroganova et al. (2006)	44 participants Age: (M = 19.6 weeks; SD= 2.59) Theta (3.6–5.2 Hz)	The attention tasks (ExA, EnA, and CA). Spectral analysis (Fast Fourier Transform). Spectral theta peak amplitude (SA). Tests: Bayley-2 scale	At five months of age, children who maintain attention on a hidden object demonstrate the emergence of a highly synchronized theta rhythm compared to their child stages.
Gomarus et al. (2006)	18 participants (4 F/15 M) Age: (M = 10.4 years, SD = 1.5) ERPs Theta (4–8 Hz) Alpha (8–12 Hz).	Selective memory search task ERD, ERS.	Theta ERS was most evident during the most challenging task condition in the recognition set, while alpha ERD exhibited a load effect only during memorization.
Spironelli et al. (2010)	70 participants: 28 children (14 F/14 M) Age: (M = 10.01 years, SD = 0.18) 22 young adults (11 F/11 M) Age: (M = 22.59 years, SD = 3.67) 20 middle-aged adults (7 F/13 M) Age: (M = 59.10 years, SD = 7.11) Delta (1–4 Hz) Theta (4–8 Hz) Alpha (8–13 Hz Beta (β1: 13–20 Hz, β2: 21–28 Hz)	Three linguistic tasks (orthographic, phonological, semantic). Spectral FFT analysis.	In adults, task-dependent right lateralization was observed in theta and alpha distributions, whereas children exhibited differences primarily between linguistic and non-linguistic tasks without clear lateralization patterns.
Jiang et al. (2017)	37 participants (18 F/19 M; 10 excluded) Age: (4.5–5.5 years) Delta (1–3 Hz) Theta (3–8 Hz) Alpha (8–13 Hz) Beta (13–30 Hz) Gamma (30–70 Hz)	Event-related spectral perturbation (ERSP).	Negative stimuli, compared to neutral ones, induced greater theta ERS in children who showed improved cognitive efficiency with negative emotional content.
Morasch et al. (2009)	48 participants (16 F/32 M) Age: (M = 10 months, 6 days) alpha (6–9 Hz)	Discrete Fourier transform (DFT) Tasks: The activity-matched baseline Encoding (modeling) Recall Test.	Infants who demonstrated successful ordered recall exhibited an increase in brain activity from the baseline to the task at the front temporal scalp areas. In contrast, infants who did not show ordered recall had no changes in brain activity from baseline to task.
Schneider et al. (2018)	39 participants: 23 adults (14 F/9 M) (M = 24 years, SD = 4.3) 16 children (8 F/8 M) (M = 10.88 years, SD = 0.96 years) Theta, alpha, and beta = 3–30 Hz	Event-related spectral perturbation.	The increase in children's theta power after 2400 ms over the frontal and right frontocentral areas during the processing of grammatically and semantically correct sentences significantly differed from that of adults.

Table 1. *Cont.*

	Development Stages and Focused Oscillations	Modality and Paradigms	Inference
Attaheri et al. (2022)	60 participants Age: (longitudinal cohort: (M = 212.2 days SD = 7.2 to M = 333.1 days SD = 5.6)) Delta Theta	(mTRF) method in delta, theta, and alpha bands. Periodogram (PSD) analysis. Phase-amplitude coupling. (PAC) for delta–beta, delta–gamma, theta–beta, and theta–gamma coupling.	(a) PAC was significantly present at all ages, with both delta and theta bands driving the coupling. (b) Delta and theta were significant carrier phases, linking with higher-frequency amplitudes in the infant brain.
Meyer et al. (2023)	23 infants (10 F/13 M) Age: 15.5–16.5 months (M = 15.9) Theta (4–5 Hz)	EEG recorded during demonstration video clips at one of the three conditions (normal, high, and variable) includes five goal-directed movement actions (acting on balls, cups, and rings). After EEG recording, 1 min. exploration phase started and infants had the opportunity to act on the objects themselves. Time–frequency analysis of Fz, FCz, Cz.	Infants with increased frontal theta activity were prone to investigating new objects within the present task framework. Variability has an important role in drawing infants' attention to and thereby fostering their learning from infant-directed actions (IDAs).
Meyer et al. (2019)	29 children (9 F/10 M) Age: 4 years (M = 52.48, SD = 1.94 months) Theta (3–6 Hz)	Tasks: No task, a color-naming task, an imitation task. Theta band (3–6 Hz) power was calculated using a Fast Fourier Transform with multitaper method.	Frontocentral theta power was found higher when children engaged in tasks than when they were not involved in tasks. Higher theta power over left frontotemporal electrode sites when being engaged in the language-related task compared to the motor-related task Theta oscillations have a role in top-down control.
Grossmann et al. (2007)	12 infants (5 girls), 4-month-olds Gamma oscillations (ERPs)		
Köster et al. (2019)	38 infants (14 F/24 M) Age: nine-month-olds (M = 9.4 months, SD = 7 days) Theta (3–5 Hz) Alpha (5–7 Hz)	Four classical violations of expectation paradigms. The picture sequences were visually flickered at a theta (4-Hz) or an alpha (6-Hz) frequency. Steady-state visually evoked potentials (SSVEPs) of posterior electrodes (O1, Oz, O2, P3, P4, Pz, P7, P8) were analyzed.	At unexpected events, the visually entrained theta oscillations increased. However, entrained alpha oscillations did not differ between outcomes. Expected and unexpected events are factors that cause a frequency-specific SSVEP response.

Table 1. *Cont.*

	Development Stages and Focused Oscillations	Modality and Paradigms	Inference
Tang et al. (2019)	201 participants (124 F/77 M) 58 children (38 F/20 M) Age: 10–12 years (M = 10.83, SD = 0.82) 64 adolescents (39 F/25 M) Age: 14–17 years (M = 15.02, SD = 0.98) 79 young adults (47 F/32 M) Age: 18–28 (M = 19.42, SD = 1.92) Theta (4–7 Hz)	Social Exclusion Task (participants played the online ball-toss game Cyberball). ERSP between 200 to 600 ms at F3, FCz, P4. ITC from 100 to 400 ms. Event-related theta EEG power and phase coherence.	Adolescents and children showed the highest spectral power at 400–600 ms on the rejection part. However, adults' spectral power was the highest theta spectral power at the not-my-turn event. Increased theta spectral power responses to both social exclusion and threats of exclusion shown in adolescent development (ITC); adolescents showed lower levels of left frontal theta synchrony in response to rejection compared to children and adults.
Orekhova et al. (1999)	60 infants Age: 7–12 months born after 33.5–41 weeks gestation (M: 37.3, SD = 1.96) Theta 1 (3.6–4.8 Hz) Theta 2 (5.2–6.0 Hz)	EEG recorded in three attention conditions: looking without overt emotional reactions, the peek-a-boo game, and attention to the adult's reappearance.	EEG theta rhythm (3.6–6 Hz) synchronized during internally controlled (anticipatory) attention. As infants become older, the attention-related slow theta synchronization tends to decrease. The fast theta rhythm appeared only during effortful concentration of attention.
Bazhenova et al. (2007)	16 infants (9 F/7 M) Age: 20 weeks (M = 19.7; SD = 0.89) Theta (3.6–5.6 Hz) Alpha (6.0–9.0 Hz)	EEG recorded an adult's smiling (SF) and a blank face (BF) in a face-to-face setting.	Looking at the adult's BF causes increases in theta activity in the 4.8 Hz narrow band, observed over the majority of anterior and right posterior temporal scalp areas. The right hemisphere is more dominant to changes in a partner's facial expressivity.
Yordanova et al. (2009)	70 participants (37 F/33 M) 54 children (25 F/29 M) 16 young Adults (12 F/4 M) Age: M = 306.4, SD = 55.7 Theta (4–7 Hz) Alpha (8–14 Hz)	Three task conditions: passive listening condition (PLC), simple reaction task (SRT), serial learning reaction task (SLRT). EROs.	Results show that fast alpha oscillations are stronger in SLRT condition and more synchronized in children than adults. In the slower theta-to-alpha range, the phase synchronization increases with age. Developmental dynamics specific for theta, alpha-1, and alpha-2 frequencies differ in memory activation tasks.
Nanova et al. (2018)	36 children (13 F/20 M) Age: 7–10 years Delta (0.5–4 Hz) Theta (4–7 Hz) Slow alpha (7–10 Hz) Fast alpha (10–14 Hz)	Sensorimotor task with fixed stimulus sequences. ERPs N1, P2, N2 and P3 amplitudes. Phase-synchronization of EROs.	Proactive processing mode was marked by increased pre-stimulus theta activity. Notable reduction in the temporal synchronization of event-related theta/alpha oscillations within 300 milliseconds after the stimulus.

Table 1. *Cont.*

	Development Stages and Focused Oscillations	Modality and Paradigms	Inference
Nanova et al. (2011)	36 healthy children (18 F/18 M): Age: 7–8 years old (M = 7.7) (9 F/9 M) Age: 9–10 years old (M = 9.2) (9 F/9 M) Delta (0.5–4 Hz), Theta (4–7 Hz), Slow Alpha(7–10 Hz) Fast Alpha (10–14 Hz)	EEG recorded 3 different conditions: auditory stimulation, a passive listening condition (PLC), a simple reaction task (SRT), and a serial learning reaction task (SLRT). Phase-locking analysis. Auditory ERP.	Accelerated functional maturation of processing networks is frequency-specific and shown in increased girls' phase-locking of the slower (delta, theta, and slow alpha) but not for faster (fast alpha) frequency bands. Between 7 and 10 years of age, the development of neurophysiological mechanisms crucial for early auditory information processing varies based on gender.
Ashkinazi et al. (2010)	30 healthy children (M = 6.05, SD = 0.75): Age: 5–6 years old (n = 14) Age:6–7 years old (n = 16) Set: a plastic set (n = 14), a rigid set(n = 16) Theta (4–7 Hz) Alpha (8–13 Hz)	Resting-state EEG. Visual set paradigm. Spectral power analysis using FFT.	Higher alpha band spectral power (SP) has been shown at the plastic set rather than the rigid set. Both hemsipheres at occipital lobes showed increased alpha activity.
Chung et al. (2022)	49 full-term infants (25 F/24 M): Age: 21 infants, 9-month-olds (M = 9.21, SD = 10.0 days) (8 F/13 M) Age: 28 infants, 12-month-olds (M = 12.21, SD = 17.8 days) (17 F/11 M) Mu (6- to 9-Hz)	EEG recorded during action observation conditions: grasp condition, cane-use condition. Action execution period: Allow the infant to attempt trials in grasping and using a cane. Interchannel phase coherence (ICPC).	At the 6–9 Hz frequency range phase, coherence of central–ocipital regions is higher than shorter distances such as central–parietal and central–frontal regions while observing grasping action.
Vollebregt et al. (2015)	27 children (16 F/9 M): Age: 7–10 years (M = 9.11, SD = 1.29) Alpha (8–12 Hz)	Attention paradigm (Posner's cueing paradigm). EEG recorded 2 min EO, 2 min EC, and during the task.	Healthy children exhibited adult-like alpha lateralization patterns during covert attention. Children who were less affected by spatial cueing showed stronger left alpha modulation.

3.3. Alpha Activity: Markers of Attention and Rest in Developing Minds

The alpha rhythm in early childhood functions as an alpha rhythm in adulthood but at a lower frequency. The 6–9 Hz band is suggested to be similar to the adult alpha rhythm for infants and toddlers [103]. Different frequency bands identified in infant EEG studies are slightly lower than those reported in adults but may support similar cognitive processes such as attention, memory, and emotion [54]. Alpha activity is increased in adults at rest, while decreased alpha is triggered during cognitively demanding tasks [104]. It is stated that alpha rhythms reflect activity in the thalamocortical feedback loop, as well as memory processes [29,105]. In a study using spectral analysis to examine longitudinal baseline data, 6–9 Hz was found to be the predominant frequency in most scalp electrode sites from infancy to 4 years of age. Therefore, the infant/toddler's 6–9 Hz rhythm during silent wakefulness should be considered just like the adult's 8–13 Hz alpha rhythm [103]. The findings from Yordanova et al.'s study suggest that the alpha response system is functionally active in children aged 6 to 11 years, although its development remains incomplete by age 11. The study indicates that the magnitude and phase-locking parameters may be associated with various functional aspects of the alpha response system. Consequently, while younger children can generate alpha responses during information processing, they do not engage this system as effectively as older children and adults [106]. Another study focused on alpha responses discusses that stimulus-locked alpha responses noted that when spontaneous alpha rhythms are absent, as seen in 3-year-old children, stimulus-locked alpha responses to visual stimuli are not recorded [107]. One of the studies explaining the complex nature of alpha oscillations mentions that alpha phase locking depends on age, topography and pre-stimulus alpha power in a way that is different from alpha amplitude. The ability to reorganize alpha activity after auditory stimulation and produce repeatable alpha patterns is emphasized in relation to brain development processes with age [108].

Alpha oscillations can be modulated both by internal rhythms and external stimuli, and studies on the mechanism of the relevant system continue. Considering the ability of interneurons specialized to work in the communication of certain cell groups to produce alpha oscillations, or the contribution of structures such as the thalamus to the production of relevant oscillations that are reflected in all brain regions, understanding the rhythm of the developing brain will also contribute significantly to interpreting the nature of the oscillations [109]. It is mentioned that low-frequency rhythms, such as alpha oscillations, provide synchronization characterized by a larger spatial volume [53,110,111]. As the time it takes for signals to travel between linked regions usually increases depending on the distance of spatial separation, the observed developmental rise in lower-frequency EEG patterns like mu and alpha might stem from improved signal speed across extended (anterior to posterior) white matter pathways. This enhancement could be due to the growth and development of axons and the myelination process, which may extend into early adulthood [21,112].

In an experiment involving maintenance and shifting of attention in 7–10-year-old children, researchers were able to describe the pattern of alpha oscillations modulated by attention. They explain that children tend to focus on the right side because the alpha oscillations in the left hemisphere are not strong enough to break the focus of attention. Considering that the ventral network, controlled by structures around the right temporal parietal junction (TPJ), is required for extrinsically directed attention switches, they explain why this effect applies only to the right side. It is argued that lateralized alpha modulation can be used as a tool to examine attention processes in groups such as children with attention disorders since alpha modulation is already present at these ages [113]. At the source level, seed-based connectivity analyses show that sustained attention induces a reduction in the alpha frequency band (localized within dorsal attention and default mode networks) [114].

Resting state activity continues to change during childhood and adolescence, during which the amplitude of delta and theta frequency activities decreases, while faster frequency band activities (alpha, beta, and gamma) increase [115,116]. This dynamic is a sign of central nervous system (CSN) maturation [69,117]. Alpha oscillations with a 6–9 Hz frequency are related to visual attention, emotional expression, working memory, and inhibitory control in infants [85,118]. When time–frequency (TF) analyses were applied to EEG data collected from 50 infants at different times (8 months and four years old) using a working memory task, the working memory process was found to be associated with increased alpha power (6–9 Hz) at an average of all electrodes at eight months, and for medial anterior electrodes, it was determined to occur at four years [70]. This more focused activity was interpreted as an increased functional specialization through development [119,120]. In the previous study, we discussed the increase in theta and alpha power and phase locking in the parietal–occipital regions to the objects remembered of 6–7-year-old children's visual and auditory memory tasks [121]. Moreover, while beta and low-frequency components work together for encoding, it is mentioned that alpha oscillations sharpen this process. In a study with children involving memory tasks, reaction time is used as a marker of behavioral responses and has the most consistent relationship with the alpha (8.5–10 Hz) frequency range [93]. Alpha phase cycles contribute to the simultaneous functioning of brain regions by creating a time window for excitation or inhibition [122,123], which is thought to be important in attention and learning functions. In a study comparing sensory-evoked potentials in children and adults, auditory-evoked potentials were detected and considered to be a superposition of rhythms in the delta and theta ranges. It is also mentioned that 3-year-old children do not produce an alpha resonance response to sensory stimuli [124].

Upon reviewing all articles listed in Table 1 within the scope of this study, spectral analysis of the alpha frequency range emerges as a crucial method for detecting and interpreting shifts in frequency during neurodevelopmental processes. The phase and power data concerning the event are believed to be instrumental in delineating the link between functional development and brain oscillations in children. Given that infants in their early life stages need to interpret environmental signals critical for survival and managing physiological needs, the gradual transition from slower oscillations indicative of autonomic regulation to faster oscillations that underlie higher cognitive abilities is natural. It is thought that as infants develop skills, the evolution in functions and corresponding brain oscillations occur concurrently.

3.4. Mu Rhythms: Mirroring Movement and Social Cognition

Over the years following its initial identification, the mu rhythm in adults has been defined by three primary characteristics: specific regional brain distribution, particular frequency range, and a reliance on behavioral context for its activation. In practical terms, the mu rhythm is notably recognized for its reduced synchronization during deliberate muscle activity [21,125]. Differing from the posterior alpha rhythm, the mu rhythm remains relatively stable and is not significantly altered by shifts in environmental lighting conditions [126]. Due to the significant role of this rhythm, motor execution has emerged as the definitive functional criterion for defining the mu rhythm [21].

The characterization of mu oscillations acknowledges a distinction between the upper (10–13 Hz) and lower (8–10 Hz) mu band frequencies. Pfurtscheller et al. (2000) advocate that these bands exhibit distinct functional and spatial characteristics, where the upper band is associated with a more localized, movement-specific ERD pattern, unlike the lower band, which displays a widespread and more generalized ERD pattern [21,127]. The link between movement perception and execution in infants has been explored through EEG research. Substantial event-related desynchronization was noted in the central regions when infants engaged in intentional activities, and a similar desynchronization occurred in these regions while observing the same actions. Task-related studies suggest that the central rhythm in infants, occurring at 6–9 Hz, differs from the alpha range rhythmic activity observed in other areas of the scalp. It is mentioned that this research supports the notion that the infant rhythm exhibits functional similarities with the adult mu rhythm [128].

Motor related mu supression develops early in children. The power of mu rhythm (8–14 Hz) decreased significantly during movement, and it returned to baseline after movement for both children in 7-year-olds [129]. Morover, the reduced mu power is shown in infants with task-dependent activity (congruent to incongruent stimuli). Central electrode sides (C3–C4) might be evaluated as primary channels of interest in studying mu rhythms (6–9 Hz). It is suggested that occipital channels are unavailable because of the widespread posterior alpha effect [130]. The parietal cortex is recognized for supporting the exact kind of visuomotor integration necessary for performing reaching and grasping tasks, which is frequently used in EEG investigations of the mu rhythm [131,132]. It has also been demonstrated that mu suppression correlates with changes in the BOLD (Blood Oxygen Level-Dependent) signal within the premotor and parietal regions [133]. Central–parietal desynchronization, which is thought of as representitive of the neural mirroring, was found in a study using a paradigm based on active and observational learning in children. They found 3- to 6-year-old children exhibit neural mirroring within the mu ryhthm [134]. Given that the mu rhythm responds to both the execution and observation of motor actions, researchers have advocated for the utilization of mu rhythm modulations as an effective instrument for investigating the human mirror neuron system [135].

The mu rhythm exhibits a comb-like appearance, indicating it consists of two primary frequency components that have a nearly harmonic relationship, specifically the alpha (10 Hz) and beta (20 Hz) frequency bands. These two components of the mu rhythm are subsequently referred to as "mu alpha" and "mu beta," based on their spectral properties. The two frequency components of the mu rhythm can be influenced by various stimuli or tasks, such as somatosensory stimulations, imagined movements, observed movements, the redirection of spatial attention, and the anticipation of focused stimuli [21,136]. Greater occipital alpha (8–10 Hz) ERD is found during the perception of the active training task in children. This suggests that occipital alpha may be facilitating the processing of attended stimuli while suppressing attention to irrelevant information [134]. ERD of the mu rhythm (including both mu alpha and mu beta) indicates sensorimotor network activation, while ERS or the rebound observed after movement is considered to represent processes of top-down inhibitory control [137]. Morover, it is found that anticipatory ERD of the mu rhythm is related to executive function skills, whereas it is not associated with the accuracy of children's behavioral responses to tactile stimulation [137].

ERD and ERS can be altered by different parameters of movement, including its strength, duration, frequency, and complexity [138]. Observing a real action situation led to greater power desynchronization in the μ range in the medial frontal and medial parietal loci compared to observing a fake action situation. Children with higher levels of speech comprehension showed greater EEG power desynchronization in the μ range when observing a real action in the frontal and parietal loci of the left hemisphere [139]. The results suggested that children with higher levels of speech comprehension have higher activation of the mirror neuron system, which has a key role in forming and perceiving speech [139,140].

From the moment of birth, children start to develop motor skills. The organization of attention and limb movement is crucial for mastering a motor skill, and analyzing the event-related mu rhythm in children offers valuable insights into the construction of movements. Within this framework, the mu alpha rhythm contributes to cognitive reserve, while the mu beta rhythm delivers significant details about the proficiency level of the executed movement. While language development primarily involves slower frequencies, the spatial proximity of the mu rhythm to language regions and the required coordination of jaw muscles for speech exemplify the mu rhythm's significance for understanding brain dynamics.

3.5. Beta Oscillations: The Rhythms of Active Thinking and Problem-Solving

The beta frequency band comprises small-amplitude fast oscillations, and these characteristics, along with short time spectra, may facilitate identification. It exhibits activation patterns during action observation in adults, and similar changes in brain activity are observed during developmental processes. In research involving 3-year-old participants, variations in beta-range power were associated with time-locked motor responses to a partner's actions, whereas changes in mu power suggested a broader engagement of the motor system during a cooperative task. Even in early childhood, the way the motor system integrates the actions of others varies depending on whether children are involved in a joint action. This context-dependent activation of the motor system may have significant implications for the development of successful cooperative behaviors in joint activities. Furthermore, a more pronounced time-locked increase in beta power correlated with fewer mistakes made by children in acting jointly [141]. Köster et al. (2000) showed an increase in motor cortex activity within the 7–10 Hz range during the observation of actions, which anticipated the imitation of actions in 20-month-old infants. Infants at ten months of age, who were not yet consistently imitating the actions of others, exhibited a neural activity pattern remarkably similar during the observation of actions [142]. Liao et al. (2015) identified the sources of the mu rhythm in toddlers as being located in the somatomotor cortex, likely in proximity to the hand/arm homunculus within the central sulcus. This finding underscores the recent progress in signal analysis techniques, which have significantly enhanced the precision of EEG source localization [143].

In a study focused on developmental training investigation of the sensorimotor beta rhythm, they found a significant central–paeriatal ERD during action execution but not action observation. They have also added that there are no training effects on the beta rhythm during action perception. The decreased beta ERD has been associated with conceptual knowledge of actions [134]. Moreover, whereas conceptual training decreases the beta ERD, motor training increases activity [144]. It was argued that the beta rhythm might be more reactive to the observation of complex actions that need relatively long-term training, and numerous factors can influence beta ERD; thus, further research is required to identify the specific contexts and variables that modulate this activity [134]. With another perspective, beta power (15–30 Hz) has different dynamics between children and adults. The movement induce beta ERD in both, but post-movement beta power rebound (PMBR) is observed only in adults. PMBR might be associated with a more prolonged motor development process [129].

In adults, the power of somatomotor mu rhythms and their beta harmonics is reduced during the execution of intentional actions or while observing the actions of others [145]. Liao et al. (2005) observed a reduction in beta power in the left hemisphere during action execution, suggesting this phenomenon as a result of hemispheric lateralization and the design of unilateral tasks. However, they also emphasized the need for further studies to explore the connections between mu and beta rhythms in upcoming research [143].

A study evaluating the family difference between family-raised children and orphans' performance during the speech perception task showed that during direct speech perception, beta and gamma oscillations were synchronized. However, in the institutionalized group, the synchronization of these rhythms was less pronounced. The presence of a pronounced beta rhythm in EEG data is interpreted as indicating an activated state of both cortical and cortico-subcortical neuronal networks [146]. Vrobel et al. (2000) suggested that beta-frequency oscillations in cortico-thalamic pathways are associated with a reduced activation threshold of thalamic relay neurons. This reduction improves the specificity of information flow to targeted neocortical areas during tasks that require attention [147]. Angel and Fries (2010) put forward a theory regarding the significance of the mechanisms behind high-frequency EEG rhythms (beta and gamma) production. They suggested that during motor and cognitive tasks, the beta rhythm facilitates the maintenance of the current motor state and aids in the sustained and routine execution of cognitive activities [148].

A reduction in beta power, which previous studies have linked to efficient syntactic processing, was observed [149,150]. Research exploring the developmental evolution of neural oscillations in auditory sentence comprehension identified developmental distinctions through neural oscillations and ERP dynamics. Findings revealed that children display a distinct pattern of involvement, showing an absence of the beta power reduction in response to grammatical errors observed in adults [146]. It was demonstrated in another study investigating the processing of auditory stimuli that time–frequency analyses of the EEG data that are collected from infants' data indicate that the stimulus evoked 1/f pattern neural synchrony like adults. The stimulus onset shows synchronous bursts of theta, beta, and gamma power in the brain's right and left auditory regions. This article is on the concept of technical details and approach for EEG data, and they have suggested the creation of a model to understand the integrity of early auditory processing mechanisms for future studies [30].

Children encounter numerous novel experiences and develop various skills from an early age. As previously detailed, this involves activities like focusing attention, processing data, or monitoring their surroundings. Here, it becomes apparent that the beta rhythm plays a crucial role among the rapid frequencies that escalate as children age. While predominantly linked to motor activities in adults, beta oscillations in children indicate functional capabilities across various domains, including language development, movement observation, and the interpretation of social signals.

3.6. Gamma Band: High-Frequency Insights into Advanced Cognitive Integration

The production of the gamma rhythm is associated with the functioning of pacemaker neurons located in the specific and intralaminar nuclei of the thalamus, where these cells produce bursts of spikes at a frequency of approximately 40 Hz. This activity enhances and synchronizes neuronal activity in select cortical areas, providing the neurophysiological foundation for the efficient integration of incoming information and consciousness [146,151]. When a person detects a significant external stimulus demanding extra cognitive work, gamma oscillations are among the rhythms that become more pronounced. The synchronization of EEG oscillations in the gamma-band frequency can quickly connect cell ensembles spread across different locations, thereby offering the required speed for cognitive processing [148]. Gamma oscillations have been associated with various sensory and higher cognitive abilities, including feature binding, attentional processes, and working memory capacity [152–154]. Furthermore, it has been proposed that these oscillations be generated by several cellular and synaptic mechanisms [110]. As a corollary of this, these oscillations are sensitive to the maturational changes in the brain. Therefore, understanding the characteristics of these oscillations across developmental stages will provide further information about the development of sensory and cognitive abilities in humans. However, when compared to studies on slower frequencies, the number of studies on gamma event/task-related oscillations in children is limited.

Consistent with the previous findings on the effects of social gaze on infants' EEG activity, 4-month-old infants had larger gamma band responses to faces with direct gaze compared to faces with averted gaze, pointing to the importance of this ability in social interactions even in the earlier stages of the development [155,156]. Furthermore, object permanence ability in infants has been shown to be developing earlier than previously thought. Specifically, 6–8-month-old infants had higher gamma band activity in response to maintaining visual representations of two occluded objects versus one [157]. Also, in order to eliminate the potential effects of object memory on this greater gamma response over the right temporal region, researchers compared different types of object disappearance (occlusion vs. disintegration) and concluded that the detected increase was linked with the continued presence of the object, not the object memory [158,159]. On the other hand, the occlusion of face stimuli did not elicit this response, implying the presence of distinct mechanisms underlying these processes. In addition to the processes mentioned above, violating infants' expectations regarding completing a goal-directed chain of events elicited higher power in gamma responses in the left frontal area [160,161]. Also, infants demonstrated a preference for their native language as early as six months, which was evident by the increase in their induced gamma band responses [80].

Similar to adults, children demonstrated increased gamma band power in response to moving grating stimuli as well, which was predominantly observed over the occipital area [162]. Also, the peak frequency of their gamma oscillations was increased in response to the increasing velocity of the moving gratings. Moreover, compared to institutionalized orphans, children who their families raise had higher synchronization in the gamma frequency band in response to direct speech [146]. This difference was linked to developmental delays in language and cognitive abilities of the institutionalized children. Furthermore, target auditory stimuli elicited larger gamma band responses (within 0–120 ms) with stronger phase-locking in children, which was interpreted as being related to earlier attentional mechanisms [163].

4. Concluding Remarks

This review has provided a detailed examination of the developmental trajectories of brain oscillations from infancy through childhood, highlighting the significance of delta, theta, alpha, beta, mu, and gamma rhythms in various cognitive and sensory processes. Each oscillation band presents unique characteristics and developmental patterns that contribute to the overall understanding of brain maturation.

Delta oscillations are crucial in early development, particularly in language acquisition and cognitive processing. Their decrease with age signifies the maturation of cortical structures. Theta rhythms bridge perception and cognition, with their role in attention and memory becoming more specialized as children grow. Alpha activity, reflective of thalamocortical interactions, shows increased complexity and is essential for inhibitory control and memory. Beta rhythms are vital for active thinking and problem-solving, showing distinct patterns during motor and cognitive tasks. Mu rhythms, which reflect motor control and social cognition, highlight the importance of the mirror neuron system in learning and social interactions. Gamma oscillations are associated with high-level cognitive functions such as attention, memory, and perceptual binding, showing significant developmental increases in response to sensory and cognitive stimuli. Understanding these oscillatory patterns and their developmental trajectories provides a foundation for further research into the neurobiological underpinnings of cognitive and sensory development. This knowledge is essential for identifying biomarkers for cognitive and developmental disorders, offering potential for early diagnosis and intervention strategies. Future studies should focus on longitudinal analyses to capture the dynamic changes in brain oscillations and explore their implications for clinical practice.

The integration of advanced neuroimaging techniques and computational models will enhance our ability to interpret the complex interactions between different oscillation bands. Additionally, examining the impact of environmental factors, such as early education and social experiences, on the development of brain oscillations will provide valuable insights into optimizing developmental outcomes. In summary, the intricate patterns of brain oscillations from infancy to childhood reflect the profound changes occurring in the developing brain. These oscillations serve as vital indicators of neural maturation and cognitive development, offering a window into the functional architecture of the brain. By continuing to unravel the complexities of brain oscillations, we can pave the way for innovative approaches to support healthy development and address neurodevelopmental challenges.

Author Contributions: Conceptualization: E.Ü. and B.G.; writing—original draft: E.Ü. methodology: E.Ü., R.D., E.B., and İ.Y.; investigation: R.D., E.B., and İ.Y.; writing—review and editing: E.Ü. and B.G.; supervision: Ö.C. All authors have read and agreed to the published version of the manuscript.

Funding: This research received no external funding.

Conflicts of Interest: The authors declare no conflicts of interest.

References

1. Higginbotham, H.; Yokota, Y.; Anton, E. Strategies for analyzing neuronal progenitor development and neuronal migration in the developing cerebral cortex. *Cereb. Cortex* **2011**, *21*, 1465–1474. [CrossRef]
2. Ackerman, S. *Discovering the Brain*; National Academies Press (US): Washington, DC, USA, 1992.
3. Alberts, B. *Molecular Biology of the Cell*; Garland Science: New York, NY, USA, 2017.
4. Goodman, C.S.; Shatz, C.J. Developmental mechanisms that generate precise patterns of neuronal connectivity. *Cell* **1993**, *72*, 77–98. [CrossRef] [PubMed]
5. Penn, A.A.; Shatz, C.J. Brain waves and brain wiring: The role of endogenous and sensory-driven neural activity in development. *Pediatr. Res.* **1999**, *45*, 447–458. [CrossRef]
6. Casey, B.J.; Tottenham, N.; Liston, C.; Durston, S. Imaging the developing brain: What have we learned about cognitive development? *Trends Cogn. Sci.* **2005**, *9*, 104–110. [CrossRef] [PubMed]
7. Gogtay, N.; Giedd, J.N.; Lusk, L.; Hayashi, K.M.; Greenstein, D.; Vaituzis, A.C.; Nugent, T.F., III; Herman, D.H.; Clasen, L.S.; Toga, A.W.; et al. Dynamic mapping of human cortical development during childhood through early adulthood. *Proc. Natl. Acad. Sci. USA* **2004**, *101*, 8174–8179. [CrossRef]
8. Sowell, E.R.; Thompson, P.M.; Leonard, C.M.; Welcome, S.E.; Kan, E.; Toga, A.W. Longitudinal mapping of cortical thickness and brain growth in normal children. *J. Neurosci.* **2004**, *24*, 8223–8231. [CrossRef] [PubMed]
9. Sowell, E.R.; Peterson, B.S.; Thompson, P.M.; Welcome, S.E.; Henkenius, A.L.; Toga, A.W. Mapping cortical change across the human life span. *Nat. Neurosci.* **2003**, *6*, 309–315. [CrossRef]
10. Dreyfus-Brisac, C. Ontogenesis of brain bioelectrical activity and sleep organization in neonates and infants. In *Human Growth: Volume 3 Neurobiology and Nutrition*; Springer: Berlin/Heidelberg, Germany, 1979; pp. 157–182.
11. Vanhatalo, S.; Palva, J.M.; Andersson, S.; Rivera, C.; Voipio, J.; Kaila, K. Slow endogenous activity transients and developmental expression of K+–Cl− cotransporter 2 in the immature human cortex. *Eur. J. Neurosci.* **2005**, *22*, 2799–2804. [CrossRef]

12. Ehlers, C.L.; Wills, D.N.; Phillips, E.; Havstad, J. Low voltage alpha EEG phenotype is associated with reduced amplitudes of alpha event-related oscillations, increased cortical phase synchrony, and a low level of response to alcohol. *Int. J. Psychophysiol.* **2015**, *98*, 65–75. [CrossRef]
13. Eisermann, M.; Kaminska, A.; Moutard, M.-L.; Soufflet, C.; Plouin, P. Normal EEG in childhood: From neonates to adolescents. *Neurophysiol. Clin./Clin. Neurophysiol.* **2013**, *43*, 35–65. [CrossRef]
14. Bhavnani, S.; Lockwood Estrin, G.; Haartsen, R.; Jensen, S.K.; Gliga, T.; Patel, V.; Johnson, M.H. EEG signatures of cognitive and social development of preschool children—A systematic review. *PLoS ONE* **2021**, *16*, e0247223. [CrossRef] [PubMed]
15. Başar, E. *Brain-Body-Mind in the Nebulous Cartesian System: A Holistic Approach by Oscillations*; Springer: Berlin/Heidelberg, Germany, 2010.
16. Whitford, T.J.; Rennie, C.J.; Grieve, S.M.; Clark, C.R.; Gordon, E.; Williams, L.M. Brain maturation in adolescence: Concurrent changes in neuroanatomy and neurophysiology. *Hum. Brain Mapp.* **2007**, *28*, 228–237. [CrossRef]
17. Shaw, P.; Kabani, N.J.; Lerch, J.P.; Eckstrand, K.; Lenroot, R.; Gogtay, N.; Greenstein, D.; Clasen, L.; Evans, A.; Rapoport, J.L. Neurodevelopmental trajectories of the human cerebral cortex. *J. Neurosci.* **2008**, *28*, 3586–3594. [CrossRef] [PubMed]
18. Feinberg, I.; Campbell, I.G. Sleep EEG changes during adolescence: An index of a fundamental brain reorganization. *Brain Cogn.* **2010**, *72*, 56–65. [CrossRef] [PubMed]
19. Segalowitz, S.J.; Santesso, D.L.; Jetha, M.K. Electrophysiological changes during adolescence: A review. *Brain Cogn.* **2010**, *72*, 86–100. [CrossRef] [PubMed]
20. Barriga-Paulino, C.I.; Flores, A.B.; Gómez, C.M. Developmental changes in the eeg rhythms of children and young adults. *J. Psychophysiol.* **2011**, *25*, 143–158. [CrossRef]
21. Thorpe, S.G.; Cannon, E.N.; Fox, N.A. Spectral and source structural development of mu and alpha rhythms from infancy through adulthood. *Clin. Neurophysiol.* **2016**, *127*, 254–269. [CrossRef]
22. Rodríguez-Martínez, E.; Ruiz-Martínez, F.; Barriga Paulino, C.; Gómez, C.M. Frequency shift in topography of spontaneous brain rhythms from childhood to adulthood. *Cogn. Neurodynamics* **2017**, *11*, 23–33. [CrossRef] [PubMed]
23. Britto, P.R.; Lye, S.J.; Proulx, K.; Yousafzai, A.K.; Matthews, S.G.; Vaivada, T.; Perez-Escamilla, R.; Rao, N.; Ip, P.; Fernald, L.C. Nurturing care: Promoting early childhood development. *Lancet* **2017**, *389*, 91–102. [CrossRef] [PubMed]
24. Green, J.; Charman, T.; Pickles, A.; Wan, M.W.; Elsabbagh, M.; Slonims, V.; Taylor, C.; McNally, J.; Booth, R.; Gliga, T. Parent-mediated intervention versus no intervention for infants at high risk of autism: A parallel, single-blind, randomised trial. *Lancet Psychiatry* **2015**, *2*, 133–140. [CrossRef] [PubMed]
25. Basar, E. *EEG-Brain Dynamics: Relation between EEG and Brain Evoked Potentials*; Elsevier: Amsterdam, The Netherlands, 1980.
26. Başar, E. A review of alpha activity in integrative brain function: Fundamental physiology, sensory coding, cognition and pathology. *Int. J. Psychophysiol.* **2012**, *86*, 1–24. [CrossRef] [PubMed]
27. Başar, E.; Güntekin, B. A short review of alpha activity in cognitive processes and in cognitive impairment. *Int. J. Psychophysiol.* **2012**, *86*, 25–38. [CrossRef] [PubMed]
28. Bullock, T.; McClune, M.; Achimowicz, J.; Iragui-Madoz, V.; Duckrow, R.; Spencer, S. Temporal fluctuations in coherence of brain waves. *Proc. Natl. Acad. Sci. USA* **1995**, *92*, 11568–11572. [CrossRef] [PubMed]
29. Klimesch, W.; Doppelmayr, M.; Schwaiger, J.; Auinger, P.; Winkler, T. Paradoxical'alpha synchronization in a memory task. *Cogn. Brain Res.* **1999**, *7*, 493–501. [CrossRef]
30. Musacchia, G.; Ortiz-Mantilla, S.; Realpe-Bonilla, T.; Roesler, C.P.; Benasich, A.A. Infant auditory processing and event-related brain oscillations. *JoVE (J. Vis. Exp.)* **2015**, *101*, e52420.
31. Pfurtscheller, G.; Da Silva, F.L. Event-related EEG/MEG synchronization and desynchronization: Basic principles. *Clin. Neurophysiol.* **1999**, *110*, 1842–1857. [CrossRef]
32. Singer, W. Synchronization of cortical activity and its putative role in information processing and learning. *Annu. Rev. Physiol.* **1993**, *55*, 349–374. [CrossRef] [PubMed]
33. Başar, E.; Başar-Eroglu, C.; Röschke, J.; Schütt, A. The EEG is a quasi-deterministic signal anticipating sensory-cognitive tasks. In *Brain Dynamics: Progress and Perspectives*; Springer: Berlin/Heidelberg, Germany, 1989; pp. 43–71.
34. Başar-Eroglu, C.; Strüber, D.; Schürmann, M.; Stadler, M.; Başar, E. Gamma-band responses in the brain: A short review of psychophysiological correlates and functional significance. *Int. J. Psychophysiol.* **1996**, *24*, 101–112. [CrossRef] [PubMed]
35. Ward, L.M. Synchronous neural oscillations and cognitive processes. *Trends Cogn. Sci.* **2003**, *7*, 553–559. [CrossRef] [PubMed]
36. Page, M.J.; McKenzie, J.E.; Bossuyt, P.M.; Boutron, I.; Hoffmann, T.C.; Mulrow, C.D.; Shamseer, L.; Tetzlaff, J.M.; Akl, E.A.; Brennan, S.E. The PRISMA 2020 statement: An updated guideline for reporting systematic reviews. *Int. J. Surg.* **2021**, *88*, 105906. [CrossRef] [PubMed]
37. Ouzzani, M.; Hammady, H.; Fedorowicz, Z.; Elmagarmid, A. Rayyan—A web and mobile app for systematic reviews. *Syst. Rev.* **2016**, *5*, 210. [CrossRef] [PubMed]
38. Steriade, M. *Brain Electrical Activity and Sensory Processing during Waking and Sleep States, Principles and Practice of Sleep Medicine*; Kryger, M., Roth, T., Dement, W., Eds.; WB Saunders: Philadelphia, PA, USA, 2000.
39. Babiloni, C.; Frisoni, G.; Steriade, M.; Bresciani, L.; Binetti, G.; Del Percio, C.; Geroldi, C.; Miniussi, C.; Nobili, F.; Rodriguez, G. Frontal white matter volume and delta EEG sources negatively correlate in awake subjects with mild cognitive impairment and Alzheimer's disease. *Clin. Neurophysiol.* **2006**, *117*, 1113–1129. [CrossRef] [PubMed]

40. Penolazzi, B.; Spironelli, C.; Angrilli, A. Delta EEG activity as a marker of dysfunctional linguistic processing in developmental dyslexia. *Psychophysiology* **2008**, *45*, 1025–1033. [CrossRef] [PubMed]
41. Spironelli, C.; Angrilli, A. Developmental aspects of language lateralization in delta, theta, alpha and beta EEG bands. *Biol. Psychol.* **2010**, *85*, 258–267. [CrossRef] [PubMed]
42. Güntekin, B.; Başar, E. Review of evoked and event-related delta responses in the human brain. *Int. J. Psychophysiol.* **2016**, *103*, 43–52. [CrossRef] [PubMed]
43. Schürmann, M.; Başar-Eroglu, C.; Kolev, V.; Başar, E. Delta responses and cognitive processing: Single-trial evaluations of human visual P300. *Int. J. Psychophysiol.* **2001**, *39*, 229–239. [CrossRef] [PubMed]
44. Başar, E.; Parnefjord, R.; Başar, E. Evoked delta oscillations on the hearing threshold. In *Brain Function and Oscillations: Integrative Brain Function. Neurophysiology and Cognitive Processes*; Springer Science & Business Media: Berlin/Heidelberg, Germany, 1999; pp. 161–175.
45. Knyazev, G.; Slobodskoj-Plusnin, J.Y.; Bocharov, A. Event-related delta and theta synchronization during explicit and implicit emotion processing. *Neuroscience* **2009**, *164*, 1588–1600. [CrossRef] [PubMed]
46. Moreira, J.F.G.; Silvers, J.A. In due time: Neurodevelopmental considerations in the study of emotion regulation. In *Emotion Regulation*; Routledge: Abingdon, UK, 2018; pp. 93–116.
47. Dehaene-Lambertz, G.; Hertz-Pannier, L.; Dubois, J. Nature and nurture in language acquisition: Anatomical and functional brain-imaging studies in infants. *Trends Neurosci.* **2006**, *29*, 367–373. [CrossRef]
48. Obleser, J.; Kayser, C. Neural entrainment and attentional selection in the listening brain. *Trends Cogn. Sci.* **2019**, *23*, 913–926. [CrossRef] [PubMed]
49. Kolev, V.; Başar-Eroglu, C.; Aksu, F.; Başar, E. EEG rhythmicities evoked by visual stimuli in three-year-old children. *Int. J. Neurosci.* **1994**, *75*, 257–270. [CrossRef] [PubMed]
50. Ehlers, J.; Strüber, D.; Basar-Eroglu, C. Multistable perception in children: Prefrontal delta oscillations in the developing brain. *Int. J. Psychophysiol.* **2016**, *103*, 129–134. [CrossRef] [PubMed]
51. Attaheri, A.; Choisdealbha, Á.N.; Di Liberto, G.M.; Rocha, S.; Brusini, P.; Mead, N.; Olawole-Scott, H.; Boutris, P.; Gibbon, S.; Williams, I. Delta-and theta-band cortical tracking and phase-amplitude coupling to sung speech by infants. *NeuroImage* **2022**, *247*, 118698. [CrossRef] [PubMed]
52. Howard, M.F.; Poeppel, D. Discrimination of speech stimuli based on neuronal response phase patterns depends on acoustics but not comprehension. *J. Neurophysiol.* **2010**, *104*, 2500–2511. [CrossRef] [PubMed]
53. Von Stein, A.; Sarnthein, J. Different frequencies for different scales of cortical integration: From local gamma to long range alpha/theta synchronization. *Int. J. Psychophysiol.* **2000**, *38*, 301–313. [CrossRef] [PubMed]
54. Saby, J.N.; Marshall, P.J. The utility of EEG band power analysis in the study of infancy and early childhood. *Dev. Neuropsychol.* **2012**, *37*, 253–273. [CrossRef] [PubMed]
55. Wang, X.-J. Neurophysiological and computational principles of cortical rhythms in cognition. *Physiol. Rev.* **2010**, *90*, 1195–1268. [CrossRef] [PubMed]
56. Nanova, P.; Kolev, V.; Yordanova, J. Developmental gender differences in the synchronization of auditory event-related oscillations. *Clin. Neurophysiol.* **2011**, *122*, 907–915. [CrossRef] [PubMed]
57. Anaya, B.; Ostlund, B.; LoBue, V.; Buss, K.; Pérez-Edgar, K. Psychometric properties of infant EEG: Developmental stability, reliability, and construct validity of frontal alpha asymmetry and delta-beta coupling. *Dev. Psychobiol.* **2021**, *63*, e22178. [CrossRef]
58. Anaya, B.; Vallorani, A.M.; Pérez-Edgar, K. Individual dynamics of delta–beta coupling: Using a multilevel framework to examine inter-and intraindividual differences in relation to social anxiety and behavioral inhibition. *J. Child Psychol. Psychiatry* **2021**, *62*, 771–779. [CrossRef]
59. Basar, E.; Baar-Eroglu, C.; Rahn, E.; Schürmann, M. Sensory and cognitive components of brain resonance responses: An analysis of responsiveness in human and cat brain upon visual and auditory stimulation. *Acta Oto-Laryngol.* **1991**, *111*, 25–35. [CrossRef]
60. Demiralp, T.; Başar-Eroglu, C.; Rahn, E.; Başar, E. Event-related theta rhythms in cat hippocampus and prefrontal cortex during an omitted stimulus paradigm. *Int. J. Psychophysiol.* **1994**, *18*, 35–48. [CrossRef]
61. Harris, J.J.; Jolivet, R.; Attwell, D. Synaptic energy use and supply. *Neuron* **2012**, *75*, 762–777. [CrossRef] [PubMed]
62. Köster, M.; Haese, A.; Czernochowski, D. Neuronal oscillations reveal the processes underlying intentional compared to incidental learning in children and young adults. *PLoS ONE* **2017**, *12*, e0182540. [CrossRef] [PubMed]
63. Köster, M.; Langeloh, M.; Hoehl, S. Visually entrained theta oscillations increase for unexpected events in the infant brain. *Psychol. Sci.* **2019**, *30*, 1656–1663. [CrossRef] [PubMed]
64. Xie, W.; Mallin, B.M.; Richards, J.E. Development of infant sustained attention and its relation to EEG oscillations: An EEG and cortical source analysis study. *Dev. Sci.* **2018**, *21*, e12562. [CrossRef] [PubMed]
65. Orekhova, E.V.; Stroganova, T.A.; Posikera, I.N. Theta synchronization during sustained anticipatory attention in infants over the second half of the first year of life. *Int. J. Psychophysiol.* **1999**, *32*, 151–172. [CrossRef] [PubMed]
66. Stroganova, T.A.; Orekhova, E.V.; Posikera, I.N. Externally and internally controlled attention in infants: An EEG study. *Int. J. Psychophysiol.* **1998**, *30*, 339–351. [CrossRef] [PubMed]
67. van Noordt, S.; Heffer, T.; Willoughby, T. A developmental examination of medial frontal theta dynamics and inhibitory control. *NeuroImage* **2022**, *246*, 118765. [CrossRef] [PubMed]

68. Meyer, M.; van Schaik, J.; Poli, F.; Hunnius, S. How infant-directed actions enhance infants' attention, learning, and exploration. *Dev. Sci.* **2022**, *26*, e13259. [CrossRef] [PubMed]
69. Clarke, A.R.; Barry, R.J.; McCarthy, R.; Selikowitz, M. Age and sex effects in the EEG: Development of the normal child. *Clin. Neurophysiol.* **2001**, *112*, 806–814. [CrossRef] [PubMed]
70. Bell, M.A.; Wolfe, C.D. Changes in brain functioning from infancy to early childhood: Evidence from EEG power and coherence during working memory tasks. *Dev. Neuropsychol.* **2007**, *31*, 21–38. [CrossRef]
71. Adam, N.; Blaye, A.; Gulbinaite, R.; Delorme, A.; Farrer, C. The role of midfrontal theta oscillations across the development of cognitive control in preschoolers and school-age children. *Dev. Sci.* **2020**, *23*, e12936. [CrossRef] [PubMed]
72. Liu, Z.-X.; Woltering, S.; Lewis, M.D. Developmental change in EEG theta activity in the medial prefrontal cortex during response control. *NeuroImage* **2014**, *85*, 873–887. [CrossRef]
73. Orekhova, E.; Stroganova, T.; Posikera, I.; Elam, M. EEG theta rhythm in infants and preschool children. *Clin. Neurophysiol.* **2006**, *117*, 1047–1062. [CrossRef]
74. Bosseler, A.N.; Taulu, S.; Pihko, E.; Mäkelä, J.P.; Imada, T.; Ahonen, A.; Kuhl, P.K. Theta brain rhythms index perceptual narrowing in infant speech perception. *Front. Psychol.* **2013**, *4*, 690. [CrossRef]
75. Cavanagh, J.F.; Figueroa, C.M.; Cohen, M.X.; Frank, M.J. Frontal theta reflects uncertainty and unexpectedness during exploration and exploitation. *Cereb. Cortex* **2012**, *22*, 2575–2586. [CrossRef]
76. Posner, M.I.; Tang, Y.-Y.; Lynch, G. Mechanisms of white matter change induced by meditation training. *Front. Psychol.* **2014**, *5*, 115924. [CrossRef] [PubMed]
77. Conejero, Á.; Guerra, S.; Abundis-Gutiérrez, A.; Rueda, M.R. Frontal theta activation associated with error detection in toddlers: Influence of familial socioeconomic status. *Dev. Sci.* **2018**, *21*, e12494. [CrossRef] [PubMed]
78. Köster, M.; Langeloh, M.; Michel, C.; Hoehl, S. Young infants process prediction errors at the theta rhythm. *NeuroImage* **2021**, *236*, 118074. [CrossRef] [PubMed]
79. Meyer, M.; Endedijk, H.M.; Van Ede, F.; Hunnius, S. Theta oscillations in 4-year-olds are sensitive to task engagement and task demands. *Sci. Rep.* **2019**, *9*, 6049. [CrossRef]
80. Ortiz-Mantilla, S.; Hämäläinen, J.A.; Musacchia, G.; Benasich, A.A. Enhancement of gamma oscillations indicates preferential processing of native over foreign phonemic contrasts in infants. *J. Neurosci.* **2013**, *33*, 18746–18754. [CrossRef]
81. Fuentemilla, L.; Penny, W.D.; Cashdollar, N.; Bunzeck, N.; Düzel, E. Theta-coupled periodic replay in working memory. *Curr. Biol.* **2010**, *20*, 606–612. [CrossRef]
82. Rutishauser, U.; Ross, I.B.; Mamelak, A.N.; Schuman, E.M. Human memory strength is predicted by theta-frequency phase-locking of single neurons. *Nature* **2010**, *464*, 903–907. [CrossRef] [PubMed]
83. Siegel, M.; Warden, M.R.; Miller, E.K. Phase-dependent neuronal coding of objects in short-term memory. *Proc. Natl. Acad. Sci. USA* **2009**, *106*, 21341–21346. [CrossRef] [PubMed]
84. Bell, M.A. Brain electrical activity associated with cognitive processing during a looking version of the A-not-B task. *Infancy* **2001**, *2*, 311–330. [CrossRef] [PubMed]
85. Morasch, K.C.; Bell, M.A. Patterns of brain-electrical activity during declarative memory performance in 10-month-old infants. *Brain Cogn.* **2009**, *71*, 215–222. [CrossRef]
86. Ishii, R.; Shinosaki, K.; Ukai, S.; Inouye, T.; Ishihara, T.; Yoshimine, T.; Hirabuki, N.; Asada, H.; Kihara, T.; Robinson, S.E. Medial prefrontal cortex generates frontal midline theta rhythm. *Neuroreport* **1999**, *10*, 675–679. [CrossRef]
87. Ellis, C.T.; Skalaban, L.J.; Yates, T.S.; Bejjanki, V.R.; Córdova, N.I.; Turk-Browne, N.B. Evidence of hippocampal learning in human infants. *Curr. Biol.* **2021**, *31*, 3358–3364.e4. [CrossRef]
88. Matsuzawa, J.; Matsui, M.; Konishi, T.; Noguchi, K.; Gur, R.C.; Bilker, W.; Miyawaki, T. Age-related volumetric changes of brain gray and white matter in healthy infants and children. *Cereb. Cortex* **2001**, *11*, 335–342. [CrossRef]
89. Dehaene-Lambertz, G. The human infant brain: A neural architecture able to learn language. *Psychon. Bull. Rev.* **2017**, *24*, 48–55. [CrossRef]
90. Raz, G.; Saxe, R. Learning in infancy is active, endogenously motivated, and depends on the prefrontal cortices. *Annu. Rev. Dev. Psychol.* **2020**, *2*, 247–268. [CrossRef]
91. Schlichting, M.L.; Mumford, J.A.; Preston, A.R. Learning-related representational changes reveal dissociable integration and separation signatures in the hippocampus and prefrontal cortex. *Nat. Commun.* **2015**, *6*, 8151. [CrossRef]
92. Nanova, P.; Kolev, V.; Yordanova, J. Effect of Proactive Mode of Processing on Event-related Oscillatory Brain Responses in Children. *Int. J. Bioautomation* **2018**, *22*, 253–262. [CrossRef]
93. Gómez, C.M.; Munoz, V.; Rodríguez-Martínez, E.I.; Arjona, A.; Barriga-Paulino, C.I.; Pelegrina, S. Child and adolescent development of the brain oscillatory activity during a working memory task. *Brain Cogn.* **2023**, *167*, 105969. [CrossRef]
94. Rajan, V.; Bell, M.A. Developmental changes in fact and source recall: Contributions from executive function and brain electrical activity. *Dev. Cogn. Neurosci.* **2015**, *12*, 1–11. [CrossRef] [PubMed]
95. Kuperberg, G.R.; Jaeger, T.F. What do we mean by prediction in language comprehension? *Lang. Cogn. Neurosci.* **2016**, *31*, 32–59. [CrossRef] [PubMed]
96. Daube, C.; Ince, R.A.; Gross, J. Simple acoustic features can explain phoneme-based predictions of cortical responses to speech. *Curr. Biol.* **2019**, *29*, 1924–1937.e9. [CrossRef]

97. Maulsby, R.L. An illustration of emotionally evoked theta rhythm in infancy: Hedonic hypersynchrony. *Electroencephalogr. Clin. Neurophysiol.* **1971**, *31*, 157–165. [CrossRef]
98. Cuevas, K.; Raj, V.; Bell, M.A. A frequency band analysis of two-year-olds' memory processes. *Int. J. Psychophysiol.* **2012**, *83*, 315–322. [CrossRef]
99. Bazhenova, O.V.; Stroganova, T.A.; Doussard-Roosevelt, J.A.; Posikera, I.A.; Porges, S.W. Physiological responses of 5-month-old infants to smiling and blank faces. *Int. J. Psychophysiol.* **2007**, *63*, 64–76. [CrossRef] [PubMed]
100. Porges, S.W. The polyvagal theory: Phylogenetic contributions to social behavior. *Physiol. Behav.* **2003**, *79*, 503–513. [CrossRef]
101. Bentley, P.; Vuilleumier, P.; Thiel, C.M.; Driver, J.; Dolan, R.J. Cholinergic enhancement modulates neural correlates of selective attention and emotional processing. *Neuroimage* **2003**, *20*, 58–70. [CrossRef]
102. Wang, J.; Gao, D.; Li, D.; Desroches, A.S.; Liu, L.; Li, X. Theta–gamma coupling reflects the interaction of bottom-up and top-down processes in speech perception in children. *Neuroimage* **2014**, *102*, 637–645. [CrossRef] [PubMed]
103. Marshall, P.J.; Bar-Haim, Y.; Fox, N.A. Development of the EEG from 5 months to 4 years of age. *Clin. Neurophysiol.* **2002**, *113*, 1199–1208. [CrossRef]
104. Klimesch, W. EEG alpha and theta oscillations reflect cognitive and memory performance: A review and analysis. *Brain Res. Rev.* **1999**, *29*, 169–195. [CrossRef] [PubMed]
105. Babiloni, C.; Babiloni, F.; Carducci, F.; Cappa, S.; Cincotti, F.; Del Percio, C.; Miniussi, C.; Moretti, D.V.; Pasqualetti, P.; Rossi, S. Human cortical EEG rhythms during long-term episodic memory task. A high-resolution EEG study of the HERA model. *Neuroimage* **2004**, *21*, 1576–1584. [CrossRef]
106. Yordanova, J.; Kolev, V. Alpha response system in children: Changes with age. *Int. J. Psychophysiol.* **1997**, *26*, 411–430. [CrossRef] [PubMed]
107. Yordanova, J.; Kolev, V.; Bas¸ar-Eroglu, C. Is the alpha rhythm a control parameter for brain responses? *Biol. Cybern.* **1997**, *76*, 471–480.
108. Yordanova, J.Y.; Kolev, V.N. Developmental changes in the alpha response system. *Electroencephalogr. Clin. Neurophysiol.* **1996**, *99*, 527–538. [CrossRef]
109. Chapeton, J.I.; Haque, R.; Wittig, J.H.; Inati, S.K.; Zaghloul, K.A. Large-scale communication in the human brain is rhythmically modulated through alpha coherence. *Curr. Biol.* **2019**, *29*, 2801–2811.e5. [CrossRef]
110. Buzsáki, G.; Wang, X.-J. Mechanisms of gamma oscillations. *Annu. Rev. Neurosci.* **2012**, *35*, 203–225. [CrossRef] [PubMed]
111. Halgren, M.; Fabó, D.; Ulbert, I.; Madsen, J.R.; Erőss, L.; Doyle, W.K.; Devinsky, O.; Schomer, D.; Cash, S.S.; Halgren, E. Superficial slow rhythms integrate cortical processing in humans. *Sci. Rep.* **2018**, *8*, 2055. [CrossRef]
112. Yakovlev, P.I.; Lecours, A.R. The myelogenetic cycles of regional maturation of the brain. In *Regional Development of the Brain in Early Life*; Minkowski, A., Ed.; Blackwell: Oxford, UK, 1967; pp. 3–70.
113. Vollebregt, M.A.; Zumer, J.M.; Ter Huurne, N.; Castricum, J.; Buitelaar, J.K.; Jensen, O. Lateralized modulation of posterior alpha oscillations in children. *NeuroImage* **2015**, *123*, 245–252. [CrossRef]
114. Vinck, M.; Oostenveld, R.; Van Wingerden, M.; Battaglia, F.; Pennartz, C.M. An improved index of phase-synchronization for electrophysiological data in the presence of volume-conduction, noise and sample-size bias. *Neuroimage* **2011**, *55*, 1548–1565. [CrossRef]
115. John, E.; Ahn, H.; Prichep, L.; Trepetin, M.; Brown, D.; Kaye, H. Developmental equations for the electroencephalogram. *Science* **1980**, *210*, 1255–1258. [CrossRef]
116. Matoušek, M.; Petersén, I. Automatic evaluation of EEG background activity by means of age-dependent EEG quotients. *Electroencephalogr. Clin. Neurophysiol.* **1973**, *35*, 603–612. [CrossRef]
117. Knyazev, G.G. EEG delta oscillations as a correlate of basic homeostatic and motivational processes. *Neurosci. Biobehav. Rev.* **2012**, *36*, 677–695. [CrossRef] [PubMed]
118. Orekhova, E.V.; Stroganova, T.A.; Posikera, I.N. Alpha activity as an index of cortical inhibition during sustained internally controlled attention in infants. *Clin. Neurophysiol.* **2001**, *112*, 740–749. [CrossRef]
119. Cuevas, K.; Cannon, E.N.; Yoo, K.; Fox, N.A. The infant EEG mu rhythm: Methodological considerations and best practices. *Dev. Rev.* **2014**, *34*, 26–43. [CrossRef]
120. Debnath, R.; Salo, V.C.; Buzzell, G.A.; Yoo, K.H.; Fox, N.A. Mu rhythm desynchronization is specific to action execution and observation: Evidence from time-frequency and connectivity analysis. *NeuroImage* **2019**, *184*, 496–507. [CrossRef] [PubMed]
121. Güntekin, B.; Uzunlar, H.; Çalışoğlu, P.; Eroğlu-Ada, F.; Yıldırım, E.; Aktürk, T.; Atay, E.; Ceran, Ö. Theta and alpha oscillatory responses differentiate between six-to seven-year-old children and adults during successful visual and auditory memory encoding. *Brain Res.* **2020**, *1747*, 147042. [CrossRef]
122. Hahn, G.; Ponce-Alvarez, A.; Deco, G.; Aertsen, A.; Kumar, A. Portraits of communication in neuronal networks. *Nat. Rev. Neurosci.* **2019**, *20*, 117–127. [CrossRef]
123. Fries, P. Rhythms for cognition: Communication through coherence. *Neuron* **2015**, *88*, 220–235. [CrossRef] [PubMed]
124. Başar-Eroglu, C.; Kolev, V.; Ritter, B.; Aksu, F.; Başar, E. EEG, auditory evoked potentials and evoked rhythmicities in three-year-old children. *Int. J. Neurosci.* **1994**, *75*, 239–255. [CrossRef]
125. Chatrian, G.E.; Petersen, M.C.; Lazarte, J.A. The blocking of the rolandic wicket rhythm and some central changes related to movement. *Electroencephalogr. Clin. Neurophysiol.* **1959**, *11*, 497–510. [CrossRef] [PubMed]

126. Kuhlman, W.N. Functional topography of the human mu rhythm. *Electroencephalogr. Clin. Neurophysiol.* **1978**, *44*, 83–93. [CrossRef] [PubMed]
127. Pfurtscheller, G.; Neuper, C.; Krausz, G. Functional dissociation of lower and upper frequency mu rhythms in relation to voluntary limb movement. *Clin. Neurophysiol.* **2000**, *111*, 1873–1879. [CrossRef] [PubMed]
128. Marshall, P.J.; Young, T.; Meltzoff, A.N. Neural correlates of action observation and execution in 14-month-old infants: An event-related EEG desynchronization study. *Dev. Sci.* **2011**, *14*, 474–480. [CrossRef]
129. Hao, J.; Feng, W.; Zhang, L.; Liao, Y. The post-movement beta rebound and motor-related Mu suppression in children. *J. Mot. Behav.* **2020**, *52*, 590–600. [CrossRef]
130. Patzwald, C.; Matthes, D.; Elsner, B. Eighteen-month-olds integrate verbal cues into their action processing: Evidence from ERPs and mu power. *Infant Behav. Dev.* **2020**, *58*, 101414. [CrossRef]
131. Culham, J.C.; Valyear, K.F. Human parietal cortex in action. *Curr. Opin. Neurobiol.* **2006**, *16*, 205–212. [CrossRef]
132. Iacoboni, M. Visuo-motor integration and control in the human posterior parietal cortex: Evidence from TMS and fMRI. *Neuropsychologia* **2006**, *44*, 2691–2699. [CrossRef]
133. Arnstein, D.; Cui, F.; Keysers, C.; Maurits, N.M.; Gazzola, V. μ-suppression during action observation and execution correlates with BOLD in dorsal premotor, inferior parietal, and SI cortices. *J. Neurosci.* **2011**, *31*, 14243–14249. [CrossRef] [PubMed]
134. Bryant, L.J.; Cuevas, K. Effects of active and observational experience on EEG activity during early childhood. *Psychophysiology* **2019**, *56*, e13360. [CrossRef]
135. Fox, N.A.; Bakermans-Kranenburg, M.J.; Yoo, K.H.; Bowman, L.C.; Cannon, E.N.; Vanderwert, R.E.; Ferrari, P.F.; Van IJzendoorn, M.H. Assessing human mirror activity with EEG mu rhythm: A meta-analysis. *Psychol. Bull.* **2016**, *142*, 291. [CrossRef]
136. Hari, R.; Salmelin, R. Human cortical oscillations: A neuromagnetic view through the skull. *Trends Neurosci.* **1997**, *20*, 44–49. [CrossRef]
137. Klimesch, W.; Sauseng, P.; Hanslmayr, S. EEG alpha oscillations: The inhibition–timing hypothesis. *Brain Res. Rev.* **2007**, *53*, 63–88. [CrossRef]
138. van Wijk, B.C.; Beek, P.J.; Daffertshofer, A. Neural synchrony within the motor system: What have we learned so far? *Front. Hum. Neurosci.* **2012**, *6*, 252. [CrossRef]
139. Mikhailova, A.; Orekhova, L.; Dyagileva, Y.O.; Mukhtarimova, T.; Pavlenko, V. Reactivity of the EEG μ rhythm on observing and performing actions in young children with different levels of receptive speech development. *Neurosci. Behav. Physiol.* **2021**, *51*, 85–92. [CrossRef]
140. Ramachandran, V.S. *The Tell-Tale Brain: A Neuroscientist's Quest for What Makes Us Human*; WW Norton & Company: New York, NY, USA, 2012.
141. Meyer, M.; Hunnius, S.; Van Elk, M.; Van Ede, F.; Bekkering, H. Joint action modulates motor system involvement during action observation in 3-year-olds. *Exp. Brain Res.* **2011**, *211*, 581–592. [CrossRef] [PubMed]
142. Köster, M.; Langeloh, M.; Kliesch, C.; Kanngiesser, P.; Hoehl, S. Motor cortex activity during action observation predicts subsequent action imitation in human infants. *NeuroImage* **2020**, *218*, 116958. [CrossRef] [PubMed]
143. Liao, Y.; Acar, Z.A.; Makeig, S.; Deak, G. EEG imaging of toddlers during dyadic turn-taking: Mu-rhythm modulation while producing or observing social actions. *NeuroImage* **2015**, *112*, 52–60. [CrossRef]
144. Gerson, S.; Meyer, M.; Hunnius, S.; Bekkering, H. Unravelling the contributions of motor experience and conceptual knowledge in action perception: A training study. *Sci. Rep.* **2017**, *7*, 46761. [CrossRef] [PubMed]
145. Pineda, J.A. The functional significance of mu rhythms: Translating "seeing" and "hearing" into "doing". *Brain Res. Rev.* **2005**, *50*, 57–68. [CrossRef]
146. Belalov, V.; Dyagileva, Y.O.; Pavlenko, V.; Kochukhova, O.M. Neurophysiological analysis of speech perception in 2.5 to 3.5-year-old orphans and children raised in a family. *Neurophysiology* **2014**, *46*, 79–87. [CrossRef]
147. Wróbel, A. Beta activity: A carrier for visual attention. *Acta Neurobiol. Exp.* **2000**, *60*, 247–260. [CrossRef]
148. Engel, A.K.; Fries, P. Beta-band oscillations—Signalling the status quo? *Curr. Opin. Neurobiol.* **2010**, *20*, 156–165. [CrossRef]
149. Hahne, A.; Eckstein, K.; Friederici, A.D. Brain signatures of syntactic and semantic processes during children's language development. *J. Cogn. Neurosci.* **2004**, *16*, 1302–1318. [CrossRef]
150. Davidson, D.J.; Indefrey, P. An inverse relation between event-related and time–frequency violation responses in sentence processing. *Brain Res.* **2007**, *1158*, 81–92. [CrossRef]
151. Zeman, A. Consciousness. *Brain* **2001**, *124 Pt 7*, 1263–1289. [CrossRef]
152. Csibra, G.; Davis, G.; Spratling, M.; Johnson, M. Gamma oscillations and object processing in the infant brain. *Science* **2000**, *290*, 1582–1585. [CrossRef] [PubMed]
153. Jensen, O.; Kaiser, J.; Lachaux, J.-P. Human gamma-frequency oscillations associated with attention and memory. *Trends Neurosci.* **2007**, *30*, 317–324. [CrossRef] [PubMed]
154. Axmacher, N.; Henseler, M.M.; Jensen, O.; Weinreich, I.; Elger, C.E.; Fell, J. Cross-frequency coupling supports multi-item working memory in the human hippocampus. *Proc. Natl. Acad. Sci. USA* **2010**, *107*, 3228–3233. [CrossRef] [PubMed]
155. Farroni, T.; Csibra, G.; Simion, F.; Johnson, M.H. Eye contact detection in humans from birth. *Proc. Natl. Acad. Sci. USA* **2002**, *99*, 9602–9605. [CrossRef] [PubMed]
156. Grossmann, T.; Johnson, M.H.; Farroni, T.; Csibra, G. Social perception in the infant brain: Gamma oscillatory activity in response to eye gaze. *Soc. Cogn. Affect. Neurosci.* **2007**, *2*, 284–291. [CrossRef] [PubMed]

157. Leung, S.; Mareschal, D.; Rowsell, R.; Simpson, D.; Iaria, L.; Grbic, A.; Kaufman, J. Oscillatory activity in the infant brain and the representation of small numbers. *Front. Syst. Neurosci.* **2016**, *10*, 4. [CrossRef] [PubMed]
158. Kaufman, J.; Csibra, G.; Johnson, M.H. Representing occluded objects in the human infant brain. *Proc. R. Soc. Lond. Ser. B Biol. Sci.* **2003**, *270*, S140–S143. [CrossRef] [PubMed]
159. Kaufman, J.; Csibra, G.; Johnson, M.H. Oscillatory activity in the infant brain reflects object maintenance. *Proc. Natl. Acad. Sci. USA* **2005**, *102*, 15271–15274. [CrossRef] [PubMed]
160. Southgate, V.; Csibra, G.; Kaufman, J.; Johnson, M.H. Distinct processing of objects and faces in the infant brain. *J. Cogn. Neurosci.* **2008**, *20*, 741–749. [CrossRef] [PubMed]
161. Reid, V.M.; Csibra, G.; Belsky, J.; Johnson, M.H. Neural correlates of the perception of goal-directed action in infants. *Acta Psychol.* **2007**, *124*, 129–138. [CrossRef] [PubMed]
162. Orekhova, E.V.; Butorina, A.V.; Sysoeva, O.V.; Prokofyev, A.O.; Nikolaeva, A.Y.; Stroganova, T.A. Frequency of gamma oscillations in humans is modulated by velocity of visual motion. *J. Neurophysiol.* **2015**, *114*, 244–255. [CrossRef] [PubMed]
163. Yordanova, J.; Kolev, V.; Heinrich, H.; Woerner, W.; Banaschewski, T.; Rothenberger, A. Developmental event-related gamma oscillations: Effects of auditory attention. *Eur. J. Neurosci.* **2002**, *16*, 2214–2224. [CrossRef]

brain sciences

MDPI

Article

Socioeconomic Inequalities Affect Brain Responses of Infants Growing Up in Germany

Annika Susann Wienke and Birgit Mathes *

Bremer Initiative to Foster Early Childhood Development (BRISE), Faculty for Human and Health Sciences, University of Bremen, 28359 Bremen, Germany; annika.wienke@uni-bremen.de
* Correspondence: birgit.mathes@uni-bremen.de

Abstract: Developmental changes in functional neural networks are sensitive to environmental influences. This EEG study investigated how infant brain responses relate to the social context that their families live in. Event-related potentials of 255 healthy, awake infants between six and fourteen months were measured during a passive auditory oddball paradigm. Infants were presented with 200 standard tones and 48 randomly distributed deviants. All infants are part of a longitudinal study focusing on families with socioeconomic and/or cultural challenges (Bremen Initiative to Foster Early Childhood Development; BRISE; Germany). As part of their familial socioeconomic status (SES), parental level of education and infant's migration background were assessed with questionnaires. For 30.6% of the infants both parents had a low level of education ($\leq$10 years of schooling) and for 43.1% of the infants at least one parent was born abroad. The N2–P3a complex is associated with unintentional directing of attention to deviant stimuli and was analysed in frontocentral brain regions. Age was utilised as a control variable. Our results show that tone deviations in infants trigger an immature N2–P3a complex. Contrary to studies with older children or adults, the N2 amplitude was more positive for deviants than for standards. This may be related to an immature superposition of the N2 with the P3a. For infants whose parents had no high-school degree and were born abroad, this tendency was increased, indicating that facing multiple challenges as a young family impacts on the infant's early neural development. As such, attending to unexpected stimulus changes may be important for early learning processes. Variations of the infant N2–P3a complex may, thus, relate to early changes in attentional capacity and learning experiences due to familial challenges. This points towards the importance of early prevention programs.

Keywords: brain development; infant; EEG; ERP; passive auditory oddball; SES; parental education; migration background; risk factors

1. Introduction

Infancy represents a formative phase in life in which, environmental conditions and experience shape emotional, social, cognitive and learning related developmental processes.

1.1. Impact of Socioeconomic Inequalities on Family Dynamics

A growing body of research highlights the impact of familial socioeconomic status (SES) and being part of a cultural minority on child development [1–3]. SES is typically characterised by several factors including parental educational attainment, family income, and parental occupation [4]. Together with migration background, SES reflects how a child is positioned within a society. These markers exert a broad influence on family life, from access to resources to parental stress levels and contribute to developmental inequalities [5]. Disparities in access to resources include healthcare, nutrition, and stimulating learning environments [6,7]. As such, disparities are often part of a transgenerational cycle and, thereby, affect parents as well as their offspring [8,9].

As a consequence, low educational background has been associated with limited strategies and knowledge about child development and child-rearing, both affecting secure attachment and healthy socioemotional development of the infant. Lower educational background is also being discussed in connection to reduced lexical richness and the grammar complexity of parents affecting children's language abilities [10]. Disparities in parental education further contribute to differences in mental and physical health and relate to learning opportunities or academic expectations [6,11–13].

Families with a migration background often face a range of challenges associated with acculturation, language barriers, social integration, and acceptance [14]. These challenges can affect family dynamics, parental roles, and access to health care and support networks, leading to increased levels of stress and uncertainty within the family unit [9], as well as families feeling isolated and vulnerable [15]. The situation may be more difficult when families have few socioeconomic resources. Migrating from countries with lower SES may correlate with increased financial burdens, adding an additional risk factor as it may further reduce access to essential resources, heighten stress levels, and lead to fewer opportunities for social and educational advancement [10,16]. Overall, migration may lead to reduced parental well-being and increased stress levels affecting parent–child interactions and the infant's emotional security. The home learning environment may comprise of reduced educational stimulation and increased disquiet affecting the infant's cognitive development [9].

Multiple factors interact within the social context and may lead to an overall tenser family life and increased levels of parental stress. Studies have shown that parental stress can adversely affect parent–child interactions, resulting in less sensitive and responsive caregiving behaviours [8]. The accumulation of socioeconomic challenges, such as those resulting from low education and migration with few resources, may amplify detrimental effects on child development and perpetuate cycles of disadvantage. Thus, understanding the complexity of social and cultural characteristics in families and child development is important for addressing inequalities and promoting positive outcomes for all children.

1.2. Impact of Socioeconomic Inequalities on Cognitive and Neural Development

Extensive research highlights the profound impact of socioeconomic disadvantage on various domains of child development, including cognitive, socioemotional, and neurobiological processes [3,17]. A lower socioeconomic status (SES) is known to detrimentally impact language development [18,19], executive functions, including memory, cognitive control [20–24], and attention [25]; thereby, also influencing later academic outcomes [11].

Disparities in cognitive and social skills emerge early in life, as shown by the U.S. Early Childhood Longitudinal Study—Birth Cohort (ECLS-B). Significant differences based on family background were evident by nine months and increased by age two [26]. In Germany, longitudinal data from a nationwide cohort [27] reveal migration-related disparities in scientific skills at ages four to six and social disparities in language skills by age two [28].

In addition, selective attention is important for learning, encompassing the ability to attend to relevant, while ignoring irrelevant information or distractors. A handful of studies revealed a correlation between SES and children's selective attention, linking a lower SES to a poorer ability of supressing unattended stimuli for children of school age [29].

Measuring neural processing during tasks may uncover nuanced processing distinctions in children's selective attention that may not be apparent through behavioural assessments, either because the task is too easy [30,31], or because the children are very young.

The infant brain undergoes rapid growth and cellular changes during development, due to neurogenesis, neuronal linking, pruning and myelinization. Building new neuronal connections is vital for cognitive development, learning, and adaptation to the environment [32]. Consequently, this development is heavily influenced by early experiences and environmental conditions. Emerging evidence suggests that socioeconomic disparities in early life experiences can have lasting effects on brain structure and function [17,33,34].

Lawson et al. for example, demonstrated an influence of parental education on cortical thickness in frontal brain structures of 11-year-olds [35].

Most studies on SES and neurocognitive development focus on children and adolescents, with limited research on infancy and early childhood [17,34,36]. Early experiences, parenting and home environments have been shown to effect executive skill development from a young age [24,37–40], yet few studies explore how SES impacts brain function in early infancy. Longitudinal studies have demonstrated associations between familial income and grey matter growth in frontal and parietal lobes [41]. Furthermore, activation of the executive attention network was diminished upon observation of incorrect puzzle solving in toddlers from low-SES families [42]. Cantiani et al. demonstrated that higher SES (Hollingshead scale including educational and occupational background) was associated with increased left central gamma power at 6 months of age, which subsequently predicted improved language scores at 24 months. Their findings suggest that SES-related differences in brain activity may emerge early in life and influence later language development [43].

Disentangling maturational effects based on single SES factors is, to date, still difficult as the literature is scarce and diverse as further addressed in the limitations. Taken together, the studies suggest that environmental factors influence maturation of the frontal lobe early in life, potentially leading to significant individual differences in developing attentional abilities.

1.3. Infant N2–P3a Complex following Unexpected Deviants

We investigated event-related infant brain responses during auditory attention by utilising a passive auditory oddball paradigm while measuring an electroencephalogram (EEG) [44]. During the measurement we presented a sequence of uniform tones (standard), occasionally interspersed with rare deviant tones (deviant). The paradigm utilised simple piano tones with salient pitch differences between deviant and standard, and a sufficient interstimulus interval (ISI) to elicit an automated orientation response [45] and to minimise distortions of the results due to familial languages in our multicultural sample [46–49].

The passive auditory oddball paradigm is widely used in developmental, neurological and psychiatric settings and can be repeated easily over the entire lifespan [50–54]. It taps into the basic and partly automated processes enabling stimulus matching and directing attention towards unexpected events. These skills can already be observed during the earliest stages of life [55–57] and may be early precursors for later cognitive abilities and language acquisition [58].

Deviant, i.e., unexpected, novel or rule-breaking stimuli, elicit an enhanced frontal N2–P3a complex in adults. The N2–P3a complex is a sequence of event-related potentials (ERPs) occurring approximately between 150 and 400 ms [59,60]. The enhancement of the N2–P3a complex for rare deviant, in comparison to standard stimuli, reflects (1) automatic and pre-attentive neuronal processes elicited by an unexpected change (N2) and (2) the subsequent unintentional shift of attention towards the deviant and the orientation response (P3a) [60–62].

Understanding timing, polarity and topography of infant ERPs during passive auditory oddball paradigms is still part of ongoing research. Developmental changes occurring rapidly over the first year of life and immature coordination within infant brain networks makes detection of ERPs more difficult. Further, infant studies are rare in comparison to adult studies, more difficult to conduct, often incorporate only a moderate number of infants, and utilise varying task designs. It seems clear that faster and more efficient neuronal processes in adults cannot be directly extrapolated to how the infant brain processes similar tasks. Development may even give rise to specialised components specific to infancy [44]. This also leads to variations in ERP nomenclature in the literature (see discussion below).

ERPs likely resembling the N2–P3a complex emerge during the first two years of life, as indicated for varying stimulus attributes, paradigm design and age ranges [44,63]. The ERPs show some characteristics similar to ERPs observed in adults but are not entirely matching [44,64,65]. Thus, we refer to an "infant N2–P3a complex" in the following.

Infant oddball studies often investigate the Mismatch Negativity (MMN), which is computed by subtracting the deviant from the standard ERP, subsequently, accentuating the difference in processing of the stimulus types [52]. The direction of the shift is less consistent in early developmental studies and, thus, referred to as the Mismatch Response (MMR) [63,66]. The MMR may comprise time-windows overlapping with the N2 and early P3a [60,64]. To reflect the infant N2–P3a complex, we analyse the amplitude around the N2 ("N2") and P3a peak ("late P3a") as well as an intermediate time-window centring around the maximum of the difference wave ("early P3a").

Notwithstanding all variations, ERP components elicited during the passive auditory oddball paradigm within the infant N2–P3a complex, reflect early cognitive processes training the infant brain to emphasise relevant information inherent to unexpected events over expected stimulus repetitions. Thus, this can be regarded as underlying early learning experiences and capabilities. Studying ERPs separately for each stimulus class provides a direct approach of how the brain responds to both auditory stimuli.

1.4. Aim of the Study

This paper aims to further elucidate the complex interplay between familial socioeconomic and cultural challenges and neural development. In general, studies exploring infant brain development in the context of these challenges still remain limited. Utilising an oddball paradigm offers the opportunity to identify neuronal markers with minimal cultural dependency, providing greater precision in understanding cognitive processes and their timing; adding to the field of infant studies mainly addressing spontaneous ongoing EEG baseline activity. As a novel approach, we investigated the influence of parental education and migration background on the infant N2–P3a complex elicited during a passive auditory oddball paradigm in infants. The infant N2–P3a complex reflects the immature neural response of automated attentional processes to stimulus change. Differential responses to standard and deviant stimuli shown in previous studies indicate that infants already discriminate unexpected stimulus changes. This study, enables us to better understand the possible early stages of developmental disparities by investigating the infantile neural processes important for discovering and learning about one's surroundings. We assumed that accumulation of disadvantages particularly affects early development. A long-term goal is to better define vulnerabilities and, thereby, substantiate targeted interventions to address disparities and promote equitable opportunities for all children [37,67].

2. Materials and Methods

2.1. Study Design

The current study is part of the Bremer Initiative to Foster Early Child Development (BRISE, German: Bremer Initiative zur Stärkung frühkindlicher Entwicklung) [68]. BRISE is a longitudinal intervention project investigating early development of children using a multi-method approach. All families were recruited from pre-selected urban neighbourhoods with structural disadvantages of the mid-sized city Bremen, Germany. Families further met at least one of three inclusion criteria: (1) low educational attainment of the parents—at least one of the parents did not complete a vocational training or at least one of the parents holds, at most, a school-leaving certificate after 10 years (German: Mittlere Reife, similar to an American GED), (2) migration background—at least one of the parents or grandparents was born outside of Germany, (3) precarious family income—due to at least one parent being unemployed or a low-wage earner or both parents working part-time. The BRISE sample comprises N = 550 children in total. The project started in 2017 and is currently funded until 2025. Families are scientifically accompanied from pregnancy or shortly after the child's birth up to the child's first year of primary school. Several measurement time points enable observation of a broad range of aspects of the family environment, behavioural abilities of the child as well as neurophysiological development. BRISE aims to identify strategies supporting health and early education during child development.

The ethics commission of the German Psychological Society approved BRISE and all legal guardians gave written consent (OK 092013_rev, 10/2013).

2.2. Participants

The data in this study is based on the whole BRISE sample. A total of 255 healthy infants (117 females) are included in the current analysis. At EEG measurement they were 6.3 to 14.1 months old (Ø 8.3 ± 1.4) and achieved sufficient data quality (see below). A total of 38 data sets had to be excluded due to early termination, too much movement or too high impedances of the EEG electrodes. Two datasets were excluded due to missing data in the SES background. Of the remaining families, EEG data was not recorded for this paradigm. Sixteen datasets were from siblings, of which four were twins.

2.3. Demographic Group Variables

Table 1 lists the demographic variables. Parental education and migration background were assessed at first contact with the family during a screening interview which ensured eligibility for the longitudinal study. The telephone interview took place when the mother was still pregnant, or their first enrolled infant was only a couple of weeks old. Education level and migration background was assessed separately for both parents.

Table 1. Family and demographic characteristics of infants.

N	**255**	
Mean age, months (SD) Gender (% female)	8.3 (1.2) 45.9	
	Parental Education	
	Higher (university entrance degree of at least one parent)	Lower (neither parent with university entrance degree)
N	177	78
	Infant Migration Background	
	None or 2nd grade	1st grade
N	145	110

Education level was dichotomously rated as one for neither parent having a university entrance degree ("low") or zero for at least one of the parents having a university entrance degree ("high"). Educational qualifications obtained abroad were assessed according to their official recognition in Germany (German: Abitur) to reflect their socioeconomic significance within the country in which the families currently live. In 21 cases, information regarding the father's education was not available and, thus, only the mother's information was considered during analysis.

Parental migrant background was categorised as zero for both parents being born in Germany ("None or 2nd grade") and as one for at least one parent being born abroad ("1st grade"). Since all infants were born in Germany, a rating of one was the equivalent of a 1st generation migration background of the child. In five cases, information of paternal birthplace was not known and only the mother's information was considered. In N = 55 infants, both parents were born abroad, and in N = 55 one parent (father: N = 32).

Parents migrated from 56 countries, various cultural backgrounds and all continents. It is neither possible, nor the aim, to differentiate between specific sub-cultures, as has been done in other studies [69]. The diversity of the countries of origin between the participating families rather gives a realistic picture of the suburban areas in which the families live, and which determines the work of family support programs, schools or other areas of public life. According to a 2024 four-level categorisation of the German Federal Ministry of Finance, 90% of the parents migrated from countries with lower costs of living than Germany (level 1 country). A total of 80% of the parents migrated from countries categorised as level 3 and 4.

Thus, parents dominantly migrated from countries with a low socioeconomic background, which may increase difficulties in acculturation and financial burdens.

To assess education and migration background as well as its intersection, four groups were defined as: EDU0/MIG0 (low risk regarding both parental education and migration background); EDU1/MIG0 (high risk regarding parental education background); EDU0/MIG1 (high risk regarding parental migration background) and EDU1/MIG1 (high risk regarding both demographic variables).

2.4. Experimental Procedure

The BRISE study design encompasses a sequence of measurement time points at home and in the laboratory. EEG data was exclusively obtained at the laboratory during the first visit. The laboratory (a general guideline for the EEG testing as implemented in the project BRISE is provided in the German publication [46] provided a sound-attenuated and electromagnetically shielded room, divided into an area with toys to acclimatise and EEG-preparation, and a distraction-poor area separated by a screen for the experimental part. During the experiments, the child was held by a parent on its lap in a sitting position and both were positioned in front of a monitor and two sound boxes. The experimenter was seated behind the family operating the computers for presentation and measurement. The experimenter had sight of the family via a camera. Electrically shielded ceiling lights were dimmed during measurements. Spontaneous EEG as well as two short visual tasks were presented before the oddball paradigm. In total, infants spent about 20 min in front of the screen.

For the passive auditory oddball, 200 standard stimuli interspersed by 48 deviants were presented to the awake infant in four runs using the stimulus delivery software Presentation® version 20.1 (Neurobehavioral Systems, Berkeley, CA, USA). The appearance rate of 24% deviants was kept equal in each run. Order of stimulus sequence was pseudo-randomised with the constraint that each deviant was followed by a standard tone. An inter-stimulus interval (ISI) varied randomly between 1400 and 2400 ms. As stimuli, two naturally complex piano tones were used for standards with 446.14 Hz (Bb4) and for deviants with 554.37 Hz (Db5; difference three semitones; ~20% deviance). Each tone consisted of a 200 ms tone length including 50 ms fade-out presented at approximately 70 dB at ear distance from the speaker. The frequencies of the tones were chosen based on the literature [70–72]. They were intended to match the pitch of a women's voice of childbearing age plus the tendency towards higher pitch similar to infant-directed speech in order to attract the infant's attention. A still face of a woman was presented throughout the trials to raise attention during the task, taking advantage of infants' natural interest in faces. Between runs, a short age-appropriate and calm video was presented. These breaks were incorporated to attend to the infants needs with minimal influence on the task, although it was possible to interrupt stimulus presentation at all times if the infant grew restless. Total duration of the task was about 9 min.

All stimuli were presented on a 24″ monitor with loudspeakers next to the monitor. Parents were seated 1 m in front of the monitor and the speakers. They were asked not to talk to the infant. Calm interaction and examination of quiet toys to reduce fussiness was possible.

2.5. EEG Measurement

The EEG was recorded at 500 Hz with band limits of 0.01–250 Hz by means of a 32-channel BrainAmp System (Brain Products®, Gilching, Germany). For EEG measurements a fitting electrode actiCAP with 32 Ag–AgCl electrodes (Fp1, Fp2, F7, F3, Fz, F4, F8, FT9, FC5, FC1, FC2, FC6, FT10, T7, C3, Cz, C4, T8, TP9, CP5, CP1, CP2, CP6, TP10, P7, P3, Pz, P4, P8, O1, O2; Easycap, Falk Minow Services) was placed onto the infant's head according to the international 10-10 system [73]. Cz was utilised as reference during recording.

2.6. EEG Pre-Processing

Analysis of the EEG data was performed in MATLAB R2022a® (MathWorks, Natick, USA) utilising several features of the package eeglab version 2024.0 [74]. Continuous data was filtered by the eeglab bandpass filter from 1 to 20 Hz (function pop_eegfiltnew()). If necessary, up to a maximum of three electrodes were interpolated per infant (N = 149 infants in total). The data was re-referenced to a common average. Subsequently, the data was segmented into epochs around the stimulus onset. All epochs were manually scanned for artefacts caused by eye or other movements or technical disturbances occurring between 500 ms before, to 1000 ms after stimulus onset. Epochs containing artefacts were rejected. A minimum of 20 artefact-free epochs per stimulus class was set as a limit for further analysis. On average, 28.7 (±6.6) deviants were included in the analysis. For statistical comparison, the number of standard epochs was randomly reduced to match the number of deviant epochs, while the first and each standard tone after a break were always excluded.

2.7. Analysis of the Stimulus-Locked N2 and P3a in the Time Domain

For each infant, the artefact-free epochs were baseline-corrected by the mean amplitude between 150 to 0 ms prior to stimulus onset and averaged for each stimulus class, i.e., deviant and standard. The resulting ERPs were averaged across all infants to inspect time course and topography of the grand average for both stimulus classes (see Figure 1).

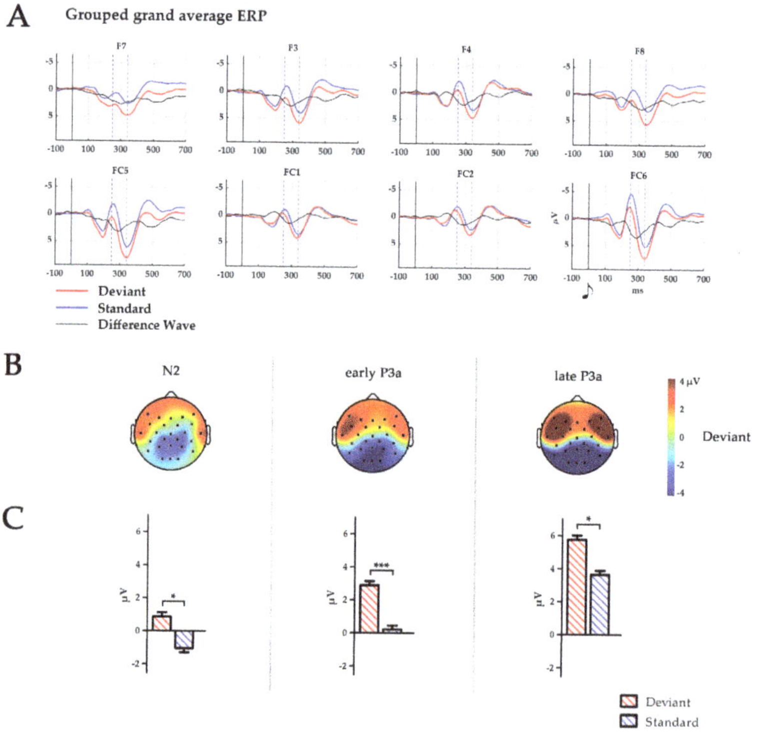

Figure 1. (**A**) Temporal ERP courses for six fronto-central electrodes averaged across all 255 infants for standard (blue), deviant stimuli (red) and difference wave (black). Time points 0 ms indicate the presentation of the respective tone. Dashed lines indicate peaks of N2 at 250 ms and P3a at 340 ms. (**B**) Topography of the deviant stimulus in the time windows of N2: 224–274 ms, early P3a: 264–314 ms and late P3a: 314–364 ms. Black dots depict electrode placement. (**C**) Boxplots of the mean amplitude over all six electrodes (F7, F8, F3, F4, FC5, and FC6) for the time window N2, early P3a and late P3a. * $p < 0.05$; *** $p \leq 0.001$.

The infant N2–P3a complex was defined as an anterior stimulus response encompassing F7, F8, F3, F4, FC5, and FC6, which is in accordance with the literature. Electrode sites were pooled into three ROIs: frontal 1 (F7/F8), frontal 2 (F3/F4) and fronto-central (FC5/FC6). Within these electrodes, the infant P3a peak was defined as the largest positive-going peak of the ERP time course, while the infant N2 was defined as the negative deflection preceding the P3a, in analogy to [75]. Peak amplitudes of the grand average for the infant N2–P3a complex were maximal at FC6. For the statistical analysis, 50 ms time windows were defined centred around the peaks at FC6 for deviants (for the infant N2 at 246 ms and for the infant P3a at 340 ms). A time window encompassing the peak of the difference wave at 290 ms (calculated as deviant minus standard) was similarly defined. The time windows are, thus, defined as N2 (224–274 ms), early P3a (264–314 ms) and late P3a (314–364 ms). Amplitudes for all ERPs were calculated as the mean area under the curve within each of the time windows.

2.8. Statistical Analysis

ANOVAs were conducted separately for each time window (N2, early P3a, late P3a) to cover the entire N2–P3a complex. They included the within-subject factors condition (deviant and standard), ROI (frontal1, frontal2 and fronto-central) and hemisphere (left, right), and the between-subject factor group (EDU0/MIG0, EDU1/MIG0, EDU0/MIG1 and EDU1/MIG1). ROIs were chosen according to local maxima of the ERP. Groups reflect either no, single or combined occurrence of risk factors to illustrate specificity as well as accumulation of occurring risk factors. Age at the time of EEG assessment was incorporated as a control variable, since preliminary results indicated that the age range may affect the N2–P3a complex. Huynh–Feldt corrected probabilities are reported to correct for any violations of the assumption of sphericity. Partial eta square values are reported as an estimate of effect size. Since group variables are nominally distributed, the assumption of homogeneity of regression slopes was tested. Post-hoc comparisons using multiple comparisons were applied following significance in the overall ANOVA. All post hoc tests were corrected using the Sidak procedure. The significance level was set to $p < 0.05$ for all statistical tests.

3. Results

The time course, topographical distribution and mean amplitudes of the ERPs averaged for all infants are depicted in Figure 1. Figure 2 displays similar information; however, separated for the four subject factor groups. A separate boxplot is shown in Figure 3 for the statistical effect of group on standard processing at electrode FC6.

3.1. Characterisation of the Infant N2–P3a Complex

Mean amplitude during the N2 time window was larger for deviants than standards ($F(1250) = 7.3$, $p = 0.008$, $\eta2 = 0.03$). This trend became more pronounced during the early P3a time window ($F(1250) = 14.4$, $p = 0.001$, $\eta2 = 0.05$) and remained until the late P3a time window ($F(1250) = 4.1$, $p = 0.045$, $\eta2 = 0.02$). Amplitude maximum was found at F7/F8 electrode sites during the early P3a time window ($F(2500) = 3.9$, $p = 0.021$, $\eta2 = 0.02$) and at FC5/FC6 electrode sites during the late P3a time window ($F(2500) = 6.3$, $p = 0.036$, $\eta2 = 0.03$, $p < 0.001$ for both post-hoc comparisons).

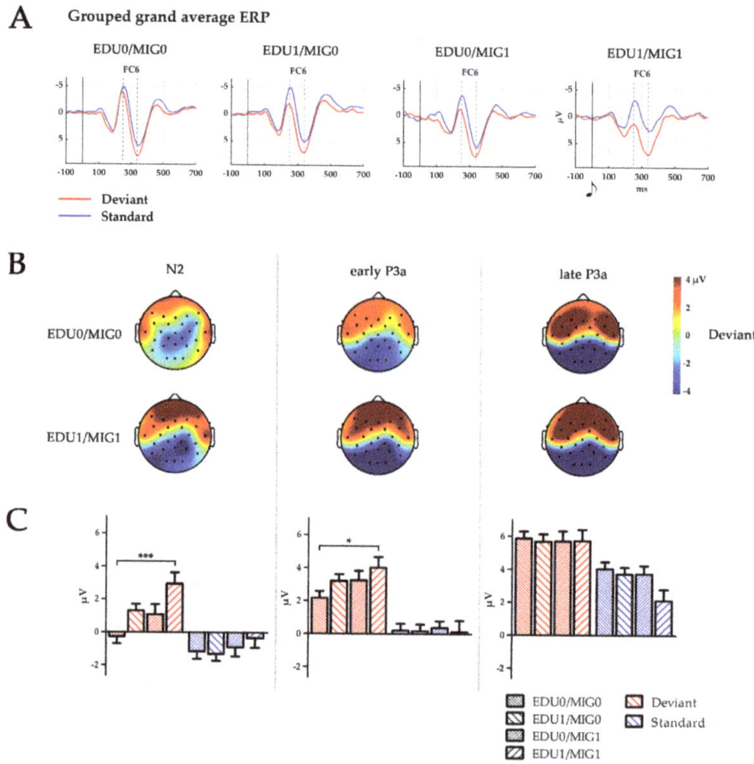

Figure 2. (**A**) Temporal ERP courses for electrode FC6 for standards left, and deviants right, separated by subject factor groups (EDU0/MIG0, EDU1/MIG0, EDU0/MIG1 and EDU1/MIG1; EDU = education level, MIG = migration background; 0 low risk, 1 high risk). Time points 0 ms indicate the presentation of the respective tone. Dashed lines indicate peaks of N2 at 250 ms and P3a at 340 ms. (**B**) Topography of the deviant stimulus in the time windows of N2: 224–274 ms, early P3a: 264–314 ms and late P3a: 314–364 ms for groups EDU0/MIG0 and EDU1/MIG1. Black dots depict electrode placement. (**C**) Boxplot depicting mean amplitude of standard and deviant processing over all six electrodes (F7, F8, F3, F4, FC5, and FC6) for the time window N2, early P3a and late P3a separated for all four subject factor groups (EDU0/MIG0, EDU1/MIG0, EDU0/MIG1 and EDU1/MIG1; EDU = education level, MIG = migration background; 0 = low risk, 1 = high risk). * $p < 0.05$; *** $p \leq 0.001$.

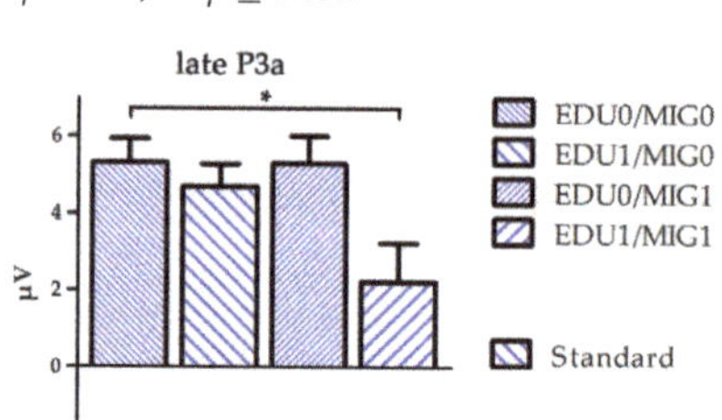

Figure 3. Boxplot depicting mean amplitude of standard processing over electrode FC6 for the time window late P3a separated for all four subject factor groups (EDU0/MIG0, EDU1/MIG0, EDU0/MIG1 and EDU1/MIG1; EDU = education level, MIG = migration background; 0 = low risk, 1 = high risk). * $p < 0.05$.

3.2. Group Differences

A group (F(3250) = 2.7, p = 0.049, η2 = 0.03) and group × stimulus interaction effect (F(3250) = 5.0, p = 0.002, η2 = 0.06) indicated increasing positive values with the occurrence of risk factors during the N2 time window. Post-hoc comparisons indicated an overall significant difference between the EDU0/MIG0 and EDU1/MIG1 group (p = 0.037), which was pronounced for deviants (p = 0.001). A group x stimulus interaction during the early P3a time window indicated that this trend persevered to later processing stages (F(3250) = 2.7, p = 0.047, η2 = 0.03). Additionally, multiple comparisons conducted following a group × stimulus × ROI × hemisphere interaction effect (F(6488) = 2.3, p = 0.036, η2 = 0.03) indicated that the late P3a amplitude for standards was reduced for the EDU1/MIG1 in comparison to the EDU0/MIG0 group at F7 and FC6.

4. Discussion

We investigated the anterior infant N2–P3a complex elicited by auditory stimulus changes in infants between 6 to 14 months of age. Amplitudes for deviants were more positive than for standards during all time windows encompassing the infant N2–P3a complex. This indicates an immature N2 component, since the response pattern for standards and deviants during infancy differs from those described in older children and adults. This effect was pronounced when parents were born abroad and had no higher education, indicating a neurodevelopmental delay in these infants. Furthermore, the amplitude during the P3a time window following standard stimuli was reduced in these infants. The results indicate that social living conditions impact early neural dynamics and may affect early attentional capacities and learning abilities.

4.1. Immature Processing of Stimulus Changes—The Infant N2–P3a Complex

The utilised passive oddball paradigm evoked an infant frontal N2–P3a complex, which differed between standards and unexpected deviants. This suggests early development of rapid neural processes underlying stimulus discrimination, which automatically draws attention towards unexpected, deviating stimuli [44,55]. These abilities are fundamental for cognitive processes, learning, and adaptive behaviour throughout life. Finely tuned attention processing during infancy is pivotal for healthy and robust development [76].

The infant N2 exhibited a more positive maximum for deviant than standard stimuli. Most adult studies report an opposite effect [62]. This may result from the immaturity of the N2 component at a young age. The amplitude difference between standards and deviants was most pronounced during the early P3a time window, which enclosed the initial phases of the P3a wave preceding its peak. Infant studies report for the MMR both negative e.g., [71,77–82] and positive deflections e.g., [83–86] for the difference wave. Interestingly, some infant studies describe both, a negative deflection followed by a positive one in our time range [45,55,66]. Thus, the observed pronounced amplitude difference between both stimulus types during the early P3a time window seems to be related to the late deflection of the MMR.

Infant P3a responses are described as following the MMR and, in accordance with our results, are more positive for deviants than standards [44,56]. This might indicate that within the infant N2–P3a complex, late responses eliciting an orientation response seem to mature or at least resemble adult-like responses earlier in life than initial processing stages following an unexpected event. This assumption is in line with a longitudinal study investigating deviant processing in infants by Kushnerenko et al. [87]. They showed that a N250 and a P350 peak in the ERP become apparent in the group average at about 6 months of age, i.e., within the age range of our infants, and likely mirroring the infant N2–P3a complex we have described. Due to their longitudinal measurements, they showed that prior to 6 months of age the ERP time course is dominated by a single slow positive wave, which is subdivided into an early and late positive wave by the emerging N250. Their results indicate differential time courses of development for infantile ERP components. Taken together, our results indicate a lack of fully tuned neural dynamics underlying stimulus matching during the N2 and early P3a time window and the involuntary attentional shift during the P3a time window (orientation response) [64].

Future studies might profit from analysing event-related oscillations (EROs). ERPs may result from the superposition of EROs within different frequency ranges [88–91]. The adult N2 is dominantly driven by concurrent theta oscillations [92–94], while the adult P3a is dominated by the slower delta oscillations [61,95–97].

Spontaneous activity in the infant brain is dominated by slow, high amplitude oscillatory activity in the delta range. Increasingly faster waves in the 6–9 Hz range emerge over the first year of life [98–100]. Thus, it is possible that faster oscillations are still diminished in the infant ERP, resulting in a dominating slow-wave emerging during the early time-windows and peaking concurrent to the P3a. Superposition of a large slow-wave and age-expected still emerging faster oscillations might explain the positive shift for deviants during the N2 time window. In accordance, Kolev et al. and Başar-Eroglu et al. showed that adult-like alpha responses are diminished in both visual and auditory evoked potentials as well as spontaneous activity in 3-year-olds [101,102]. Their studies indicate that the existence of a given frequency component in the ERP depends on the presence of the corresponding rhythm in the spontaneous EEG.

This suggestion aligns with the approach proposed e.g., by Mathes et al. and Wienke et al. for investigating maturational processes of ERP components on an oscillatory level [103,104]. It is also in line with the general idea of neuronal network communication being implemented by oscillatory activity patterns during development.

4.2. Changes of the Infant N2–P3a Complex due to Social and Cultural Challenges

Lower education, a migration background, and not being fluent in the local language are barriers to health care [7,9] and cultural participation [14,105]. Both challenges often co-vary with a higher risk of lower income and unemployment, as well as increased social burdens, everyday emotional stress, and fewer possibilities of coping with these challenges [106,107]. In accordance, our group-defining factors of education level and migration background should be understood as multifaceted indicators of early developmental risks due to socioeconomic inequalities.

We defined four groups, mirroring the possibilities with which educational- (low/high) or migration-related risks (parents born abroad/parents born in country they currently live) were reported. In accordance with our assumption that the occurrence of both risk factors put the children at a severe disadvantage, we found particularly high differences between the EDU0/MIG0 and the EDU1/MIG1 group, with both groups reporting one risk factor (EDU1/MIG0 and EDU0/MIG0) being intermediate.

Our results show that parental education and migration background impact on neuronal processing reflected to the infant N2–P3a complex. The N2 amplitude is diminished when either one of the risk factors defined in this study was apparent. The effect is most pronounced between the EDU0/MIG0 group and the double burdened group EDU1/MIG1 when processing deviants. It further prolongs into the early P3a time window. During the late P3a time window the positive deflection elicited by standard stimuli is minimal for the EDU1/MIG1 group. This indicates that the accumulation of risk factors affects infant brain development and processing of information within a very simple passive auditory paradigm. The results may specifically indicate that the automatic process of discriminating deviants from standards and attentional orientation towards standards is altered in these infants.

We argued above, that during infancy the N2 component is still emerging and not yet matured enough to mirror the N2 increase for deviants observed in older children and adults. Kushnerenko et al. have further shown that N2 maturation during the first year of life is highly individual and dynamic [87]. Variations of the N2, thus, seem to tap into a process that specifically indexes brain maturation and brain plasticity in this age range. Its pronounced reduction in the EDU1/MIG1 group may indicate a developmental delay or developmental variations, which may pose a risk for future gains from learning experiences.

Studies investigating social and cultural challenges on infant brain development, though rare, are increasingly acknowledged to be integral to understand lifetime social disparities [32,37,108]. They often focus on changes of the oscillatory patterns of the spontaneous EEG and show that SES disadvantages are not apparent at birth [109]; however, they may emerge over the first year of life. Tomalski et al. and Cantinai et al. related lower SES to decreased amplitudes of frontal [110] and of left central gamma oscillations at rest and related the latter to higher language scores at age 24 months [43]. Otero et al. observed increased delta absolute power for toddlers of lower SES, while high SES related to increased alpha absolute power and increased alpha and beta relative power [111]. Tarullo et al. assessed highly disadvantaged 48-month-olds from rural Pakistan showing associations between an increase in absolute gamma power and better executive functional skills [112].

Two studies combined markers of SES and perceived maternal stress with infant EEG markers. Pierce et al. found that during rest, maternal stress indirectly related to 2-month-old infants' gamma and beta oscillations, while maternal education seemed to additionally impact directly on infants' alpha, beta and gamma oscillations [113]. The findings indicate opposing as well as unique contributions of maternal stress and education on infants' neurodevelopment. A recent study by Troller-Renfree et al. indicated that spontaneous oscillations of infants living below the poverty threshold are slower when parents report increased stress levels [114].

These studies suggest a maturational lag reflected by a slower shift towards higher oscillatory frequency patterns at rest in high-risk children. Taken in accordance with our results, it assumes increased developmental risks may delay emergence of faster spontaneous oscillations and the infant N2 emerges concurrently with them (see also [102]).

Even fewer studies investigated task related brain activity in infants in relation to SES. In a three-stimulus auditory oddball, Katus et al. observed a shift from an intensity-based response at 1 month to a novelty-based response at 5 months of age, accompanied by increased habituation to stimulus intensity in a UK cohort. Conversely, a cohort from Gambia did not show this shift or significant changes in habituation. The degree of transition to a novelty-based response correlated with early language scores at 5 months, while growth between 1 and 5 months did not. This implies that infants who exhibit a greater shift towards novelty-based responses may have more advanced language skills at 5 months of age, in turn, indicating heightened risk of adverse neurodevelopmental outcome in infants of lower SES background [69]. Conejero et al. indicated diminished stimulus differentiation, indicated through theta oscillations, in infants with disadvantages in SES during an error detection task [42].

Importantly, brain maturation is an adaptive response to the specific environment in which the infant is raised to aid themselves in navigating and thriving within their social surroundings [32]. Adaptational processes may activate resources matching to everyday challenges in all children. Family support programs, however, may be important to prevent maturational risks and long-term disadvantages. Understanding the nuanced interplay between brain maturation and socioeconomic factors is essential for devising effective interventions to support children's development early on.

The pathways by which SES affects early brain development are multifaceted, given the diverse impact SES has on family dynamics. One potential mechanism is an increase in perceived stress, which could play a significant role in influencing developmental outcomes (see e.g., [114]). However, SES encompasses a broad spectrum of risk factors, each with its own challenges on family life and its own unique impact on infant development. The convergence of multiple risk factors likely results in complex interactions, amplifying the overall burden, as evidenced by our findings in the EDU1/MIG1 group. Longitudinal studies involving cohorts of diverse family background are essential for a more nuanced understanding of these risk factors and their specific effects. Our study, BRISE, offers initial insights into several of these factors, as outlined below.

4.3. Embedding Results into the BRISE Longitudinal Study

Risk factors in families participating in BRISE can be differentiated in families with a low- and a high-risk profile, with the latter having cumulating risks for different family based variables [115]. Cumulating risks seem related to early stages of a more difficult temperament in infants during the second half of their first year of life, i.e., concurrent to the presented EEG data. Future studies may better show how differing brain responses are related to the children's abilities and behaviour.

Future studies may further reveal if not only the accumulation of risk but certain combinations ("intersections") are of particular importance for a developing child [116]. We have already shown that education attainment and migration background may have varying impacts on a mother's behaviour during their pregnancy. For example, mothers with a low education level reported more often that they smoked while being pregnant, which we interpreted in relation to heightened pre-pregnancy burdens and fewer resources for coping strategies. On the other hand, an Islamic cultural background often prevented drinking during pregnancies [117].

An important factor in addition to the socioeconomic background of the child seems to be maternal behaviour. One of these factors is maternal self-efficacy, i.e., the confidence about one's own effectiveness as a mother. Lower self-efficacy is associated with the adoption of more distant soothing strategies, which in turn predicts reported regulatory problems of the children [118]. Additionally, maternal self-efficacy is positively associated with social support, mitigating adverse effects of low confidence. However, migration background is linked to reduced benefits from urban family support [119]. In addition, video-rated maternal sensitivity is directly related to SES, particularly maternal education, in our sample [120]. Taken together, these findings underscore how lower parental education and migration background can impede access to support and information and impact on parenting, which may consequently adversely affect infant development through a broad range of aspects.

4.4. Classification of Socioeconomic Backgrounds

Many developmental studies do not report the familial socioeconomic and cultural background of participating children, or that the recruitment primarily reaches families with an above-average SES [121,122]. Our inclusion criteria ensured inclusion of families with socioeconomic and/or cultural challenges. Heterogeneity within the sample still enabled the study of circumstantial effects during early upbringing on brain development. Data analysis thus provided important information from populations that are not-well studied and who may profit from suitable support regarding their everyday challenges.

Group effects were approximately mid-sized, which may concur with studies arguing that developmental disadvantages become more apparent over time [123]. This prospect is encouraging, as it suggests that developmental outcomes may still be influenced by compensatory strategies, such as support programs targeting families of low SES. Detrimental effects related to SES, however, may intensify and have an increasing effect on cognitive and social abilities with increasing age. This study is part of a longitudinal initiative which scientifically accompanies children and their families for approximately the first seven years of life. BRISE is also an interventional study, where some families attend early and longitudinal family support programs (see [68]). BRISE is prospectively planned out as a multi-method study to link together different aspects of development at multiple time-points of young children growing up in challenged families. This includes longitudinal effects of early occurring environmental risk factors, supporting resources received, and stages of brain maturation. In the future, analysis may elucidate the capacity of such an intervention chain to effectively mitigate the manifestation of SES disparities often observed at later stages of development (for further discussion see e.g., [123,124]).

The families participating in BRISE live in a mid-sized city in Germany, a welfare state in Western Europe with a worldwide above-average living standard. Despite the apparent familial challenges, infants in this study may have had on average more resources than infants raised in poorer countries. Germany supports families of low to no income through financial support for housing, living and education, as well as regular medical check-ups free of charge throughout pregnancy and for the child. Thus, it could be assumed that group differences would be pronounced when comparing populations with more extreme SES differences. This raises the question to what extent brain development can be impacted by even lower SES levels and/or less supportive environments for families to live in. Tarullo et al. recently demonstrated how in underprivileged countries, birth weight is related to early brain maturation [125]. This stresses the limitation of our understanding of children growing up in different societies as research remains scarce (see [126] for review). In the German social context, the findings here highlight an impact of parental education and migration on the infant's brain already within the first year of life.

4.5. Limitations and Directions for Future Studies

Neurophysiological studies with infants and challenged families are difficult and entail limitations as well as a need for future studies.

EEG data from infants often exhibits high levels of noise due to increased movement artefacts and variability of the brain response, which may partially stem from the immature structure of the developing brain [58,121,127]. This study investigated a rather large sample, data was high-pass filtered at 1 Hz and infants with less than 20 artefact-free epochs for each stimulus category were excluded to overcome these challenges. Rigorous filtering may, however, diminish the naturally occurring dominance of slow oscillations in infants. More research is needed to strike a balance between noise reduction and preservation of slow oscillations in typical infant brain responses to ensure optimal data quality.

Timing and topography of ERPs varies in infant studies and makes component classification and nomenclature difficult and manifold [44,62,64]. Due to fewer studies than other age groups, varying paradigms (stimuli, stimulus difference, length of the interstimulus interval, presentation rate, etc.) and analytical approaches pinpointing results to a specific ERP seems not entirely clear [56,58,62,64,65,70]. Neuronal networks are still immature, as several structural changes occur during development and individual differences in development might further obscure ERP investigation [87]. Hence, relating infant components to well-known adult components is likely not directly translatable. The components of the N2–P3a complex in infants may just be beginning to manifest, and potentially be related to distinct underlying processes, or specific components may be unique to infancy [87]. Further research is warranted to elucidate the developmental trajectories and functional significance of the neural responses to an oddball paradigm during early infancy, further development and relation to later cognitive abilities.

Socioeconomic challenges can be versatile and often lead to increased stress and decreased access to information, social institutions, and goods [20,128]. Low educational attainment and migration background, especially when migration ensues from countries of lower SES, as utilised in this study, relate to social exclusion and marginalisation [9,129]. The underlying processes linking challenges to detrimental effects on child development may be differential, and family income seems of particular relevance [10]. Thus, integrating information on different aspects of family life is important. To date, many of the developmental EEG studies investigate the aspect of poverty which can be defined in various ways and encompasses different SES factors, as discussed by D'Angiulli et al. and Farah et al. [17,130]. Some studies, such as Katus et al., compared cohorts from different SES backgrounds [69]. SES factors are often not completely independent. They relate to an individual's access to resources, opportunities, and social environments. The home learning environment and family stress seem to be the key aspects through which SES acts upon child development. Therefore, including short questionnaires to assess key aspects of SES in any EEG study would enhance comparability and provide more nuanced insights into the impact of SES on cognition. Understanding the interplay between different familial challenges and their relation to the home learning environment, emotional climate, and stress, is essential for elucidating the mechanisms underlying early brain development [121]. It is of further importance to link early changes in brain development to (later) cognitive abilities, e.g., language acquisition [18] and attention regulation as well as to temperament in infancy [115,131]. It has already been shown that infant N2–P3a-like responses may predict pre-school reading and language abilities [58,132].

In addition, future studies should reflect upon the inclusion barriers. Recruitment may not address all families to the same degree [121]. EEG measurement techniques are often particularly suitable for certain hair styles and hair structures, which often decreases equal representation of all ethnic and racial groups [133].

BRISE is aimed at identifying the potential of long-term interventions and support for child development. Future studies including later developmental stages need to consider the degree of participation in familial support programs. No formal analysis of these effects was conducted at this early stage, since the intervention has just begun or may have not been started.

5. Conclusions

The increased positive response for deviants as opposed to standards, even during the N2 time window of the infant N2–P3a complex, may be characteristic of the immature neural response of infants. This response, nevertheless, indicates neural reactivity towards stimulus changes within a passive oddball paradigm. This study further demonstrates differences in the maturing N2–P3a complex in relation to socioeconomic inequalities. The results indicate that the socioeconomic position, as marked by parental educational attainment and migration background, may influence brain development during infancy and that accumulation of challenges may have detrimental effects. The infant N2–P3a complex may reflect early attentional capacity and learning experiences, which emphasises the early emergence of developmental risks due to social disparities. Attention to unexpected stimulus changes likely plays a crucial role in early learning processes. Thus, family support programs should be implemented early, preceding possible manifestations of developmental disadvantages. The current research indicates that policies and early interventions need easy access to target diverse groups. They should aim at reducing sociocultural inequalities, especially if several factors coincide. The longitudinal nature of BRISE should prospectively allow the combined analysis of early risk factors and later consequences for cognitive development up to children's school age.

Author Contributions: A.S.W. ran EEG data analysis and wrote the first draft of the manuscript. A.S.W. and B.M. cooperated on all stages in defining the statistical analysis and writing the manuscript. All authors have read and agreed to the published version of the manuscript.

Funding: This manuscript is based on a project that is funded by the German Federal Ministry of Education and Research (grant no. 01NV2001A–I). The contents are the responsibility of the authors. Additional support was provided by the City of Bremen and the Jacobs Foundation.

Institutional Review Board Statement: The study was conducted in accordance with the Declaration of Helsinki, and approved by the Ethics Committee of the German Psychological Society and all legal guardians gave written consent (OK 092013_rev, 9 October 2013).

Informed Consent Statement: Informed consent was obtained from all families involved in the study.

Data Availability Statement: The data presented in this article are not readily available because of an ongoing study. Request to access the datasets should be directed to the corresponding author and will be made available upon request via a scientific repository in the future.

Acknowledgments: We are thankful to the BRISE Research Alliance for support, particularly to all undertaking the EEG measurements. We thank Robin Kemmerich for valuable discussions and insight on EEG data analysis and Foroogh Razavi for support implementing data analysis routines. We thank all families for participating.

Conflicts of Interest: The authors declare no conflicts of interest.

References

1. Berger, L.M.; Paxson, C.; Waldfogel, J. Income and Child Development. *Child. Youth Serv. Rev.* **2009**, *31*, 978–989. [CrossRef] [PubMed]
2. Coll, C.G.; Lamberty, G.; Jenkins, R.; McAdoo, H.P.; Crnic, K.; Wasik, B.H.; García, H.V. An Integrative Model for the Study of Developmental Competencies in Minority Children. *Child. Dev.* **1996**, *67*, 1891–1914. [CrossRef]
3. Letourneau, N.L.; Linda, D.L.; Leah, L.; Barry, W.; Catherine, Y.M. Socioeconomic Status and Child Development: A Meta-Analysis. *J. Emot. Behav. Disord.* **2013**, *21*, 211–224. [CrossRef]
4. McLoyd, V.C. Socioeconomic disadvantage and child development. *Am. Psychol.* **1998**, *53*, 185–204. [CrossRef]
5. Conger, R.D.; Conger, K.J.; Martin, M.J. Socioeconomic Status, Family Processes, and Individual Development. *J. Marriage Fam.* **2010**, *72*, 685–704. [CrossRef] [PubMed]
6. Bradley, R.H.; Corwyn, R.F. Socioeconomic status and child development. *Annu. Rev. Psychol.* **2002**, *53*, 371–399. [CrossRef] [PubMed]
7. Duncan, G.J.; Magnuson, K. Investing in Preschool Programs. *J. Econ. Perspect.* **2013**, *27*, 109–132. [CrossRef] [PubMed]
8. Conger, R.D.; Donnellan, M.B. An interactionist perspective on the socioeconomic context of human development. *Annu. Rev. Psychol.* **2007**, *58*, 175–199. [CrossRef] [PubMed]
9. Suárez-Orozco, C.; Motti-Stefanidi, F.; Marks, A.; Katsiaficas, D.A.-O. An integrative risk and resilience model for understanding the adaptation of immigrant-origin children and youth. *Am. Psychol.* **2018**, *73*, 781–796. [CrossRef]
10. Schneider, J.M.; Behboudi, M.H.; Maguire, M.J. The Necessity of Taking Culture and Context into Account When Studying the Relationship between Socioeconomic Status and Brain Development. *Brain Sci.* **2024**, *14*, 392. [CrossRef]
11. Sirin, S.R. Socioeconomic Status and Academic Achievement: A Meta-Analytic Review of Research. *Rev. Educ. Res.* **2005**, *75*, 417–453. [CrossRef]
12. Heineck, G.; Riphahn, R.T. Intergenerational Transmission of Educational Attainment in Germany—The Last Five Decades. *Jahrbücher Natl. Stat.* **2009**, *229*, 36–60. [CrossRef]
13. Yamamoto, Y.; Holloway, S.D. Parental Expectations and Children's Academic Performance in Sociocultural Context. *Educ. Psychol. Rev.* **2010**, *22*, 189–214. [CrossRef]
14. Berry, J.W. Acculturation: Living successfully in two cultures. *Int. J. Intercult. Relat.* **2005**, *29*, 697–712. [CrossRef]
15. Portes, A. *Legacies: The Story of the Immigrant Second Generation*; Rumbaut, R.G., Ed.; University of California Press: Berkeley, CA, USA, 2001.
16. Johnson, S.B.; Riis, J.L.; Noble, K.G. State of the Art Review: Poverty and the Developing Brain. *Pediatrics* **2016**, *137*, e20153075. [CrossRef] [PubMed]
17. Farah, M.J. The Neuroscience of Socioeconomic Status: Correlates, Causes, and Consequences. *Neuron* **2017**, *96*, 56–71. [CrossRef] [PubMed]
18. Perkins, S.C.; Finegood, E.D.; Swain, J.E. Poverty and language development: Roles of parenting and stress. *Innov. Clin. Neurosci.* **2013**, *10*, 10–19. [PubMed]
19. Hirsh-Pasek, K.; Adamson, L.B.; Bakeman, R.; Owen, M.T.; Golinkoff, R.M.; Pace, A.; Yust, P.K.; Suma, K. The Contribution of Early Communication Quality to Low-Income Children's Language Success. *Psychol. Sci.* **2015**, *26*, 1071–1083. [CrossRef] [PubMed]
20. Evans, G.W. Child Development and the Physical Environment. *Annu. Rev. Psychol.* **2006**, *57*, 423–451. [CrossRef]
21. Farah, M.J.; Shera, D.M.; Savage, J.H.; Betancourt, L.; Giannetta, J.M.; Brodsky, N.L.; Malmud, E.K.; Hurt, H. Childhood poverty: Specific associations with neurocognitive development. *Brain Res.* **2006**, *1110*, 166–174. [CrossRef]
22. Duncan, G.J.; Yeung, W.J.; Brooks-Gunn, J.; Smith, J.R. How Much Does Childhood Poverty Affect the Life Chances of Children? *Am. Sociol. Rev.* **1998**, *63*, 406–423. [CrossRef]
23. Noble, K.G.; McCandliss, B.D.; Farah, M.J. Socioeconomic gradients predict individual differences in neurocognitive abilities. *Dev. Sci.* **2007**, *10*, 464–480. [CrossRef] [PubMed]
24. Lawson, G.M.; Hook, C.J.; Farah, M.J. A meta-analysis of the relationship between socioeconomic status and executive function performance among children. *Dev. Sci.* **2018**, *21*, e12529. [CrossRef] [PubMed]
25. Stevens, C.; Bavelier, D. The role of selective attention on academic foundations: A cognitive neuroscience perspective. *Dev. Cogn. Neurosci.* **2012**, *2* (Suppl. S1), S30–S48. [CrossRef] [PubMed]
26. Halle, T.; Forry, N.D.; Hair, E.C.; Perper, K.; Wandner, L.D.; Wessel, J.S.; Vick, J. *Disparities in Early Learning and Development: Lessons from the Early Childhood Longitudinal Study—Birth Cohort (ECLS-B): (571822009-001)*; Child Trends: Washington, DC, USA, 2009.
27. Blossfeld, H.P.; Roßbach, H.G. *Education as a Lifelong Process: The German National Educational Panel Study (NEPS)*; Springer: Wiesbaden, Germany, 2019. [CrossRef]
28. Linberg, A.; Attig, M.; Weinert, S. Social disparities in the vocabulary of 2-year-old children and the mediating effect of language-stimulating interaction behavior. *J. Educ. Res. Online* **2020**, *12*, 12–35. [CrossRef] [PubMed]
29. Lupien, S.J.; King, S.; Meaney, M.J.; McEwen, B.S. Can poverty get under your skin? Basal cortisol levels and cognitive function in children from low and high socioeconomic status. *Dev. Psychopathol.* **2001**, *13*, 653–676. [CrossRef] [PubMed]

30. D'Angiulli, A.; Herdman, A.; Stapells, D.; Hertzman, C. Children's Event-Related Potentials of Auditory Selective Attention Vary With Their Socioeconomic Status. *Neuropsychology* **2008**, *22*, 293–300. [CrossRef] [PubMed]
31. Stevens, C.; Lauinger, B.; Neville, H. Differences in the neural mechanisms of selective attention in children from different socioeconomic backgrounds: An event-related brain potential study. *Dev. Sci.* **2009**, *12*, 634–646. [CrossRef]
32. Lipina, S.J.; Segretin, M.S. Strengths and weakness of neuroscientific investigations of childhood poverty: Future directions. *Front. Hum. Neurosci.* **2015**, *9*, 53. [CrossRef]
33. Ursache, A.; Noble, K.G.; Pediatric Imaging, N.; Genetics, S. Socioeconomic status, white matter, and executive function in children. *Brain Behav.* **2016**, *6*, e00531. [CrossRef]
34. Ursache, A.; Noble, K.G. Neurocognitive development in socioeconomic context: Multiple mechanisms and implications for measuring socioeconomic status. *Psychophysiology* **2016**, *53*, 71–82. [CrossRef]
35. Lawson, G.M.; Duda, J.T.; Avants, B.B.; Wu, J.; Farah, M.J. Associations between children's socioeconomic status and prefrontal cortical thickness. *Dev. Sci.* **2013**, *16*, 641–652. [CrossRef] [PubMed]
36. Pietto, M.L.; Kamienkowski, J.E.; Lipina, S.J. Electrophysiological Approaches in the Study of the Influence of Childhood Poverty on Cognition. In *Neuroscience and Social Science: The Missing Link*; Ibáñez, A., Sedeño, L., García, A.M., Eds.; Springer International Publishing: Cham, Switzerland, 2017; pp. 349–381. [CrossRef]
37. Noble, K.G.; Engelhardt, L.E.; Brito, N.H.; Mack, L.J.; Nail, E.J.; Angal, J.; Barr, R.; Fifer, W.P.; Elliott, A.J. Socioeconomic disparities in neurocognitive development in the first two years of life. *Dev. Psychobiol.* **2015**, *57*, 535–551. [CrossRef] [PubMed]
38. Hackman, D.A.; Gallop, R.; Evans, G.W.; Farah, M.J. Socioeconomic status and executive function: Developmental trajectories and mediation. *Dev. Sci.* **2015**, *18*, 686–702. [CrossRef] [PubMed]
39. Moyano, S.; Rico-Pico, J.; Conejero, A.; Hoyo, A.; Ballesteros-Duperon, M.L.A.; Rueda, M.R. Influence of the environment on the early development of attentional control. *Infant. Behav. Dev.* **2023**, *71*, 101842. [CrossRef] [PubMed]
40. Blair, C.; Granger, D.A.; Willoughby, M.; Mills-Koonce, R.; Cox, M.; Greenberg, M.T.; Kivlighan, K.T.; Fortunato, C.K.; Investigators, t.F. Salivary Cortisol Mediates Effects of Poverty and Parenting on Executive Functions in Early Childhood. *Child. Dev.* **2011**, *82*, 1970–1984. [CrossRef] [PubMed]
41. Hanson, J.L.; Hair, N.; Shen, D.G.; Shi, F.; Gilmore, J.H.; Wolfe, B.L.; Pollak, S.D. Family poverty affects the rate of human infant brain growth. *PLoS ONE* **2013**, *8*, e80954. [CrossRef] [PubMed]
42. Conejero, A.; Guerra, S.; Abundis-Gutiérrez, A.; Rueda, M.R. Frontal theta activation associated with error detection in toddlers: Influence of familial socioeconomic status. *Dev. Sci.* **2018**, *21*. [CrossRef] [PubMed]
43. Cantiani, C.; Ortiz-Mantilla, S.; Riva, V.; Piazza, C.; Bettoni, R.; Musacchia, G.; Molteni, M.; Marino, C.; Benasich, A.A. Reduced left-lateralized pattern of event-related EEG oscillations in infants at familial risk for language and learning impairment. *Neuroimage-Clin.* **2019**, *22*, 101778. [CrossRef] [PubMed]
44. Kushnerenko, E.V.; van den Bergh, B.R.H.; Winkler, I. Separating acoustic deviance from novelty during the first year of life: A review of event-related potential evidence. *Front. Psychol.* **2013**, *4*, 595. [CrossRef]
45. Kushnerenko, E.; Ceponiene, R.; Balan, P.; Fellman, V.; Näätänen, R. Maturation of the auditory change detection response in infants: A longitudinal ERP study. *Neuroreport* **2002**, *13*, 1843–1848. [CrossRef]
46. Mathes, B.; Kemmerich, R.; Wienke, A.S. Experimentelles Arbeiten mit Kindern aus verschiedenen (Sprach-)Kulturen und mit familiären Herausforderungen. *Frühe Bild.* **2024**, *13*, 116–118. [CrossRef]
47. Frischkorn, G.T.; Hilger, K.; Kretzschmar, A.; Schubert, A.-L. Intelligenzdiagnostik der Zukunft. *Psychol. Rundsch.* **2022**, *73*, 173–189. [CrossRef]
48. Bryant, G.A. Vocal communication across cultures: Theoretical and methodological issues. *Philos. Trans. R. Soc. B Biol. Sci.* **2022**, *377*, 20200387. [CrossRef]
49. Chandrasekaran, B.; Krishnan, A.; Fau-Gandour, J.T.; Gandour, J.T. Sensory processing of linguistic pitch as reflected by the mismatch negativity. *Ear Hear.* **2009**, *30*, 552–558. [CrossRef]
50. van Dinteren, R.; Arns, M.; Jongsma, M.L.A.; Kessels, R.P.C. P300 Development across the Lifespan: A Systematic Review and Meta-Analysis. *PLoS ONE* **2014**, *9*, e87347. [CrossRef]
51. Başar-Eroglu, C.; Schmiedt-Fehr, C.; Marbach, S.; Brand, A.; Mathes, B. Altered oscillatory alpha and theta networks in schizophrenia. *Brain Res.* **2008**, *1235*, 143–152. [CrossRef]
52. Justo-Guillén, E.; Ricardo-Garcell, J.; Rodríguez-Camacho, M.; Rodríguez-Agudelo, Y.; Lelo de Larrea-Mancera, E.S.; Solís-Vivanco, R. Auditory mismatch detection, distraction, and attentional reorientation (MMN-P3a-RON) in neurological and psychiatric disorders: A review. *Int. J. Psychophysiol.* **2019**, *146*, 85–100. [CrossRef] [PubMed]
53. Müller, V.; Jirsa, V.; Perdikis, D.; Sleimen-Malkoun, R.; von Oertzen, T.; Lindenberger, U. Lifespan Changes in Network Structure and Network Topology Dynamics During Rest and Auditory Oddball Performance. *Front. Aging Neurosci.* **2019**, *11*, 138. [CrossRef] [PubMed]
54. Basar, E.; Schmiedt-Fehr, C.; Mathes, B.; Femir, B.; Emek-Savas, D.D.; Tulay, E.; Tan, D.; Duzgun, A.; Güntekin, B.; Ozerdem, A.; et al. What does the broken brain say to the neuroscientist? Oscillations and connectivity in schizophrenia, Alzheimer's disease, and bipolar disorder. *Int. J. Psychophysiol.* **2016**, *103*, 135–148. [CrossRef]
55. Fellman, V.; Kushnerenko, E.; Mikkola, K.; Ceponiene, R.; Leipala, J.; Näätänen, R. Atypical auditory event-related potentials in preterm infants during the first year of life: A possible sign of cognitive dysfunction? *Pediatr. Res.* **2004**, *56*, 291–297. [CrossRef]

56. Kujala, T.; Tervaniemi, M.; Schröger, E. The mismatch negativity in cognitive and clinical neuroscience: Theoretical and methodological considerations. *Biol. Psychol.* **2007**, *74*, 1–19. [CrossRef] [PubMed]
57. Huotilainen, M.; Kujala, A.; Hotakainen, M.; Parkkonen, L.; Taulu, S.; Simola, J.; Nenonen, J.; Karjalainen, M.; Näätänen, R. Short-term memory functions of the human fetus recorded with magnetoencephalography. *Neuroreport* **2005**, *16*, 81–84. [CrossRef] [PubMed]
58. Piazza, C.; Cantiani, C.; Akalin-Acar, Z.; Miyakoshi, M.; Benasich, A.A.; Reni, G.; Bianchi, A.M.; Makeig, S. ICA-derived cortical responses indexing rapid multi-feature auditory processing in six-month-old infants. *Neuroimage* **2016**, *133*, 75–87. [CrossRef] [PubMed]
59. Tomé, D.; Barbosa, F.; Nowak, K.; Marques-Teixeira, J. The development of the N1 and N2 components in auditory oddball paradigms: A systematic review with narrative analysis and suggested normative values. *J. Neural. Trans.* **2015**, *122*, 375–391. [CrossRef] [PubMed]
60. Näätänen, R.; Kujala, T.; Winkler, I. Auditory processing that leads to conscious perception: A unique window to central auditory processing opened by the mismatch negativity and related responses. *Psychophysiology* **2011**, *48*, 4–22. [CrossRef] [PubMed]
61. Polich, J. Updating P300: An integrative theory of P3a and P3b. *Clin. Neurophysiol.* **2007**, *118*, 2128–2148. [CrossRef] [PubMed]
62. Folstein, J.R.; van Petten, C. Influence of cognitive control and mismatch on the N2 component of the ERP: A review. *Psychophysiology* **2008**, *45*, 152–170. [CrossRef] [PubMed]
63. Choudhury, N.; Benasich, A.A. Maturation of auditory evoked potentials from 6 to 48 months: Prediction to 3 and 4 year language and cognitive abilities. *Clin. Neurophysiol.* **2011**, *122*, 320–338. [CrossRef] [PubMed]
64. Wetzel, N.; Schröger, E. On the development of auditory distraction: A review. *PsyCh J.* **2014**, *3*, 72–91. [CrossRef]
65. Hämäläinen, J.A.; Ortiz-Mantilla, S.; Benasich, A.A. Source localization of event-related potentials to pitch change mapped onto age-appropriate MRIs at 6 months of age. *Neuroimage* **2011**, *54*, 1910–1918. [CrossRef]
66. He, C.; Hotson, L.; Trainor, L.J. Mismatch responses to pitch changes in early infancy. *J. Cogn. Neurosci.* **2007**, *19*, 878–892. [CrossRef] [PubMed]
67. Hackman, D.A.; Farah, M.J. Socioeconomic status and the developing brain. *Trends Cogn. Sci.* **2009**, *13*, 65–73. [CrossRef] [PubMed]
68. Schütte, K.; Hasselhorn, M.; Köller, O. Bildungsungleichheit besser verstehen und vermindern. *Frühe Bild.* **2024**, *13*, 65–70. [CrossRef]
69. Katus, L.; Mason, L.; Milosavljevic, B.; McCann, S.; Rozhko, M.; Moore, S.E.; Elwell, C.E.; Lloyd-Fox, S.; de Haan, M.; Drammeh, S.; et al. ERP markers are associated with neurodevelopmental outcomes in 1–5 month old infants in rural Africa and the UK. *NeuroImage* **2020**, *210*, 116591. [CrossRef] [PubMed]
70. He, C.; Hotson, L.; Trainor, L.J. Maturation of cortical mismatch responses to occasional pitch change in early infancy: Effects of presentation rate and magnitude of change. *Neuropsychologia* **2009**, *47*, 218–229. [CrossRef] [PubMed]
71. Marie, C.; Trainor, L.J. Development of Simultaneous Pitch Encoding: Infants Show a High Voice Superiority Effect. *Cereb. Cortex* **2013**, *23*, 660–669. [CrossRef] [PubMed]
72. Fujioka, T.; Mourad, N.; Trainor, L.J. Development of auditory-specific brain rhythm in infants. *Eur. J. Neurosci.* **2011**, *33*, 521–529. [CrossRef] [PubMed]
73. Chatrian, G.E.; Lettich, E.; Nelson, P.L. Modified nomenclature for the "10%" electrode system. *J. Clin. Neurophysiol.* **1988**, *5*, 183–186. [CrossRef]
74. Delorme, A.; Makeig, S. EEGLAB: An open source toolbox for analysis of single-trial EEG dynamics including independent component analysis. *J. Neurosci. Methods* **2004**, *134*, 9–21. [CrossRef]
75. McIsaac, H.; Polich, J. Comparison of infant and adult P300 from auditory stimuli. *J. Exp. Child. Psychol.* **1992**, *53*, 115–128. [CrossRef]
76. Posner, M.I.; Rothbart, M.K. Research on attention networks as a model for the integration of psychological science. *Annu. Rev. Psychol.* **2007**, *58*, 1–23. [CrossRef] [PubMed]
77. Marie, C.; Trainor, L.J. Early development of polyphonic sound encoding and the high voice superiority effect. *Neuropsychologia* **2014**, *57*, 50–58. [CrossRef] [PubMed]
78. Butler, B.E.; Trainor, L.J. Brief pitch-priming facilitates infants' discrimination of pitch-evoking noise: Evidence from event-related potentials. *Brain Cogn.* **2013**, *83*, 271–278. [CrossRef] [PubMed]
79. Ceponiene, R.; Rinne, T.; Näätänen, R. Maturation of cortical sound processing as indexed by event-related potentials. *Clin. Neurophysiol.* **2002**, *113*, 870–882. [CrossRef]
80. Cheour, M.; Shestakova, A.; Alku, P.; Ceponiene, R.; Näätänen, R. Mismatch negativity shows that 3-6-year-old children can learn to discriminate non-native speech sounds within two months. *Neurosci. Lett.* **2002**, *325*, 187–190. [CrossRef] [PubMed]
81. Morr, M.L.; Shafer, V.L.; Kreuzer, J.A.; Kurtzberg, D. Maturation of Mismatch Negativity in Typically Developing Infants and Preschool Children. *Ear Hear.* **2002**, *23*, 118–136. [CrossRef] [PubMed]
82. Alho, K.; Sainio, K.; Sajaniemi, N.; Reinikainen, K.; Näätänen, R. Event-related brain potential of human newborns to pitch change of an acoustic stimulus. *Electroencephalogr. Clin. Neurophysiol.* **1990**, *77*, 151–155. [CrossRef] [PubMed]
83. Trainor, L.J.; Lee, K.; Bosnyak, D.J. Cortical Plasticity in 4-Month-Old Infants: Specific Effects of Experience with Musical Timbres. *Brain Topogr.* **2011**, *24*, 192–203. [CrossRef] [PubMed]

84. Leppänen, P.H.; Lyytinen, H. Auditory event-related potentials in the study of developmental language-related disorders. *Audiol. Neurootol.* **1997**, *2*, 308–340. [CrossRef]
85. Benasich, A.A.; Choudhury, N.A.; Realpe-Bonilla, T.; Roesler, C.P. Plasticity in developing brain: Active auditory exposure impacts prelinguistic acoustic mapping. *J. Neurosci.* **2014**, *34*, 13349–13363. [CrossRef]
86. Sambeth, A.; Pakarinen, S.; Ruohio, K.; Fellman, V.; van Zuijen, T.L.; Huotilainen, M. Change detection in newborns using a multiple deviant paradigm: A study using magnetoencephalography. *Clin. Neurophysiol.* **2009**, *120*, 530–538. [CrossRef] [PubMed]
87. Kushnerenko, E.; Ceponiene, R.; Balan, P.; Fellman, V.; Huotilaine, M.; Näätänen, R. Maturation of the auditory event-related potentials during the first year of life. *Neuroreport* **2002**, *13*, 47–51. [CrossRef] [PubMed]
88. Demiralp, T.; Ademoglu, A.; Istefanopulos, Y.; Başar-Eroglu, C.; Başar, E. Wavelet analysis of oddball P300. *Int. J. Psychophysiol.* **2001**, *39*, 221–227. [CrossRef] [PubMed]
89. Başar, E.; Başar-Eroglu, C.; Karakas, S.; Schürmann, M. Gamma, alpha, delta, and theta oscillations govern cognitive processes. *Int. J. Psychophysiol.* **2001**, *39*, 241–248. [CrossRef]
90. Başar, E.; Schürmann, M.; Demiralp, T.; Başar-Eroglu, C.; Ademoglu, A. Event-related oscillations are 'real brain responses'—Wavelet analysis and new strategies. *Int. J. Psychophysiol.* **2001**, *39*, 91–127. [CrossRef] [PubMed]
91. Basar, E. The theory of the whole-brain-work. *Int. J. Psychophysiol.* **2006**, *60*, 133–138. [CrossRef]
92. Braver, T.S.; Barch, D.M.; Gray, J.R.; Molfese, D.L.; Snyder, A. Anterior cingulate cortex and response conflict: Effects of frequency, inhibition and errors. *Cereb. Cortex* **2001**, *11*, 825–836. [CrossRef] [PubMed]
93. Coderre, E.; Conklin, K.; van Heuven, W.J. Electrophysiological measures of conflict detection and resolution in the Stroop task. *Brain Res.* **2011**, *1413*, 51–59. [CrossRef] [PubMed]
94. Niendam, T.A.; Laird, A.R.; Ray, K.L.; Dean, Y.M.; Glahn, D.C.; Carter, C.S. Meta-analytic evidence for a superordinate cognitive control network subserving diverse executive functions. *Cogn. Affect. Behav. Neurosci.* **2012**, *12*, 241–268. [CrossRef]
95. Isler, J.R.; Grieve, P.G.; Czernochowski, D.; Stark, R.I.; Friedman, D. Cross-frequency phase coupling of brain rhythms during the orienting response. *Brain Res.* **2008**, *1232*, 163–172. [CrossRef]
96. Prada, L.; Barceló, F.; Herrmann, C.S.; Escera, C. EEG delta oscillations index inhibitory control of contextual novelty to both irrelevant distracters and relevant task-switch cues. *Psychophysiology* **2014**, *51*, 658–672. [CrossRef] [PubMed]
97. Kolev, V.; Demiralp, T.; Yordanova, J.; Ademoglu, A.; Isoglu-Alkac, U. Time-frequency analysis reveals multiple functional components during oddball P300. *Neuroreport* **1997**, *8*, 2061–2065. [CrossRef] [PubMed]
98. Marshall, P.J.; Bar-Haim, Y.; Fox, N.A. Development of the EEG from 5 months to 4 years of age. *Clin. Neurophysiol.* **2002**, *113*, 1199–1208. [CrossRef] [PubMed]
99. Rico-Pico, J.; Moyano, S.; Conejero, A.; Hoyo, A.; Ballesteros-Duperon, M.A.; Rueda, M.R. Early development of electrophysiological activity: Contribution of periodic and aperiodic components of the EEG signal. *Psychophysiology* **2023**, *60*, e14360. [CrossRef] [PubMed]
100. Saby, J.N.; Marshall, P.J. The utility of EEG band power analysis in the study of infancy and early childhood. *Dev. Neuropsychol.* **2012**, *37*, 253–273. [CrossRef] [PubMed]
101. Kolev, V.; Başar-Eroglu, C.; Aksu, F.; Başar, E. EEG rhythmicities evoked by visual stimuli in three-year-old children. *Int. J. Neurosci.* **1994**, *75*, 257–270. [CrossRef] [PubMed]
102. Başar-Eroglu, C.; Kolev, V.; Ritter, B.; Aksu, F.; Başar, E. EEG, auditory evoked potentials and evoked rhythmicities in three-year-old children. *Int. J. Neurosci.* **1994**, *75*, 239–255. [CrossRef]
103. Mathes, B.; Khalaidovski, K.; Wienke, A.S.; Schmiedt-Fehr, C.; Başar-Eroglu, C. Maturation of the P3 and concurrent oscillatory processes during early and late adolescence. *Clin. Neurophysiol.* **2016**, *127*, 2599–2609. [CrossRef] [PubMed]
104. Wienke, A.S.; Başar-Eroglu, C.; Schmiedt-Fehr, C.; Mathes, B. Novelty N2-P3a Complex and Theta Oscillations Reflect Improving Neural Coordination Within Frontal Brain Networks During Adolescence. *Front. Behav. Neurosci.* **2018**, *12*. [CrossRef]
105. Giovanis, E.; Akdede, S. Cultural Integration of First-Generation Immigrants: Evidence from European Union Countries. *Review of Economic Analysis* **2023**, *15*, 97–125. [CrossRef]
106. Hauge, L.J.; Torgersen, L.; Vollrath, M. Associations between maternal stress and smoking: Findings from a population-based prospective cohort study. *Addiction* **2012**, *107*, 1168–1173. [CrossRef] [PubMed]
107. Luby, J.; Belden, A.; Botteron, K.; Marrus, N.; Harms, M.P.; Babb, C.; Nishino, T.; Barch, D. The Effects of Poverty on Childhood Brain Development: The Mediating Effect of Caregiving and Stressful Life Events. *JAMA Pediatr.* **2013**, *167*, 1135–1142. [CrossRef] [PubMed]
108. Farah, M.J. Socioeconomic status and the brain: Prospects for neuroscience-informed policy. *Nat. Rev. Neurosci.* **2018**, *19*, 428–438. [CrossRef] [PubMed]
109. Brito, N.H.; Fifer, W.P.; Myers, M.M.; Elliott, A.J.; Noble, K.G. Associations among family socioeconomic status, EEG power at birth, and cognitive skills during infancy. *Dev. Cogn. Neurosci.* **2016**, *19*, 144–151. [CrossRef] [PubMed]
110. Tomalski, P.; Moore, D.G.; Ribeiro, H.; Axelsson, E.L.; Murphy, E.; Karmiloff-Smith, A.; Johnson, M.H.; Kushnerenko, E. Socioeconomic status and functional brain development—associations in early infancy. *Dev. Sci.* **2013**, *16*, 676–687. [CrossRef]
111. Otero, G.A. EEG spectral analysis in children with sociocultural handicaps. *Int. J. Neurosci.* **1994**, *79*, 213–220. [CrossRef] [PubMed]
112. Tarullo, A.R.; Obradovic, J.; Keehn, B.; Rasheed, M.A.; Siyal, S.; Nelson, C.A.; Yousafzai, A.K. Gamma power in rural Pakistani children: Links to executive function and verbal ability. *Dev. Cogn. Neurosci.* **2017**, *26*, 1–8. [CrossRef]

113. Pierce, L.J.; Thompson, B.L.; Gharib, A.; Schlueter, L.; Reilly, E.; Valdes, V.; Roberts, S.; Conroy, K.; Levitt, P.; Nelson, C.A. Association of Perceived Maternal Stress During the Perinatal Period With Electroencephalography Patterns in 2-Month-Old Infants. *Jama Pediatr.* **2019**, *173*, 561–570. [CrossRef]
114. Troller-Renfree, S.V.; Sperber, J.F.; Hart, E.R.; Costanzo, M.A.; Gennetian, L.A.; Meyer, J.S.; Fox, N.A.; Noble, K.G. Associations between maternal stress and infant resting brain activity among families residing in poverty in the U.S. *Biol. Psychol.* **2023**, *184*, 108683. [CrossRef]
115. Möwisch, D.; Wienke, A.S.; Meier-Faust, E.; Mathes, B.; Attig, M. Cumulative socioeconomic risk factors and child temperament. *Z. Für Erzieh.* **2024**. *under review*.
116. Pedrós Barnils, N.; Eurenius, E.; Gustafsson, P.A.-O. Self-rated health inequalities in the intersection of gender, social class and regional development in Spain: Exploring contributions of material and psychosocial factors. *Int. J. Equity Health* **2020**, *19*, 85. [CrossRef]
117. Römer, P.; Kemmerich, R.; Petermann, F.; Mathes, B.; Zierul, C. Alcohol and Nicotine Consumption during Pregnancy. *SUCHT* **2023**, *69*, 99–111. [CrossRef]
118. Kiel, N.; Samdan, G.; Wienke, A.S.; Reinelt, T.; Pauen, S.; Mathes, B.; Herzmann, C. From co-regulation to self-regulation: Maternal soothing strategies and self-efficacy in relation to maternal reports of infant regulation at 3 and 7 months. *Infant. Ment. Health J.* **2024**, *45*, 135–152. [CrossRef] [PubMed]
119. Samdan, G.; Reinelt, T.; Kiel, N.; Mathes, B.; Pauen, S. Maternal self-efficacy development from pregnancy to 3 months after birth. *Infant. Ment. Health J.* **2022**, *43*, 864–877. [CrossRef] [PubMed]
120. Möwisch, D.; Attig, M.; Weinert, S. Einflussfaktoren auf die frühe Mutter-Kind-Interaktion. *Frühe Bild.* **2024**, *13*, 84–92. [CrossRef]
121. Norton, E.S.; MacNeill, L.A.; Harriott, E.M.; Allen, N.; Krogh-Jespersen, S.; Smyser, C.D.; Rogers, C.E.; Smyser, T.A.; Luby, J.; Wakschlag, L. EEG/ERP as a pragmatic method to expand the reach of infant-toddler neuroimaging in HBCD: Promises and challenges. *Dev. Cogn. Neurosci.* **2021**, *51*, 100988. [CrossRef] [PubMed]
122. Dotson, V.M.; Duarte, A. The importance of diversity in cognitive neuroscience. *Ann. N. Y. Acad. Sci.* **2020**, *1464*, 181–191. [CrossRef] [PubMed]
123. Blair, C.; Raver, C.C. Poverty, Stress, and Brain Development: New Directions for Prevention and Intervention. *Acad. Pediatr.* **2016**, *16*, S30–S36. [CrossRef] [PubMed]
124. Hurt, H.; Betancourt, L.M. Effect of socioeconomic status disparity on child language and neural outcome: How early is early? *Pediatr. Res.* **2016**, *79*, 148–158. [CrossRef] [PubMed]
125. Tarullo, A.R.; Leppänen, J.M.; Evans, D.; Coetzee, L.; Lopera-Perez, D.C.; Brady, S.P.; Gabard-Durnam, L.J.; Fink, G.; Hamer, D.H.; Yousafzai, A.K.; et al. Neonatal Physical Growth Predicts Electroencephalography Power in Rural South African Children. *Brain Sci.* **2024**, *14*, 552. [CrossRef]
126. Draper, C.E.; Barnett, L.M.; Cook, C.J.; Cuartas, J.A.; Howard, S.J.; McCoy, D.C.; Merkley, R.; Molano, A.; Maldonado-Carreño, C.; Obradović, J.; et al. Publishing child development research from around the world: An unfair playing field resulting in most of the world's child population under-represented in research. *Infant. Child. Dev.* **2022**, *32*, e2375. [CrossRef]
127. Bell, M.A.; Cuevas, K. Using EEG to Study Cognitive Development: Issues and Practices. *J. Cogn. Dev.* **2012**, *13*, 281–294. [CrossRef]
128. Evans, G.W.; Kim, P. Childhood Poverty, Chronic Stress, Self-Regulation, and Coping. *Child. Dev. Perspect.* **2013**, *7*, 43–48. [CrossRef]
129. González, L.; Cortés-Sancho, R.; Murcia, M.; Ballester, F.; Rebagliato, M.; Rodríguez-Bernal, C.L. The role of parental social class, education and unemployment on child cognitive development. *Gac. Sanit.* **2020**, *34*, 51–60. [CrossRef]
130. D'Angiulli, A.; Lipina, S.J.; Olesinska, A. Explicit and implicit issues in the developmental cognitive neuroscience of social inequality. *Front. Human. Neurosci.* **2012**, *6*, 00254. [CrossRef] [PubMed]
131. Conejero, A.; Rueda, M.R. Infant temperament and family socio-economic status in relation to the emergence of attention regulation. *Sci. Rep.* **2018**, *8*, 11232. [CrossRef] [PubMed]
132. van Zuijen, T.L.; Plakas, A.; Maassen BA, M.; Been, P.; Maurits, N.M.; Krikhaar, E.; van Driel, J.; van der Leij, A. Temporal auditory processing at 17 months of age is associated with preliterate language comprehension and later word reading fluency: An ERP study. *Neurosci. Lett.* **2012**, *528*, 31–35. [CrossRef]
133. Adams, E.J.; Scott, M.E.; Amarante, M.; Ramirez, C.A.; Rowley, S.J.; Noble, K.G.; Troller-Renfree, S.V. Fostering inclusion in EEG measures of pediatric brain activity. *NPJ Sci. Learn.* **2024**, *9*, 27. [CrossRef]

brain sciences

MDPI

Article

Age-Related Aspects of Sex Differences in Event-Related Brain Oscillatory Responses: A Turkish Study

Görsev Yener [1,2,3,*], İlayda Kıyı [3], Seren Düzenli-Öztürk [4] and Deniz Yerlikaya [3]

1 Faculty of Medicine, Izmir University of Economics, 35330 Balçova, Turkey
2 Izmir Biomedicine and Genome Center, 35340 İzmir, Turkey
3 Department of Neurosciences, Health Sciences Institute, Dokuz Eylül University, 35210 İzmir, Turkey; ilaydaky@gmail.com (İ.K.); psk.denizyerlikaya@gmail.com (D.Y.)
4 Department of Speech and Language Therapy, Faculty of Health Sciences, Izmir Bakırçay University, 35665 İzmir, Turkey; seren.duzenliozturk@bakircay.edu.tr
* Correspondence: gorsev.yener@ieu.edu.tr

Abstract: Earlier research has suggested gender differences in event-related potentials/oscillations (ERPs/EROs). Yet, the alteration in event-related oscillations (EROs) in the delta and theta frequency bands have not been explored between genders across the three age groups of adulthood, i.e., 18–50, 51–65, and >65 years. Data from 155 healthy elderly participants who underwent a neurological examination, comprehensive neuropsychological assessment (including attention, memory, executive function, language, and visuospatial skills), and magnetic resonance imaging (MRI) from past studies were used. The delta and theta ERO powers across the age groups and between genders were compared and correlational analyses among the ERO power, age, and neuropsychological tests were performed. The results indicated that females displayed higher theta ERO responses than males in the frontal, central, and parietal regions but not in the occipital location between 18 and 50 years of adulthood. The declining theta power of EROs in women reached that of men after the age of 50 while the theta ERO power was more stable across the age groups in men. Our results imply that the cohorts must be recruited at specified age ranges across genders, and clinical trials using neurophysiological biomarkers as an intervention endpoint should take gender into account in the future.

Keywords: oscillations; gender; sex; aging; EEG; event-related; task-related; ERP; P300; oddball

1. Introduction

Gender differences have profound societal implications as they influence brain development, behavior, and the presentation, prevalence, and treatment of diseases. However, women have historically been underrepresented in medical research [1]. This lack of representation can negatively impact women's health outcomes. For example, from 1999 to 2000, 8 out of 10 drugs withdrawn from the market by the FDA posed higher risks for women than men [2]. More research is needed to better understand the neurobiology, behaviors, and disease vulnerabilities that may differ between sexes.

Research shows gender differences in the patterns of brain activity during memory retrieval tasks. Males tend to show greater activity than females in the prefrontal cortex, visual processing regions, parahippocampal cortex, and cerebellum [2,3], and the connectivity in brain networks associated with executive function and memory also differs between sexes and changes with age [4].

Overall, it is claimed that the brain metabolism in females is different from that in males throughout adulthood, which could impact vulnerability or resilience to neurodegenerative diseases [5]. The risk of developing AD is about twice as high for women compared to men globally [6]. However, the incidence rates appear to vary between low-to-middle-income countries, possibly due to several factors like survival rates between sexes, education levels, and genetic or hormonal influences [7]. Furthermore, increased female vulnerability to

Alzheimer's disease (AD) biomarkers in cerebrospinal fluid linked to greater hippocampal atrophy and faster age-related cognitive decline have been reported previously [8,9].

Neurophysiological signatures, measured with electroencephalography (EEG) or magnetoencephalography (MEG) techniques, have been stated as a low invasive and useful approach to investigating the progressive loss of neuronal activity in many cognitive impairments, including the Alzheimer's disease (AD) continuum [10,11]. The early pioneering work of Başar et al. on task-related potentials has shown that event-related oscillations (EROs) provide a powerful technique, with a high temporal resolution, and can be used as a tool for detecting subtle abnormalities in cognitive processes [12,13]. While ERPs can be characterized by the amplitudes and latencies of the main wave components in the time domain, these wave components represent summed-up time-varying neural activation patterns of various neural circuits [12]. Thus, a greater understanding of the simultaneous involvement of several brain networks in brain activities has resulted from the analysis of the oscillatory dynamics of the ERP signal, also known as event-related oscillations (EROs) [13]. Delta and theta ERO responses are specifically important in the cognitive process and they are the main constituents of the ERP [14–16]. Gender is an important factor in event-related potentials/oscillations (ERPs/EROs), with higher amplitudes in females [17,18]. However, there have been no ERO studies related to gender effects in healthy populations across young, middle-aged, and older adults from low–middle-income countries such as Turkey in the previous literature.

In this vein, the purpose of this study was to assess and explore whether the amplitudes of event-related oscillations (EROs) in the delta and theta frequency bands change across age groups according to gender differences. As an outcome, it would be important to consider gender across various healthy older adult age groups over 50 years in a cohort of healthy individuals in Turkey.

2. Materials and Methods

2.1. Participants

For this study, the EEG data recorded from cognitively healthy participants at the Department of Neuroscience at the Health Sciences Institute of Dokuz Eylül University were scanned retrospectively. As a result of reviewing the database, data from 155 healthy participants who underwent a neurological examination, comprehensive neuropsychological assessment (including attention, memory, executive function, language, and visuospatial skills), and magnetic resonance imaging (MRI) in past studies were used.

The following standards were used to classify the participants as healthy: (1) no history of neurological or psychiatric disorders; (2) no significant vascular lesion load on their MRI as reviewed by an expert neurologist; (3) no self-reported cognitive complaints; and (4) neuropsychological test scores within the range of age–education–gender-adjusted norms.

We had six groups in this study: [3 (age: young, middle-aged, elderly) × 2 (gender: female and male)]. The demographic and clinical characteristics, neuropsychological profiles, and age characteristics of groups are reported in Table 1.

The following were used as as exclusion criteria: (1) depression scale scores above cut off (the Beck Depression Scale (BDS) for participants younger than 55 years and the Geriatric Depression Scale (GDS) for those older than 56 years), (2) neurological and psychiatric disease history, (3) usage of medication which may affect the cognitive skills, (4) vision problems which may affect the performance in the task during the EEG recording, (5) vascular lesions and/or atrophy on brain MRI, (6) alcohol and/or drug misuse, (7) history of head trauma, and (8) the mental counting performance with more than 10% error rate during the EEG recording. Consent forms were taken from all participants according to the Declaration of Helsinki.

Table 1. Demographic and clinical characteristics of participants.

			Age Groups			
	Total Sample	**Gender**	**19–50 (n = 45)**	**51–65 (n = 48)**	**66–86 (n = 62)**	***p* Values**
Age (yr) *	55.01 ± 19.41	Female Male	24.80 ± 6.85 29.32 ± 9.46	59.28 ± 3.88 60.16 ± 4.18	71.32 ± 5.00 71.84 ± 5.07	0.197
Gender ‡	155 (86F/69M)	Female Male	20 25	29 19	37 25	0.209
Education (yr) *	12.75 ± 4.37	Female Male	14.90 ± 1.71 14.79 ± 1.50	12.45 ± 3.98 12.63 ± 3.13	9.65 ± 5.11 14.08 ± 5.26	0.005
Handedness ‡	149 R/4L/2B	Female Male	20R 24 R/1L	29R 17R/2B	36R/1L 23R/2L	0.126
Epoch Number *	28.75 ± 7.32	Female Male	26.10 ± 8.22 28.16 ± 6.25	27.86 ± 6.88 32.63 ± 7.38	28.14 ± 7.03 30.48 ± 7.60	0.604
Behavioral Data *	39.69 ± 2.20	Female Male	40.40 ± 1.50 39.24 ± 2.50	39.48 ± 2.03 39.79 ± 1.58	39.49 ± 2.91 40.04 ± 1.70	0.124
MMSE *	29.22 ± 1.04	Female Male	29.75 ± 0.55 29.70 ± 0.70	28.93 ± 1.28 29.26 ± 0.99	28.86 ± 1.12 29.16 ± 0.94	0.592
Depression ¥	4.89 ± 3.69 5.54 ± 4.51	Female Male	4.89 ± 3.70 5.58 ± 4.47	9.00 ± 6.23 4.72 ± 4.40	7.36 ± 5.00 4.20 ± 3.60	NA
OVMPT Total *	121.80 ± 12.88 120.33 ± 14.79	Female Male	128.33 ± 9.24 130.25 ± 9.31	122.38 ± 10.19 120.64 ± 13.62	118.47 ± 12.45 113.00 ± 14.70	0.467
OVMPT IR *	6.23 ± 1.97 5.99 ± 1.92	Female Male	7.42 ± 1.83 7.42 ± 1.83	6.42 ± 1.58 6.29 ± 2.02	5.25 ± 1.77 5.25 ± 1.77	0.310
OVMPT FR *	13.62 ± 1.20 13.22 ± 1.43	Female Male	14.00 ± 1.10 13.75 ± 1.06	13.65 ± 1.20 13.21 ± 1.31	13.30 ± 1.29 12.81 ± 133	0.456
OVMPT TR *	14.99 ± 0.11 15.00 ± 0.00	Female Male	15 15	15 15	15 15	0.995
Stroop *	48.09 ± 19.84 43.12 ± 18.65	Female Male	31.17 ± 5.85 40.50 ± 16.31	45.46 ± 14.80 40.57 ± 12.33	53.00 ± 20.49 49.63 ± 26.46	0.326
Categorical Fluency *	22.81 ± 4.95 24.27 ± 4.71	Female Male	25.00 ± 5.02 26.42 ± 5.14	23.69 ± 5.36 23.93 ± 4.73	22.48 ± 4.67 24.63 ± 5.18	0.419
Phonemic Fluency *	44.46 ± 14.09 44.80 ± 12.79	Female Male	61.17 ± 9.47 41.50 ± 11.94	40.42 ± 11.38 48.50 ± 16.18	38.30 ± 12.04 41.00 ± 10.68	0.007
BNT *	14.79 ± 0.47 14.94 ± 0.31	Female Male	15 15	14.96 ± 0.20 15	14.78 ± 0.42 14.81 ± 0.54	0.716

MMSE: The Mini Mental State Test, OVMPT: Öktem's Verbal Memory Performance Test, IR: Immediate Recall, FR: Free Recall, TR: Total Recognition, BNT: Boston Naming Test. ¥ Depression Scores are according to Beck Depression Scale for young group and to Yesavage's Geriatric Depression Scale for middle-aged and elderly group. * Univariate ANOVA; ‡ chi-squared test; NA: not applicable as scores were elicited by different tests.

2.2. Experimental Paradigm

In this study, EEG data were recorded during a visual oddball paradigm. All recordings with the paradigm were performed in an electrically isolated room during the morning hours. The total number of stimuli was 120 (40 target/80 nontarget). Visual stimuli were presented as light (for the target stimulus: 40 cd/m^2; for the nontarget stimulus: 10 cd/m^2) from a 22″ computer screen 120 cm in front of the participants with a refresh rate of 60 Hz in pseudorandom order. The paradigm consisted of 4 blocks and the inter-stimulus interval was varied by 3–7 s randomly. Participants were asked to do mental counting for the target stimulus. The participants were excluded with more than the 10% error rate (allowed range: 36–44).

2.3. EEG Recording and Data Processing

The EEG was recorded using 30 Ag/AgCl electrodes positioned on a standardized elastic cap according to the international 10–20 system (EasyCap; Brain Products GmbH; Gilching, Germany). The bilateral and linked earlobe electrodes (A1 + A2) were chosen as reference electrodes, and the electrooculogram (EOG) was recorded from the medial upper and lateral orbital rims of the right eye. All electrode impedances were kept below 10 kΩ. The EEG was amplified using a Brain Amp 32-channel DC amplifier with a 0.3–70 Hz band-pass filter and was digitized online with a sampling rate of 500 Hz.

Offline data preprocessing and analysis were performed using Brain Vision Analyzer 2.2 Software (Brain Products GmbH; Gilching, Germany). The 0.1 Hz high-pass filter with a zero-phase shift Butterworth filter and the 50 Hz Notch filter were applied to raw EEG data. Extended Infomax Independent Component Analysis (ICA) was used to correct horizontal and vertical eye movement artifacts. The target trials were segmented into 1500 ms epochs to include 500 ms pre-stimulus and 1000 ms post-stimulus activity. Automatic artifact rejection processing was performed with the following criteria: (a) maximum amplitude in an epoch: ±70 µV, (b) maximum allowed voltage step: 50 µV/ms, (c) maximum allowed difference in a 200 ms interval: 50 µV, (d) lowest activity in a 100 ms interval: 0.5 µV. The artifact-free epochs were averaged as time-locked to the stimulus onset for each participant. High- and low-pass filters with 8 dB/octave slope were applied to the averaged data to obtain the delta (0.5–3.5 Hz) and theta (4–7 Hz) oscillations. Frontal (F3, Fz, F4), Central (C3, Cz, C4), Parietal (P3, Pz, P4), and Occipital (O1, Oz, O2) electrodes were pooled. For each participant, the peak-to-peak activities of delta and theta oscillations were measured from the time range 0–800 ms for the delta and 0–500 ms for the theta after stimulus onset.

2.4. Statistical Analysis

Statistical analyses were carried out using IBM SPSS Statistics v. 24.0. Mixed-design ANOVA model with one 4-level within-subject factor (LOCATION: Frontal, Central, Parietal, Occipital) and two between-subject factors with 3 levels (AGE: Young, Middle-aged, Elderly) and 2 levels (GENDER: Female and Male) was performed to compare group differences for delta and theta amplitudes separately. Bonferroni correction was applied in post hoc analyses. The value of $p < 0.05$ was considered statistically significant for ANOVA analyses. Pearson's correlation analysis was used to investigate the relation between age and the amplitude values of delta and theta oscillations for each gender separately. The value of $p < 0.01$ was considered statistically significant for correlation analyses.

3. Results

The EEG analyses included 155 healthy participants. All data were analyzed for delta and theta frequency bands separately using Mixed-design ANOVA.

3.1. Delta Frequency Band Results

In the delta band responses, there was no statistically significant main AGE [$F(2, 149) = 2.369$, $p = 0.097$], main GENDER [$F(1, 149) = 0.022$, $p = 0.884$], AGE × GENDER interaction effect [$F(12, 149) = 1.073$, $p = 0.344$], and AP × AGE × GENDER interaction effect [$F(6, 447) = 1.146$, $p = 0.334$].

3.2. Theta Frequency Band Results

In the theta band responses, there is a significant AP × AGE × GENDER interaction effect [$F(6, 447) = 5.276$, $p = 0.001$]. In the post hoc analysis for the young adults, the female participants had greater theta responses than the males at the frontal, central, and parietal locations (all, $p \leq 0.001$); at the occipital location, there was no significant gender difference ($p = 0.178$). For the elderly adults, at the occipital area, the females had a greater theta response than the males ($p = 0.034$), but at the other locations, there was no significant gender effect (all, $p \geq 0.272$). In addition, for the middle-aged group, we did not find any significant gender effect for the theta responses (all, $p \geq 0.519$) (Figure 1, Table 2).

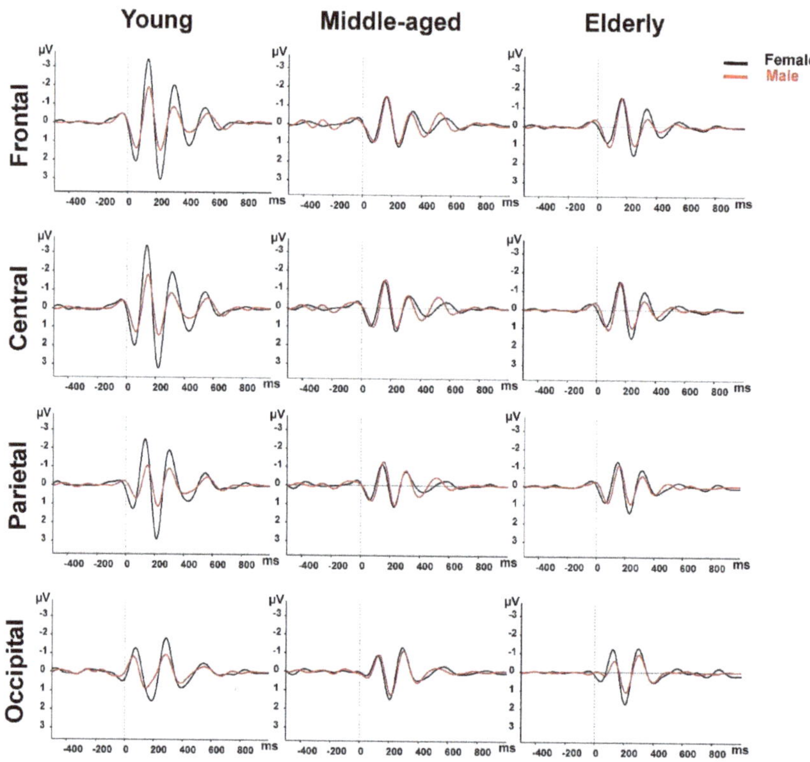

Figure 1. Grand averages of theta ERO power across age groups of both genders.

Table 2. Amplitude values (µV) of EROs in theta frequency band according to age groups and gender.

		Female ($X^{-} \pm$ SD)	Male ($X^{-} \pm$ SD)	p Values
Young (18–50 yr)	Frontal	7.41 ± 2.22	4.49 ± 1.52	**<0.001**
	Central	7.56 ± 2.15	4.70 ± 1.69	**<0.001**
	Parietal	6.28 ± 1.83	4.03 ± 1.80	**<0.001**
	Occipital	4.80 ± 1.67	3.55 ± 1.41	0.178
Middle-Aged (51–65 yr)	Frontal	4.42 ± 1.95	4.34 ± 1.59	0.875
	Central	4.35 ± 1.64	4.52 ± 1.86	0.735
	Parietal	3.94 ± 1.37	4.23 ± 2.04	0.552
	Occipital	4.48 ± 2.02	4.10 ± 1.83	0.519
Elderly (>65 yr)	Frontal	3.94 ± 1.54	4.19 ± 1.62	0.581
	Central	3.97 ± 1.57	4.01 ± 1.37	0.940
	Parietal	4.03 ± 1.67	3.55 ± 1.41	0.272
	Occipital	4.65 ± 2.36	3.58 ± 1.51	**0.034**

In the pairwise comparisons, when we compared the topographical distributions among the age and gender groups, for the young group, we found that the females had showed the greatest responses at the frontal and central locations and the lowest response at the occipital location (all, $p \leq 0.019$). On the other hand, the young males and all the participants in the middle-aged and elderly groups showed no locational significant difference (all, $p \geq 0.236$).

3.3. Correlations

3.3.1. Age

For the female participants, negative and medium correlations between the theta ERO amplitudes and the age for the frontal, central, and parietal areas (all, $p < 0.001$) were noted. We found no significant correlations between the age and ERO amplitudes in the delta or theta frequency bands for the male participants and the delta frequency band for the female participants. The significant correlations are presented as scatter plots in Figure 2.

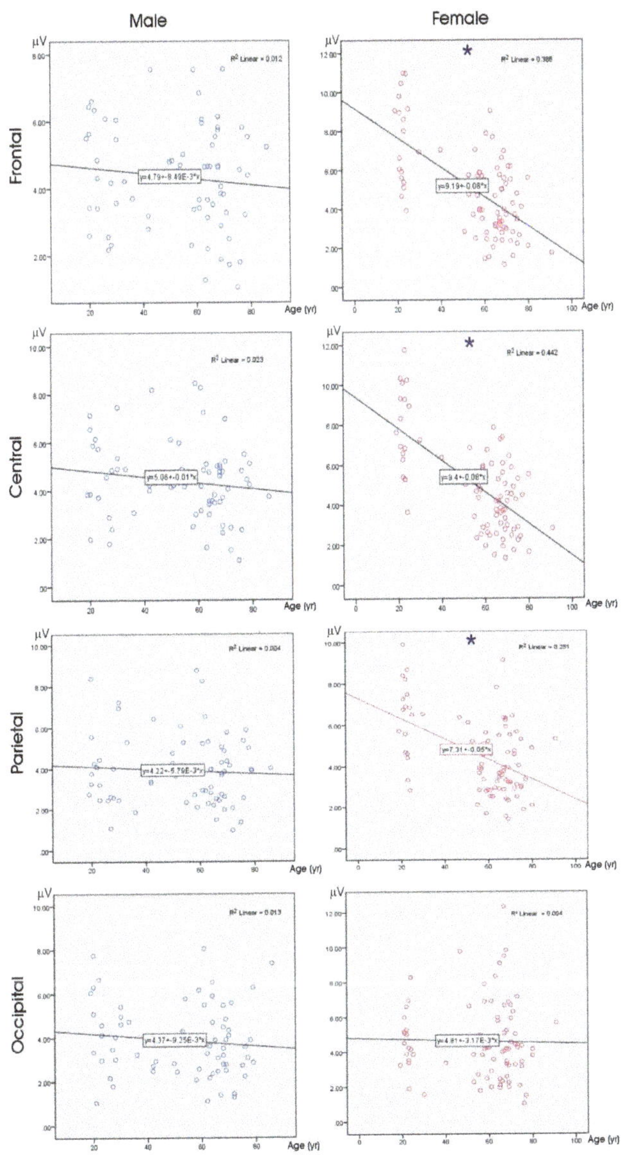

Figure 2. Correlational plots between theta ERO power and age in both genders.

3.3.2. Cognitive Functions

The correlation analyses showed significant moderate to strong negative correlations between the age and OVMPT total scores (r = −0.552, $p < 0.001$), OVMPT-IR (r = −0.383, $p = 0.001$), OVMPT-FR (r = −0.345, $p = 0.004$), digit span forward (r = −0.488, $p < 0.001$), and digit span backward (r = −0.502, $p < 0.001$) for the males. For the female participants, the correlation analysis revealed that there were moderate to strong negative correlations between the age and MMSE (r = −0.360, $p = 0.001$), OVMPT total scores (r = −0.501, $p < 0.001$), OVMPT-IR (r = −0.504, $p = 0.001$), semantic fluency (r = −0.312, $p = 0.004$), lexical fluency (r = −0.584, $p < 0.001$), digit span forward (r = −0.477, $p < 0.001$), and digit span backward (r = −0.478, $p < 0.001$). None of the correlations survived between EEG parameters and the neuropsychological tests when the statistical threshold was set at $p < 0.01$ to account for the inflating effects of testing.

4. Discussion

In the current study, we found that task-related brain responses change differently in females than males across adulthood. The three main findings of our study are as follows: (1) the higher theta EROs in females than males until middle age; (2) the declining theta power of ERO in women reaching that of men after the age groups of middle age years; and (3) stability of the theta ERO power is greater in men, in contrast with women's declining theta power with age.

Mainly, in the current study, the females displayed higher theta ERO responses than the males until the middle age. These amplitude differences were observed in the frontal, central, and parietal regions but not in the occipital region between the genders between 18 and 50 years of adulthood.

The rsEEG and task-related EEG are widely studied for cognitive disorders in clinical research [19–22]. Among the task-related EEG methods, P300 is the most commonly used event-related potential (ERP) elicited after the application of a cognitive task, mostly the oddball paradigm [18,23]. Event-related oscillations (EROs) are elicited after the spectral analysis of ERPs and provide a powerful technique with a high temporal resolution, and it can be used as a tool for detecting subtle abnormalities in cognitive impairments or processes [12,24–26]. The dynamics of EROs are different than in rsEEG. In rsEEG, the increase in the delta and theta rhythms indicates a cognitive decline or pathology [19,20,27], whereas, in ERO, the same finding means a greater brain response to the task in the post-stimulus era [24]. In this vein, the cognitive performance is negatively associated with the rsEEG rhythms in the theta and delta frequencies [19,20,27]. In contrast, the ERO responses in the same frequency bands indicate a higher cognitive performance [25] and greater brain volume [22,28]. Regarding the functional role of delta ERO responses, Demiralp et al. (2001) reported that, after the application of stimuli with decreasing intensities, the oscillatory responses occur nearly in the pure delta range when stimuli reach closer to the subjective threshold [29]. At the threshold intensity, the stimulation focuses the attention of the subject. In this context, the lower delta oscillatory responses in older subjects upon either the visual or auditory oddball paradigm are quite understandable, as healthy elderly individuals tend to show decreased attentional or decision-making focus. In previous studies, regardless of modality, delta ERO studies have shown decreased amplitudes at the frontal and central locations in Alzheimer's disease (AD) [30], Lewy body dementia (LBD), and Parkinson's disease (PD) dementia [31]. This finding indicates supramodal alterations in the anterior parts of the hemispheres in various dementia patients, differentiating them from healthy controls [31]. As expected, smaller delta and theta ERO responses were associated with lower neuropsychological test scores in either auditory or visual modality in dementia groups [31], indicating a supramodal effect.

Previously, an auditory event-related oscillation study showed higher power of the theta and beta1 responses in females, yet the connectivity in these frequency bands was greater in males [32]. This finding was explained by the thinner calvarium in women, yet, another study found the cranium thickness is larger in females compared to males [33], which must lead to fully opposite results. The thicker cranium might lead to diminished EEG transmission from source to scalp electrode. Taking into consideration the fact that, on average, females have thicker cranium, the intracranial gender EEG power differences must be even stronger. Another ERO study [34] reported increased theta and alpha power in males in the age range of 20–29 years. This contradicting result might be more related to the age range of the sample than in our cohort. In a study similar to ours [35], studying the male and female developmental trajectories of theta ERO power, the genders were significantly different in their temporal characteristics, with more rapid decreases with age in males than in females during the ages of 12 to 25. The change in the rate of decrease with age was nearly monotonic in males, with greater fluctuations in females. The influence of gender on rsEEG rhythms during physiological aging was explored to understand neurophysiological mechanisms. Investigators found higher theta power density [36] and higher rsEEG amplitudes of the delta, theta, and alpha rhythms [37] in healthy females than males. Another recent retrospective and exploratory study on rsEEG indicated gender-related effects on the spatial frequency cortical sources of rsEEG rhythms in healthy controls and the mild cognitive impairment in Alzheimer's disease (ADMCI) seniors, especially in the delta, alpha2, and alpha3 frequency bands [27].

This relationship between event-related brain oscillatory mechanisms in the delta and theta frequency bands and frontal executive functions may imply aberrant supramodal mechanisms and the impaired coordination of signal transmission [38] underlying the cognitive deficits in dementia patients [31]. Therefore, the higher theta ERO power with a declining slope in healthy women is a distinct pattern from that of healthy men during aging. This declining slope during aging may be related to the increased likelihood of women's tendency to develop [1]. The mentioned studies suggested that the task-related oscillatory responses in the theta frequency band but not in the delta frequency band may provide evidence for aging-related effects earlier. Furthermore, our group's earlier work on P300 ERP amplitudes [18], overlapping with the current study's cohort, revealed greater P300 amplitudes in women, implying that the greater P300 amplitude in women may be driven by theta rather than delta ERO responses.

Regarding the functional correlates of the theta and delta ERO responses, previous studies reported that more pronounced ERO responses in the theta frequency range were induced during the cognitive processing of auditory stimuli [39], whereas the delta frequency range was induced during the cognitive processing of visual stimuli in healthy individuals [39,40]. This finding was explained by the longer distance for the synchronization of neural networks after visual stimuli and the shorter distance after auditory stimuli. In contrast to this notion, the results of the current study using visual stimuli indicate that the greater theta ERO in women cannot be explained solely based on distance.

Furthermore, the both delta [41–45] and theta ERO responses [41,42,46–48] are impaired in cognitively impaired patients due to AD, PD, and LBD. A study by Yener et al. (2019) in AD and PD mild cognitive impairment (MCI) patients reported that both groups displayed decreased theta ERO power; however, theta phase locking was impaired only in PDMCI patients [15]. The theta ERO phase locking abnormality in the PDMCI group in contrast to the ADMCI group was explained by a more vulnerable subcortical–thalamic loop in the PD patients.

The mentioned studies suggest that a higher theta ERO in women in our healthy, relatively young participants up to the age of 50 implies greater or more effortful processing during a cognitive task [15,49]. This may be part of the subcortical ascending systems modulating the thalamocortical loop and generating ERO delta–theta responses [50]. Therefore, it can be hypothesized that EROs in slow frequency bands, i.e., delta and theta, may modulate the efficiency of neural transfer supporting various aspects of cognitive abilities including attention, executive functions, decision-making, and memory [12,51], and may regulate the cortico-striatal and thalamocortical re-entrant signals to the cerebral cortex during the processing of cognitive stimuli [31]. Within this context, the greater augmentation in the ERO theta responses to the oddball target stimuli may be associated with more integration between the prefrontal–cortical and subcortical networks in younger women. However, this augmented response disappears with a declining slope until the age of 50 in women, becoming similar to that of men.

The explanations for these distinct ERO profiles of the two genders mentioned here can be threefold: firstly, Alzheimer's disease (AD) is a greater threat for older women than men [52], and with age, the EEG patterns in elderly individuals become more similar to those of individuals in the early stages of the AD continuum [20,53]. This finding could be due to the accumulation of pathological peptides many years before any clinical manifestation [54]. In our study, in women, the decreasing theta ERO power with age may be a reflection of the increased risk of developing a neurodegenerative disorder in relation to gender. Since the AD biomarkers were not included in the current study, we cannot be sure if some individuals in our cognitively healthy elderly cohort have started accumulating amyloid in their brains, since it is a well-known fact that characteristic pathological peptide accumulation starts many years before any clinical presentation of AD [54]. Furthermore, several studies have reported gender differences regarding the progression of the neuropathological and clinical manifestations of AD [55,56]. AD disproportionally affects females; females have greater total brain atrophy, display a greater incidence rate, and perform worse in many neuropsychological tasks [57]. Secondly, these results could be related to the longer life expectancy of women in comparison to men [7]. The increased female vulnerability to cerebrospinal fluid (CSF) AD biomarkers is linked to greater hippocampal atrophy and faster age-related cognitive decline [8,9]. On the other hand, a report investigating gender-related EEG/CSF patterns in AD has shown altered rsEEG theta rhythms related to Ab-42 in men [58]. At this point, it is important to remember the different dynamics of rsEEGs and EROs in the theta frequency, where cognitive performance is negatively associated with theta rsEEG rhythms [59], whilst theta ERO responses are indicative of higher cognitive performance [16] and brain volume [22,45]. Thirdly, human brains show great plasticity during their lifespan [60], and education, gender, and sex hormones have an impact on brain plasticity [1,61]. Female-specific genes were reported in a meta-analysis on the gender effects of AD gene expression, where these genes were involved in pathways such as oxidative phosphorylation and were associated with neurodegenerative diseases [62].

Regarding the distinctively different associations between age and neuropsychological test patterns across genders, in our sample, the women tended to decline in verbal fluency while men declined in memory tests. The previous literature supports our findings. The decline in memory and visuospatial skills varied across ethnic groups, with black women displaying sharper declines in memory and visuospatial abilities than Hispanic males and non-Hispanic white women, respectively [63]. Additionally, the cognitive traits were reported to be different between genders; for example, navigation by mental maps, mental rotation, visuo-spatial skills, and working memory were stronger in men, while navigation by landmarks, verbal ability, reading and writing abilities, fine motor coordination, and perceptual skills higher in women [1]. Brain health in diverse settings, including geographical differences, age, cognition, and demographics, can affect EEG-related changes [64–67]. This study provides insights on the neuropsychological test profile between genders in a sample of the Turkish population in the West Asia region.

The current study did not display the association of the theta or delta ERO responses with cognitive tests. This finding is not in line with our expectations, as previous studies indicated theta and delta EROs are correlated with cognitive functions [31]. The reason for this discrepancy may be due to the size and the age profile of our group, and secondly, to the blunting effect of different cognitive decline patterns across genders. Another reason for the lack of correlation between neuropsychological testing and EEG parameters could be the absence of significant cognitive impairments in the study participants.

Briefly, the findings of the current exploratory study support the hypothesis that gender-related changes may be important when establishing normative data in EROs for both clinical and research applications. The expansive data offered by new technologies render new opportunities for gender-stratified clinical decision support systems available [68]. Similarly, the development of data analysis approaches integrating clinical and demographic factors, including gender and aging, is crucial, especially for electrophysiological methodologies. For performing a gender-stratified and gender-adjusted analysis of all the EEG/ERO data, the next step should be the establishment and improvement of longitudinal cohorts with repeated assessments of clinical, cognitive, and biomarker variables, such as peripheral (plasma, neuroimaging, and cerebrospinal fluid), multi-omics (proteome, and metabolome), and genetic data [69]. EEG methods show high stability across the lifespan [70] and the internal consistency of task-related oscillations in the delta and theta frequency bands is very high, with absolute Cronbach's alpha values of 0.90 and 0.94, respectively, indicating slow ERO responses are stable measurements for cognition in healthy adults [71]. Therefore, the results of our study indicate that the investigation of EROs should consider gender factors across different age groups.

The limitations of the current study include the lack of sexual hormone level data from the cohort and the relatively small number of participants for investigating gender effects. Although our healthy volunteers were neurologically intact and displayed normal limits of scores in the extensive neuropsychological tests and depression scales with no abnormality in their MRI images, they were not tested for CSF or plasma AD biomarkers. Additionally, the relatively less number of female participants in the range of 25–45 years might have accentuated the gender difference in the current paper. In future studies, adding datasets from cognitively unimpaired females of the mentioned age range and individuals with cognitive impairment to the sample might empower the statistical correlation model. Furthermore, the cohorts need to be recruited at specified age ranges, from women at pre- and post-menopausal ages to age-matched men. Then, EEG methodologies will have a continuous improvement in the diagnosis and therapy of disease.

In conclusion, in clinical trials using EEG biomarkers as an intervention endpoint, gender should be taken into account and EEG/ERO data should be enriched for performing gender-stratified analysis.

Author Contributions: Conceptualization, G.Y.; methodology, G.Y. and İ.K.; formal analysis, İ.K.; investigation, G.Y.; writing—original draft preparation, İ.K.; writing—review and editing, G.Y.; visualization, İ.K.; project administration, İ.K., S.D.-Ö., D.Y. and G.Y.; funding acquisition, S.D.-Ö., D.Y. and G.Y. All authors have read and agreed to the published version of the manuscript.

Funding: These are retrospective data elicited from projects supported by the Scientific and Technological Research Council of Turkey-TÜBITAK (project number: 112S459), Dokuz Eylül University Scientific Research Projects (Project Numbers: 2014.KB.SAG.058 on OSAS, 2018.KB.SAG.084 Longitudinal BAP), Ankara University Scientific Research Projects Coordination Unit (Project number 16L02000001).

Institutional Review Board Statement: This study was conducted in accordance with the Declaration of Helsinki, and approved by the following ethical committees: Ankara University Clinical Research Ethical Committee 31 August 2015, 13-521-15, and Dokuz Eylül University Clinical Research Ethical Committee (121-GOA, 2011/12-15 with decision number 2011/12-15 in 14 April 2011; 5119–GOA, 2019/29-01 with decision number in 2 December 2019).

Informed Consent Statement: Informed consent was obtained from all subjects.

Data Availability Statement: The data supporting the findings of this study are available through the corresponding author, upon reasonable request.

Acknowledgments: The authors thank Hilal Kula and Yağmur Özbek İşbitiren for their help in editing the manuscript.

Conflicts of Interest: The authors declare no conflicts of interest.

References

1. Zelco, A.; Wapeesittipan, P.; Joshi, A. Insights into Sex and Gender Differences in Brain and Psychopathologies Using Big Data. *Life* **2023**, *13*, 1676. [CrossRef] [PubMed]
2. Harkin, T.; Snowe, O.J.; Mikulski, B.A.; Waxman, H.A. *Drug Safety: Most Drugs Withdrawn in Recent Years Had Greater Health Risks for Women*; United States General Accounting Office: Washington, DC, USA, 2001.
3. Spets, D.S.; Slotnick, S.D. Are there sex differences in brain activity during long-term memory? A systematic review and fMRI activation likelihood estimation meta-analysis. *Cogn. Neurosci.* **2021**, *12*, 163–173. [CrossRef] [PubMed]
4. Malpetti, M.; Ballarini, T.; Presotto, L.; Garibotto, V.; Tettamanti, M.; Perani, D. Alzheimer's Disease Neuroimaging Initiative (ADNI) Database Network for Efficiency and Standardization of Dementia Diagnosis (NEST-DD) database. Gender differences in healthy aging and Alzheimer's Dementia: A 18 F-FDG-PET study of brain and cognitive reserve. *Hum. Brain Mapp.* **2017**, *38*, 4212–4227. [CrossRef] [PubMed]
5. Goyal, M.S.; Vlassenko, A.G.; Raichle, M.E. Reply to Biskup et al. and Tu et al.: Sex differences in metabolic brain aging. *Proc. Natl. Acad. Sci. USA* **2019**, *116*, 10634–10635. [CrossRef] [PubMed]
6. Perera, G.; Pedersen, L.; Ansel, D.; Alexander, M.; Arrighi, H.M.; Avillach, P.; Foskett, N.; Gini, R.; Gordon, M.F.; Gungabissoon, U.; et al. Dementia prevalence and incidence in a federation of European Electronic Health Record databases: The European Medical Informatics Framework resource. *Alzheimers Dement.* **2018**, *14*, 130–139. [CrossRef] [PubMed]
7. Masters, C.L.; Bateman, R.; Blennow, K.; Rowe, C.C.; Sperling, R.A.; Cummings, J.L. Alzheimer's disease. *Nat. Rev. Dis. Primers* **2015**, *1*, 15056. [CrossRef] [PubMed]
8. Koran, M.E.I.; Wagener, M.; Hohman, T.J. Alzheimer's Neuroimaging Initiative. Sex differences in the association between AD biomarkers and cognitive decline. *Brain Imaging Behav.* **2017**, *11*, 205–213. [CrossRef] [PubMed] [PubMed Central]
9. Li, R.; Singh, M. Sex differences in cognitive impairment and Alzheimer's disease. *Front. Neuroendocrinol.* **2014**, *35*, 385–403. [CrossRef] [PubMed] [PubMed Central]
10. Babiloni, C.; Triggiani, A.I.; Lizio, R.; Cordone, S.; Tattoli, G.; Bevilacqua, V.; Soricelli, A.; Ferri, R.; Nobili, F.; Gesualdo, L.; et al. Classification of Single Normal and Alzheimer's Disease Individuals from Cortical Sources of Resting State EEG Rhythms. *Front. Neurosci.* **2016**, *10*, 47. [CrossRef] [PubMed] [PubMed Central]
11. Maestú, F.; Cuesta, P.; Hasan, O.; Fernandéz, A.; Funke, M.; Schulz, P.E. The Importance of the Validation of M/EEG With Current Biomarkers in Alzheimer's Disease. *Front. Hum. Neurosci.* **2019**, *13*, 17. [CrossRef] [PubMed] [PubMed Central]
12. Başar-Eroglu, C.; Basar, E.; Demiralp, T.; Schürmann, M. P300-response: Possible psychophysiological correlates in delta and theta frequency channels. A review. *Int. J. Psychophysiol.* **1992**, *13*, 161–179. [CrossRef] [PubMed]
13. Başar, E.; Karakaş, S. Event-Related Oscillations in the Brain. In *Brain Function and Oscillations*; Springer Series in Synergetics; Springer: Berlin/Heidelberg, Germany, 1998. [CrossRef]
14. Başar-Eroglu, C.; Demiralp, T. Event-related theta oscillations: An integrative and comparative approach in the human and animal brain. *Int. J. Psychophysiol.* **2001**, *39*, 167–195. [CrossRef] [PubMed]
15. Yener, G.G.; Fide, E.; Özbek, Y.; Emek-Savaş, D.D.; Aktürk, T.; Çakmur, R.; Güntekin, B. The difference of mild cognitive impairment in Parkinson's disease from amnestic mild cognitive impairment: Deeper power decrement and no phase-locking in visual event-related responses. *Int. J. Psychophysiol.* **2019**, *139*, 48–58. [CrossRef] [PubMed]
16. Aktürk, T.; de Graaf, T.A.; Erdal, F.; Sack, A.T.; Güntekin, B. Oscillatory delta and theta frequencies differentially support multiple items encoding to optimize memory performance during the digit span task. *Neuroimage* **2022**, *263*, 119650. [CrossRef] [PubMed]
17. Güntekin, B.; Başar, E. Brain oscillations are highly influenced by gender differences. *Int. J. Psychophysiol.* **2007**, *65*, 294–299. [CrossRef] [PubMed]
18. Yerlikaya, D.; Hünerli-Gündüz, D.; Fide, E.; Özbek, Y.; Kıyı, İ.; Öztura, İ.; Yener, G.G. The reliability of P300 and the influence of age; gender and education variables in a 50 years and older normative sample. *Int. J. Psychophysiol.* **2022**, *181*, 1–13. [CrossRef] [PubMed]
19. Babiloni, C.; Blinowska, K.; Bonanni, L.; Cichocki, A.; De Haan, W.; Del Percio, C.; Dubois, B.; Escudero, J.; Fernández, A.; Frisoni, G.; et al. What electrophysiology tells us about Alzheimer's disease: A window into the synchronization and connectivity of brain neurons. *Neurobiol. Aging.* **2020**, *85*, 58–73. [CrossRef] [PubMed]
20. Babiloni, C.; Lopez, S.; Noce, G.; Ferri, R.; Panerai, S.; Catania, V.; Soricelli, A.; Salvatore, M.; Nobili, F.; Arnaldi, D.; et al. Resting State Alpha Electroencephalographic Rhythms Are Affected by Sex in Cognitively Unimpaired Seniors and Patients with Alzheimer's Disease and Amnesic Mild Cognitive Impairment: A Retrospective and Exploratory Study. *Cereb. Cortex.* **2022**, *32*, 2197–2215. [CrossRef] [PubMed]
21. Yener, G.G.; Güntekin, B.; Örken, D.N.; Tülay, E.; Forta, H.; Başar, E. Auditory delta event-related oscillatory responses are decreased in Alzheimer's disease. *Behav. Neurol.* **2012**, *25*, 3–11. [CrossRef] [PubMed] [PubMed Central]
22. Yener, G.G.; Emek-Savaş, D.D.; Lizio, R.; Çavuşoğlu, B.; Carducci, F.; Ada, E.; Güntekin, B.; Babiloni, C.C.; Başar, E. Frontal delta event-related oscillations relate to frontal volume in mild cognitive impairment and healthy controls. *Int. J. Psychophysiol.* **2016**, *103*, 110–117. [CrossRef] [PubMed]
23. Polich, J.; Kok, A. Cognitive and biological determinants of P300: An integrative review. *Biol. Psychol.* **1995**, *41*, 103–146. [CrossRef] [PubMed]
24. Başar, E.; Başar-Eroğlu, C.; Güntekin, B.; Yener, G.G. Brain's alpha, beta, gamma, delta, and theta oscillations in neuropsychiatric diseases: Proposal for biomarker strategies. *Suppl. Clin. Neurophysiol.* **2013**, *62*, 19–54. [CrossRef] [PubMed]

25. Güntekin, B.; O'Donnell, B.F. Special Issue: Update on Neural Oscillations in Neuropsychiatric Disorders. *Clin. EEG Neurosci.* **2023**, *54*, 347–348. [CrossRef] [PubMed]
26. Yener, G.G.; Emek-Savaş, D.D.; Güntekin, B.; Başar, E. The visual cognitive network, but not the visual sensory network, is affected in amnestic mild cognitive impairment: A study of brain oscillatory responses. *Brain Res.* **2014**, *1585*, 141–149. [CrossRef] [PubMed]
27. Babiloni, C.; Jakhar, D.; Tucci, F.; Del Percio, C.; Lopez, S.; Soricelli, A.; Salvatore, M.; Ferri, R.; Catania, V.; Massa, F.; et al. Resting state electroencephalographic alpha rhythms are sensitive to Alzheimer's disease mild cognitive impairment progression at a 6-month follow-up. *Neurobiol. Aging.* **2024**, *137*, 19–37. [CrossRef] [PubMed]
28. Hünerli, D.; Emek-Savaş, D.D.; Çavuşoğlu, B.; Dönmez Çolakoğlu, B.; Ada, E.; Yener, G.G. Mild cognitive impairment in Parkinson's disease is associated with decreased P300 amplitude and reduced putamen volume. *Clin. Neurophysiol.* **2019**, *130*, 1208–1217. [CrossRef] [PubMed]
29. Demiralp, T.; Ademoglu, A.; Istefanopulos, Y.; Başar-Eroglu, C.; Başar, E. Wavelet analysis of oddball P300. *Int. J. Psychophysiol.* **2001**, *39*, 221–227. [CrossRef] [PubMed]
30. Yener, G.; Hünerli-Gündüz, D.; Yıldırım, E.; Aktürk, T.; Başar-Eroğlu, C.; Bonanni, L.; Del Percio, C.; Farina, F.; Ferri, R.; Güntekin, B.; et al. Treatment effects on event-related EEG potentials and oscillations in Alzheimer's disease. *Int. J. Psychophysiol.* **2022**, *177*, 179–201. [CrossRef] [PubMed]
31. Yıldırım, E.; Hanoğlu, L.; Yener, G.; Yerlikaya, D.; Taylor, J.P.; Schumacher, J.; McKeith, I.; Bonanni, L.; Pantano, P.; Piervincenzi, C.; et al. Relationship between default mode network and resting-state electroencephalographic alpha rhythms in cognitively unimpaired seniors and patients with dementia due to Alzheimer's disease. *Cereb. Cortex.* **2023**, *33*, 10514–10527. [CrossRef] [PubMed] [PubMed Central]
32. Georgiev, S.; Minchev, Z.; Christova, C.; Philipova, D. Gender event-related brain oscillatory differences in normal elderly population EEG. *Int. J. Bioautomation* **2011**, *15*, 33.
33. Li, H.; Ruan, J.; Xie, Z.; Wang, H.; Liu, W. Investigation of the critical geometric characteristics of living human skulls utilising medical image analysis techniques. *Int. J. Veh. Saf.* **2007**, *2*, 345–367. [CrossRef]
34. Langrova, J.; Kremláček, J.; Kuba, M.; Kubova, Z.; Szanyi, J. Gender impact on electrophysiological activity of the brain. *Physiol. Res.* **2012**, *61*, S119–S127. [CrossRef] [PubMed]
35. Chorlian, D.B.; Rangaswamy, M.; Manz, N.; Kamarajan, C.; Pandey, A.K.; Edenberg, H.; Kuperman, S.; Porjesz, B. Gender modulates the development of theta event related oscillations in adolescents and young adults. *Behav. Brain Res.* **2015**, *292*, 342–352. [CrossRef] [PubMed]
36. Duffy, F.H.; McAnulty, G.B.; Albert, M.S. The pattern of age-related differences in electrophysiological activity of healthy males and females. *Neurobiol. Aging* **1993**, *14*, 73–84. [CrossRef] [PubMed]
37. Wada, Y.; Takizawa, Y.; Jiang, Z.Y.; Yamaguchi, N. Gender differences in quantitative EEG at rest and during photic stimulation in normal young adults. *Clin. Electroencephalogr.* **1994**, *25*, 81–85. [CrossRef] [PubMed]
38. Spagna, A.; Wu, T.; Kim, K.; Fan, J. Supramodal executive control of attention: Evidence from unimodal and crossmodal dual conflict effects. *Cortex* **2020**, *133*, 266–276. [CrossRef] [PubMed]
39. Yordanova, J.; Kolev, V.; Rosso, O.A.; Schürmann, M.; Sakowitz, O.W.; Ozgören, M.; Basar, E. Wavelet entropy analysis of event-related potentials indicates modality-independent theta dominance. *J. Neurosci. Methods.* **2002**, *117*, 99–109. [CrossRef] [PubMed]
40. Schürmann, M.; Başar-Eroglu, C.; Kolev, V.; Başar, E. Delta responses and cognitive processing: Single-trial evaluations of human visual P300. *Int. J. Psychophysiol.* **2001**, *39*, 229–239. [CrossRef] [PubMed]
41. Yener, G.G.; Güntekin, B.; Oniz, A.; Başar, E. Increased frontal phase-locking of event-related theta oscillations in Alzheimer patients treated with cholinesterase inhibitors. *Int. J. Psychophysiol.* **2007**, *64*, 46–52. [CrossRef] [PubMed]
42. Güntekin, B.; Aktürk, T.; Arakaki, X.; Bonanni, L.; Del Percio, C.; Edelmayer, R.; Farina, F.; Ferri, R.; Hanoğlu, L.; Kumar, S.; et al. Are there consistent abnormalities in event-related EEG oscillations in patients with Alzheimer's disease compared to other diseases belonging to dementia? *Psychophysiology.* **2022**, *59*, e13934. [CrossRef] [PubMed]
43. Rosenblum, Y.; Shiner, T.; Bregman, N.; Fahoum, F.; Giladi, N.; Maidan, I.; Mirelman, A. Event-related oscillations differentiate between cognitive; motor and visual impairments. *J. Neurol.* **2022**, *269*, 3529–3540. [CrossRef] [PubMed]
44. Tülay, E.E.; Güntekin, B.; Yener, G.; Bayram, A.; Başar-Eroğlu, C.; Demiralp, T. Evoked and induced EEG oscillations to visual targets reveal a differential pattern of change along the spectrum of cognitive decline in Alzheimer's Disease. *Int. J. Psychophysiol.* **2020**, *155*, 41–48. [CrossRef] [PubMed]
45. Hünerli-Gündüz, D.; Özbek İşbitiren, Y.; Uzunlar, H.; Çavuşoğlu, B.; Çolakoğlu, B.D.; Ada, E.; Güntekin, B.; Yener, G.G. Reduced power and phase-locking values were accompanied by thalamus; putamen; and hippocampus atrophy in Parkinson's disease with mild cognitive impairment: An event-related oscillation study. *Neurobiol. Aging.* **2023**, *121*, 88–106. [CrossRef] [PubMed]
46. Missonnier, P.; Gold, G.; Herrmann, F.R.; Fazio-Costa, L.; Michel, J.P.; Deiber, M.P.; Michon, A.; Giannakopoulos, P. Decreased theta event-related synchronization during working memory activation is associated with progressive mild cognitive impairment. *Dement. Geriatr. Cogn. Disord.* **2006**, *22*, 250–259. [CrossRef] [PubMed]

47. Rosenblum, Y.; Maidan, I.; Fahoum, F.; Giladi, N.; Bregman, N.; Shiner, T.; Mirelman, A. Differential changes in visual and auditory event-related oscillations in dementia with Lewy bodies. *Clin. Neurophysiol.* **2020**, *131*, 2357–2366. [CrossRef] [PubMed]
48. Schmiedt, C.; Meistrowitz, A.; Schwendemann, G.; Herrmann, M.; Basar-Eroglu, C. Theta and alpha oscillations reflect differences in memory strategy and visual discrimination performance in patients with Parkinson's disease. *Neurosci. Lett.* **2005**, *388*, 138–143. [CrossRef] [PubMed]
49. Başar, E.; Güntekin, B.; Yener, G.; Başar-Eroğlu, C. Mindful brain and EEG-neurophysiology. *Int. J. Psychophysiol.* **2016**, *103*, 1–2. [CrossRef] [PubMed]
50. Pfurtscheller, G.; Lopes da Silva, F.H. Event-related EEG/MEG synchronization and desynchronization: Basic principles. *Clin. Neurophysiol.* **1999**, *110*, 1842–1857. [CrossRef] [PubMed]
51. Harmony, T.; Fernández, T.; Silva, J.; Bernal, J.; Díaz-Comas, L.; Reyes, A.; Marosi, E.; Rodríguez, M.; Rodríguez, M. EEG delta activity: An indicator of attention to internal processing during performance of mental tasks. *Int. J. Psychophysiol.* **1996**, *24*, 161–171. [CrossRef] [PubMed]
52. *Global Status Report on the Public Health Response to Dementia*; World Health Organization: Geneva, Switzerland, 2021. Available online: https://apps.who.int/iris/handle/10665/344701 (accessed on 25 April 2023).
53. Emek-Savaş, D.D.; Özmüş, G.; Güntekin, B.; Dönmez Çolakoğlu, B.; Çakmur, R.; Başar, E.; Yener, G.G. Decrease of Delta Oscillatory Responses in Cognitively Normal Parkinson's Disease. *Clin. EEG Neurosci.* **2017**, *48*, 355–364. [CrossRef] [PubMed]
54. Moscoso, A.; Grothe, M.J.; Ashton, N.J.; Karikari, T.K.; Lantero Rodríguez, J.; Snellman, A.; Suárez-Calvet, M.; Blennow, K.; Zetterberg, H.; Schöll, M.; et al. Longitudinal Associations of Blood Phosphorylated Tau181 and Neurofilament Light Chain with Neurodegeneration in Alzheimer Disease. *JAMA Neurol.* **2021**, *78*, 396–406. [CrossRef] [PubMed]
55. Buckley, R.F.; Mormino, E.C.; Amariglio, R.E.; Properzi, M.J.; Rabin, J.S.; Lim, Y.Y.; Papp, K.V.; Jacobs, H.I.L.; Burnham, S.; Hanseeuw, B.J.; et al. Sex, amyloid, and APOE e4 and risk of cognitive decline in preclinical Alzheimer's disease: Findings from three well-characterized cohorts. *Alzheimers Dement.* **2018**, *14*, 1193–1203. [CrossRef] [PubMed]
56. Caldwell, J.Z.K.; Zhuang, X.; Leavitt, M.J.; Banks, S.J.; Cummings, J.; Cordes, D.; Papp, K.V.; Jacobs, H.I.L.; Burnham, S.; Hanseeuw, B.J.; et al. Sex moderates amyloid and apolipoprotein e4 effects on default mode network connectivity at rest. *Front. Neurol.* **2019**, *10*, 900. [CrossRef] [PubMed]
57. Williamson, J.N.; James, S.A.; Mullen, S.P.; Sutton, B.P.; Wszalek, T.; Mulyana, B.; Mukli, P.; Yabluchanskiy, A. Alzheimer's Disease Neuroimaging Initiative Consortium; Yang, Y. Sex differences in interacting genetic and functional connectivity biomarkers in Alzheimer's disease. *GeroScience* **2024**. [CrossRef] [PubMed]
58. Chino-Vilca, B.; Rodríguez-Rojo, I.C.; Torres-Simón, L.; Cuesta, P.; Vendrell, A.C.; Piñol-Ripoll, G.; Huerto, R.; Tahan, N.; Maestú, F. Sex specific EEG signatures associated with cerebrospinal fluid biomarkers in mild cognitive impairment. *Clin. Neurophysiol.* **2022**, *142*, 190–198. [CrossRef] [PubMed]
59. Babiloni, C.; Arakaki, X.; Azami, H.; Bennys, K.; Blinowska, K.; Bonanni, L.; Bujan, A.; Carrillo, M.C.; Cichocki, A.; de Frutos-Lucas, J.; et al. Measures of resting state EEG rhythms for clinical trials in Alzheimer's disease: Recommendations of an expert panel. *Alzheimers Dement.* **2021**, *17*, 1528–1553. [CrossRef] [PubMed] [PubMed Central]
60. Maguire, E.A.; Gadian, D.G.; Johnsrude, I.S.; Good, C.D.; Ashburner, J.; Frackowiak, R.S.; Frith, C.D. Navigation-related structural change in the hippocampi of taxi drivers. *Proc. Natl. Acad. Sci. USA* **2000**, *97*, 4398–4403. [CrossRef] [PubMed] [PubMed Central]
61. Hirnstein, M.; Hausmann, M. Sex/gender differences in the brain are not trivial—A commentary on Eliot et al. (2021). *Neurosci. Biobehav. Rev.* **2021**, *130*, 408–409. [CrossRef] [PubMed]
62. Moradifard, S.; Hoseinbeyki, M.; Ganji, S.M.; Minuchehr, Z. Analysis of microRNA and Gene Expression Profiles in Alzheimer's Disease: A Meta-Analysis Approach. *Sci. Rep.* **2018**, *8*, 4767. [CrossRef] [PubMed] [PubMed Central]
63. Avila, J.F.; Vonk, J.M.J.; Verney, S.P.; Witkiewitz, K.; Arce Rentería, M.; Schupf, N.; Mayeux, R.; Manly, J.J. Sex/gender differences in cognitive trajectories vary as a function of race/ethnicity. *Alzheimer's Dement.* **2019**, *15*, 1516–1523. [CrossRef] [PubMed]
64. Hernandez, H.; Baez, S.; Medel, V.; Moguilner, S.; Cuadros, J.; Santamaria-Garcia, H.; Tagliazucchi, E.; Valdes-Sosa, P.A.; Lopera, F.; OchoaGómez, J.F.; et al. Brain health in diverse settings: How age, demographics and cognition shape brain function. *Neuroimage* **2024**, *295*, 120636. [CrossRef] [PubMed]
65. Wang, K.; Umla-Runge, K.; Hofmann, J.; Ferdinand, N.K.; Chan, R.C. Cultural differences in sensitivity to the relationship between objects and contexts: Evidence from P3. *Neuroreport* **2014**, *25*, 656–660. [CrossRef] [PubMed]
66. Sonke, C.J.; Van Boxtel, G.J.; Griesel, D.R.; Poortinga, Y.H. Brain wave concomitants of cross-cultural differences in scores on simple cognitive tasks. *J. Cross-Cult. Psychol.* **2008**, *39*, 37–54. [CrossRef]
67. Lewis, R.S.; Goto, S.G.; Kong, L.L. Culture and context: East Asian American and European American differences in P3 event-related potentials and self-construal. *Personal. Soc. Psychol. Bull.* **2008**, *34*, 623–634. [CrossRef] [PubMed]
68. Kimenai, D.M.; Janssen, E.B.N.J.; Eggers, K.M.; Lindahl, B.; den Ruijter, H.M.; Bekers, O.; Appelman, Y.; Meex, S.J.R. Sex-Specific versus Overall Clinical Decision Limits for Cardiac Troponin I and T for the Diagnosis of Acute Myocardial Infarction: A Systematic Review. *Clin. Chem.* **2018**, *64*, 1034–1043. [CrossRef] [PubMed]
69. Prado, P.; Birba, A.; Cruzat, J.; Santamaría-García, H.; Parra, M.; Moguilner, S.; Tagliazucchi, E.; Ibáñez, A. Dementia ConnEEG-tome: Towards multicentric harmonization of EEG connectivity in neurodegeneration. *Int. J. Psychophysiol.* **2022**, *172*, 24–38. [CrossRef] [PubMed] [PubMed Central]

70. Lopez, K.L.; Monachino, A.D.; Vincent, K.M.; Peck, F.C.; Gabard-Durnam, L.J. Stability; change; and reliable individual differences in electroencephalography measures: A lifespan perspective on progress and opportunities. *Neuroimage* **2023**, *275*, 120116. [CrossRef] [PubMed] [PubMed Central]
71. Burgess, A.; Gruzelier, J. Individual reliability of amplitude distribution in topographical mapping of EEG. *Electroencephalogr. Clin. Neurophysiol.* **1993**, *86*, 219–223. [CrossRef] [PubMed]

MDPI AG
Grosspeteranlage 5
4052 Basel
Switzerland
Tel.: +41 61 683 77 34

Brain Sciences Editorial Office
E-mail: brainsci@mdpi.com
www.mdpi.com/journal/brainsci

www.ingramcontent.com/pod-product-compliance
Lightning Source LLC
Chambersburg PA
CBHW040814250926
51850CB00007BA/1162
* 9 7 8 3 7 2 5 8 5 1 4 5 4 *